Managing Power Electronics

MANAGING POWER ELECTRONICS

VLSI and DSP-Driven Computer Systems

Dr. Nazzareno Rossetti

A JOHN WILEY & SONS, INC., PUBLICATION

Published by John Wiley & Sons, Inc., Hoboken, New Jersey.
Published simultaneously in Canada.

Library of Congress Cataloging-in-Publication Data:

Rossetti, Nazzareno, 1951–
Managing power electronics : VLSI and DSP-driven computer systems / Nazzareno Rossetti.
p. cm.
Includes bibliographical references and index.
ISBN-13 978-0-471-70959-6 (cloth : alk. paper)
ISBN-10 0-471-70959-X (cloth : alk. paper)
1. Integrated circuits—Very large scale integration. 2. Semiconductors. 3. Signal processing—Digital techniques. I. Title.
II. Title: VLSI and DSP-driven computer systems.
TK7874.75.R67 2005
621.381'044—dc22 2005021296

Printed in the United States of America.

10 9 8 7 6 5 4 3 2 1

To Ash and Ty, my two pearls

Contents

Foreword

At $13 billion and roughly five percent of the total semiconductor market (2004 data) the power semiconductor market is big and growing fast, typically outgrowing the rest of the semiconductor market.

Modern electronic appliances, while exhibiting increasing functionality, are also expected to consume little power, for reasons of portability, thermal performance, and environmental considerations.

This book is an important contribution to the understanding of the many facets of this market, from technology to circuits, electronic appliances, and market forces at work.

The author's broad industry experience built in almost three decades of design, application, and marketing of analog and power management devices is reflected in the breadth of this book. Topics discussed range from fundamentals of semiconductor physics, to analog and digital circuit design and the complex market dynamics driving the semiconductor business. The author displays in this work a unique ability to reduce complex issues to simple concepts. The book makes good reading for the marketing engineer or business hi-tech professional wanting a quick refresh of integrated circuits and power management design, as well as the technologist wanting to expand his market horizons. The timely market and technical information also serves as excellent reference material for students interested in entering the power management field.

Seth R. Sanders, Professor
Electrical Engineering and Computer
Sciences Department
University of California, Berkeley

Preface

How to Use This Book

This book discusses state-of-the-art power management techniques of modern electronic appliances relying on such Very Large Scale Integration (VLSI) chips as CPUs and DSPs.

It also covers specific circuit design issues and their implications, including original derivation of important expressions.

This book is geared toward systems and applications, although it also gets into the specific technical aspects of discrete and integrated solutions, like the analysis of circuits within the power chips which power PCs and other modern electronics.

The first half of this book is a good complement to classic semiconductor text books because it deals with the same complex issues in a more conversational way. It avoids completely the use of complex expressions and minimizing the use of formulas to useful ones, that allow us to plug values in and get an actual result.

The second half of the book is a broad review of the modern technology landscape seen through the eyes of the power management engineer, continually challenged by the rising complexity of modern electronic appliances.

Scope

In this book, power management is covered in its many facets, including semiconductor manufacturing processes, packages, circuits, functions, and systems. The first chapter is a general overview of the semiconductor industry and gives a glimpse of its many accomplishments in a relatively short time. Semiconductor processes and packages are discussed in the second chapter. Great effort has been put here in explaining complex concepts in conversational and intuitive fashion. Chapter 3 is a guided "tour de force" in analog design building from the transistor up to higher level functions and leading to the implementation of a complete voltage regulator. In chapter 4 we discuss a number of popular DC-DC voltage regulation architectures, each responding to specific requirements demanded by the application at hand. Similarly in chapter 5 we move on to discuss AC-DC architectures for power conversion. After the technical foundation is laid with these first 5 chapters, we move to analyze some of the most popular electronic appliances. In chapter 6 we cover ultra portable appliances such as cellular telephones, Personal Digital Assistants (PDAs) and Digital Still Cameras (DSCs) and discuss the amazing success of these devices and the trend toward convergence leading to smart phones that incorporate PDAs, DSCs, Global Positioning Systems (GPS), Internet appliances and more into one small handheld device. Then in chapter 7 we cover specifically the desktop PC, a resilient device which continues to reinvent itself and defeat the many attempts by competing platforms to make it obsolete. Then we go into portable computing with the notebook PC aspiring to claim the center stage for the coming age of "computing anywhere, anytime." Finally some special power management topics are covered in chapter 8. In closure the appendix section provides more in dept information about parts discussed in the chapters.

Acknowledgments

Thanks to Fairchild Semiconductor for sponsoring this book, to Portelligent for providing some of the beautiful pictures and to Jim Holt and Steven Park for proofreading chapter 2. And finally thanks to Melissa Parker and Robert Kern of TIPS Technical Publishing for their careful editing and composition.

About the Author

Reno Rossetti is a published author of technical articles for the major electronics trade magazines, power management developer, mentor, architect, and speaker. He holds a doctorate in electrical engineering from Politecnico of Torino, Italy and a Degree in Business Administration from Bocconi University of Milan, Italy. He has more than 25 years experience in the semiconductors industry, covering integrated circuit design, semiconductor applications and marketing roles. He is currently the director of Strategy for the Integrated Circuits Group at Fairchild Semiconductor, a leading Semiconductor manufacturer providing innovative solutions for power management and power conversion.

Over the years he has designed several innovative power conversion and management solutions for Desktop and Portable System Electronics and CPUs. His patented "Valley Control" architecture (patent issued in

2000) became a leading control architecture powering many generations of voltage regulators controllers for personal computer central processing units (CPUs), He defined and released to production the first "Integrated Power Supply," LM2825, a full power supply, complete with magnetics and capacitors, confined in a standard dip 24 package and produced with standard IC manufacturing packaging technology. This resulted in a reliable and superior power supply with a mean time before failure of 20 million hours and density of 35W/cubic inch. It received several awards, including 1996 product of the year for EETimes and EDN. More recently he has been concerned with and created intellectual property (IP) for advanced power management aspects including application of micro-electro-mechanical (MEM) technologies to power supplies and untethered power distribution systems. Rossetti holds several patents in the field of voltage regulation and power management. His articles and commentaries have appeared in the main electronics magazines in the United States, Europe and Asia (EETimes, Planet Analog, PCIM, etc.).

Chapter 1

Introduction

1.1 Technology Landscape

Power management is, literally and metaphorically, the hottest area in computing and computing appliances.

In 1965, while working at Fairchild Semiconductor, Gordon Moore predicted that the number of transistors in an integrated circuit would double approximately every two years. Moore's law, as his observation has been dubbed, has so far been the foundation of the business of personal computing and its derivative applications. With its publication in Electronics magazine on April 19th, 1965, Moore's law was introduced to the world, along with its profound technological, business, and financial implications.

As long as new computers continue to deliver more performance—and Moore's law says they will—people will continue to buy them. Whether people get bored with old technology or simply outgrow it, outdated computers seem to have little value. Hence, people are only willing to pay for the *additional* value of a new product, compared to the old one, not the value of a product in its entirety. This means consumers want to pay roughly the same price or even less for the new product as for the old. In essence they want the old technology for free and are willing to pay only for the new one.

Financially, building the facilities to produce smaller and smaller transistors requires billions of dollars of investment. For every new generation of chips, the old facility is either scrapped or used to produce some electronics down the food chain. A new facility has to be built with better foundations, better concrete, and better machinery. Technologically, designing such dense chips is becoming increasingly complex, requiring new tools for simulation, production, and testing.

The combination of financial and technological constraints are such that it takes roughly two years to transition from one chip generation to the next, another interpretation of Moore's law.

Figure 1–1 shows how one function can be implemented in smaller and smaller chips as the capacity to resolve ever-smaller minimum features improves.

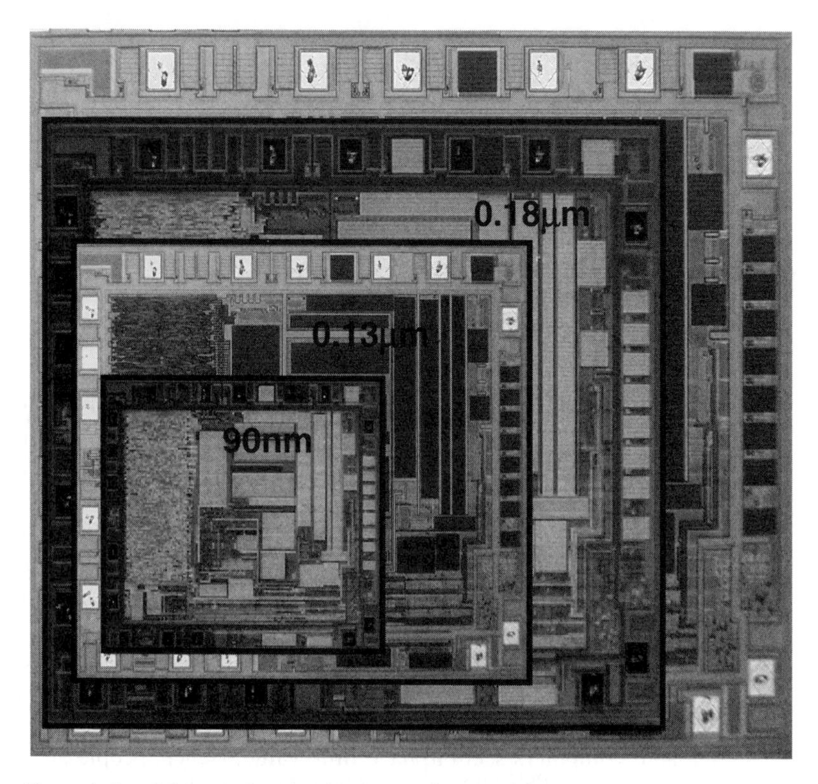

Figure 1–1 Moore's law leads to ever denser chips.

Figure 1–2 shows the progression of Pentium CPUs enabled by Moore's law. Each new CPU requires a specialized voltage regulator module (VRM), accurately specified by Intel. As chips become denser their current consumption rises steadily. With the Pentium IV, a single-phase (1Φ) voltage regulator is no longer sufficient. Recently, aggressive power management techniques inside the CPU, process enhancements like low K dielectrics, copper interconnects, strained silicon, and more recently dual-core CPUs have begun

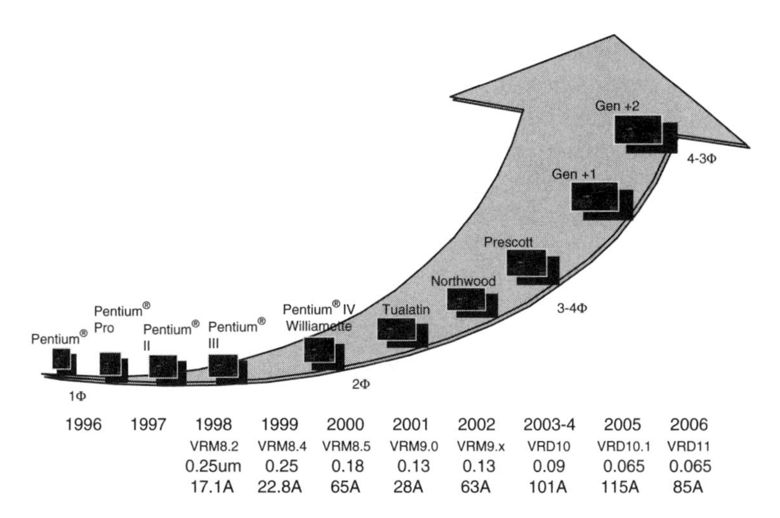

Figure 1–2 Moore's law delivers new computing platforms.

slowing down the upward spiral of power consumption. Beginning with the Centrino mobile wireless platform, even Intel has come to admit that performance can no longer be identified with clock speed (say a 3 GHz Pentium IV), but with a more global value judgment including speed of task execution, small size, wireless connectivity, and low power consumption.

The pace of such progression greatly escalates the complexity of all modern VLSI (Very Large Scale Integration) circuits, not just the PC CPU. With each transistor releasing more heat at a faster operating speed, the heat released by these complex chips is becoming difficult to handle. The heat problem is compounded by the fact that not only does the CPU get hotter, but so do the chipset, the graphics, and any other chip on the motherboard.

Power consumption containment dictates that each new generation of PC motherboards utilizes increasingly customized voltage regulators for each active load. In Figure 1–3 we show the transition from two voltage regulators for Pentium CPUs up to eight voltage regulators for the Pentium III, which power CPU periphery, the CPU, termination, the clock, memory, north bridge, AGP graphics, and stand-by.

Power management is all about feeding these power-hungry chips the energy they need to function while controlling and disposing of the heat by-product. Power management must progress faster than Moore's law in order to keep the computing business profitable.

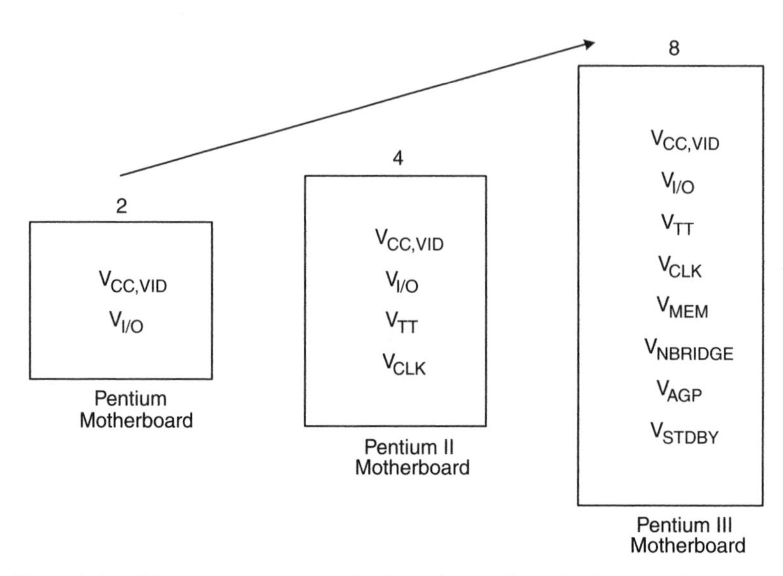

Figure 1–3 Next generation motherboards require a higher number of specialized regulators.

1.2 A Young Industry after All

Electronic gadgets are such a part of our daily lives that it is hard to believe that the electronics industry as a whole is younger than most baby boomers. This electronics revolution began in 1948 with William Shockley's invention of the solid state transistor and continues unabated at today. The first transistors were made of germanium and it was not until 1954 that silicon became a popular material. The first silicon transistors where built with a photolithographic technique known as the mesa process, a form of contact printing still conceptually at the base of any modern semiconductor process. As the name implies, these early transistors had an irregular surface like a mesa rock formation or a tiered wedding cake if you will. A fundamental step forward was Fairchild Semiconductor's invention of the planar process, in which the surface of the transistor remained flat and the various doping materials were simply diffused inside the silicon wafer surface. In the planar transistor in Figure 1–4 the smaller disk in the center is the emitter contact, lying on top of the second disk, the emitter. The bigger lopsided disk is the base and the lopsided doughnut inside it is the base contact. The collector is the entire dark square making up the rest of the picture. The creation of the planar process was a fundamental step in the creation, also by Fairchild, of the *Integrated Circuit* (IC), in which many such transistors could be "printed" on a flat silicon wafer. Figure 1–5 is the first integrated circuit—a set-reset flip-flop logic device.

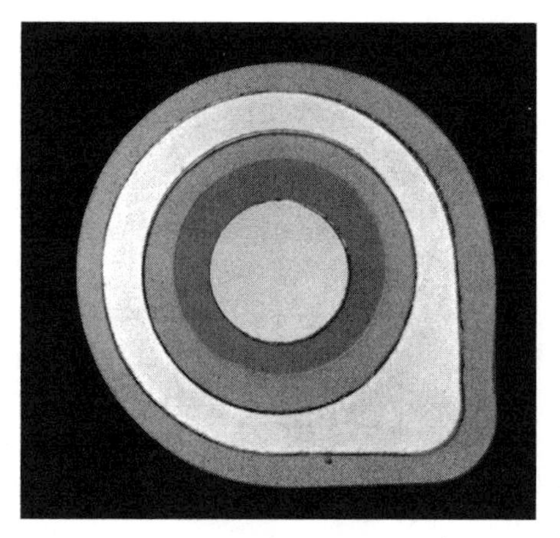

Figure 1–4 The first planar transistor (1959).

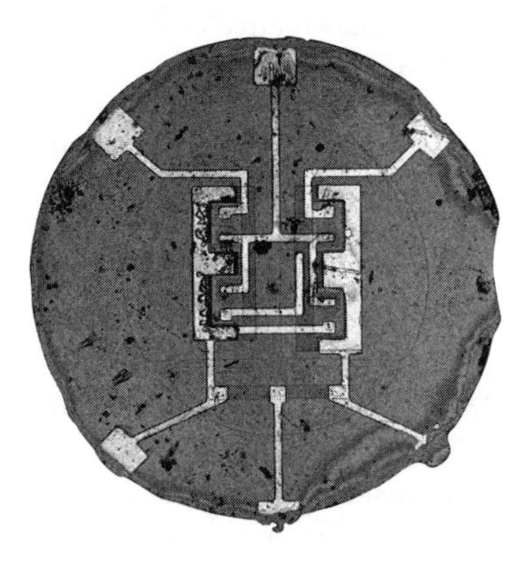

Figure 1–5 The first IC, a Set-Reset Flip-Flop (1961).

The Fairchild chip shown in Figure 1–5, vintage 1961, is 1.8 mm^2 and integrates four transistors and five resistors, barely visible under the spidery looking metal layer on the top, that make up the interconnections and contacts to the external world. Consider that in 2005 the dual core Montecito CPU integrates 1.72 billion transistors in 596 mm^2. Hence the integrated circuit process goes from a density of two transistors per square millimeter to three million transistors per square millimeter in less than fifty years. From a functional stand-point the next important step is the invention of the operational amplifier, the king of the analog world and a fundamental building block in power management integrated circuits. The first operational amplifier, the uA702, was designed at Fairchild by Robert Widlar. He subsequently designed the uA709 (Figure 1–6). This opamp

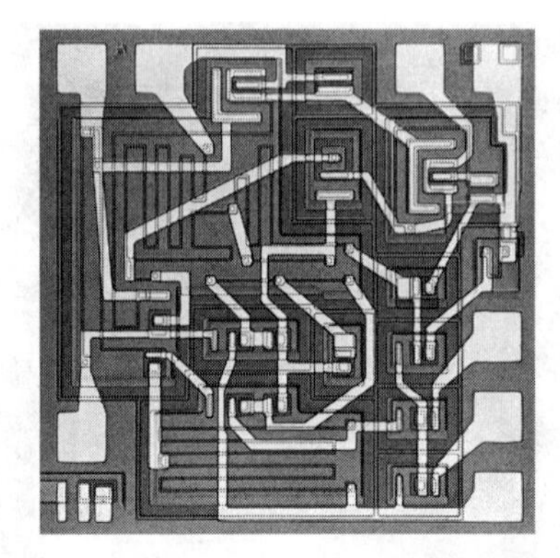

Figure 1–6 uA709 is the first operational amplifier of wide use in the industry (1965).

has 14 bipolar transistors and 15 resistors integrated in a 0.6 in^2 die and at its inception (1965) sold for one hundred dollars. Accounting for inflation, one hundred dollars in 1965 corresponds to six hundred dollars in actual value and that makes the uA709 more glamorous in its own time than a modern Pentium IV. As a corollary and as proof of the longevity of analog products, you can still buy a uA709 today but the price is a small fraction of a dollar.

Figure 1–7 shows the first planar bipolar power transistor incorporating a thin-film emitter resistor process. It was produced at Fairchild. The two identical undulated shapes show the two emitters, the square shape surrounding them is the base, and the dark surrounding area is the collector. The stubs are gold wires bonded to the two emitters and to the base and connecting to the external contact pins. Bipolar power transistors have been the workhorse of the power semiconductor industry for a long time but recently have been almost entirely supplanted by their CMOS counterparts, which are more efficient especially in static operation.

Figure 1–8 shows a modern PowerTrench™ discrete power transistor by Fairchild. This device integrates ten million cells, or elementary *MOSFET* transistors, in parallel in a small space yielding very low "on state" resistance. *Discrete power MOSFETs* like this one, in conjunction with switching regulator controllers, enable the delivery of huge amounts of power with unprecedented levels of efficiency.

Finally a true power management integrated circuit, the RC5051, is shown in Figure 1–9. In 1988 Fairchild's RC5051 pioneered the use of switching regulators in PCs, powering a Pentium II CPU by delivering 17.1 A in performance mode—a hefty amount of current at the time. This IC incorporates on a single die the equivalent of many operational amplifiers plus two driver stages that are hefty enough to drive two external MOSFET

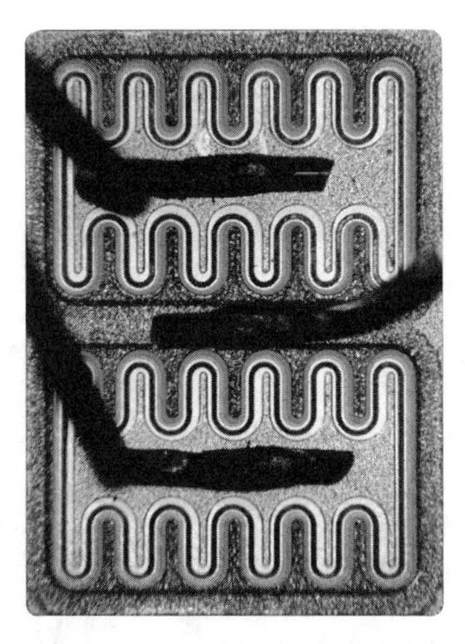

Figure 1–7 First planar power transistor incorporating a thin-film emitter resistor process (1964).

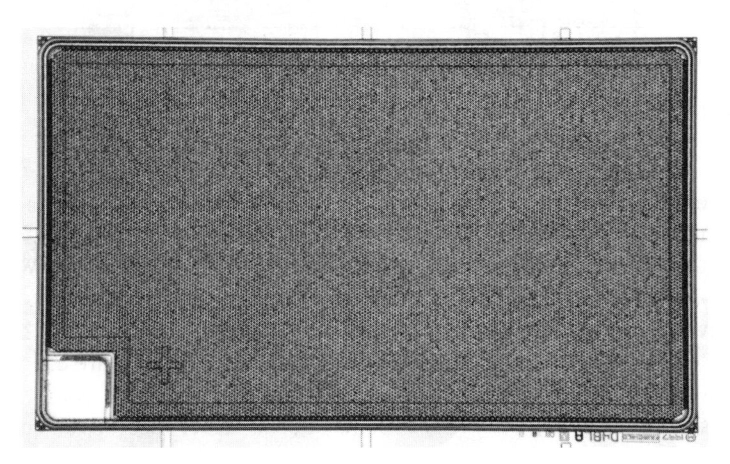

Figure 1–8 A 10 million cells per square inch PowerTrench™ MOSFET technology (1997).

transistors such as the one in Figure 1–8 in synchronous rectification mode of operation.

As these images have shown, in the last fifty years our semiconductor processes have gained tremendous efficiencies, becoming 1.5 million times denser. It is expected that in 2011 semiconductor technology will be able to resolve 12 nanometers, which is roughly the diameter of a DNA strand and only 100 times the diameter of a hydrogen atom. After that it is

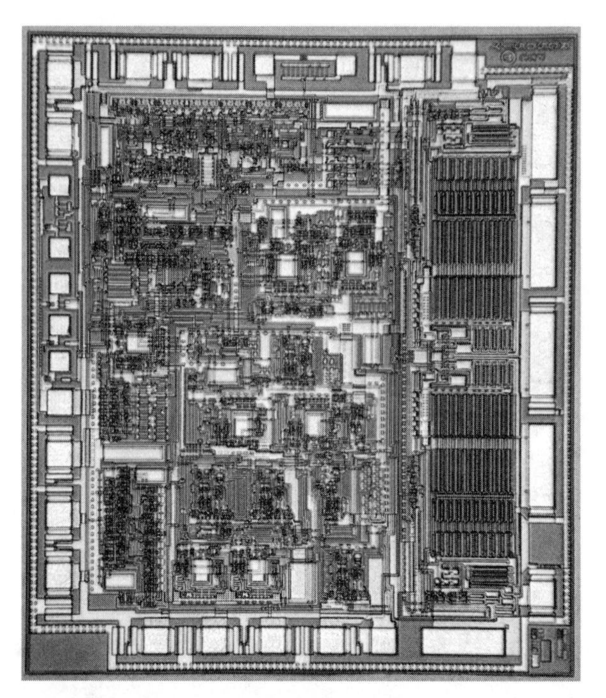

Figure 1–9 RC5051 pioneers the use of switching regulators in PCs, powering a Pentium II CPU in 1998.

widely believed that silicon will run out of steam and new materials will be necessary. A modern CPU, in the class of a Pentium IV, cranks out 3000 Million Instructions Per Second (MIPS) and consumes 100 W, an amount of power already difficult to handle even with the aid of fans and active cooling devices. On the other hand the human brain consumes 20 W and cranks out 100 million MIPS. That makes the brain more efficient than silicon by 165,000 MIPS per Watt. Perhaps this is a clue as to where we should search for a material to come after whatever succeeds silicon.

Chapter 2

Power Management Technologies

2.1 Introduction

Power management is generally accomplished by a combination of small signal transistors acting as the brain, power transistors acting as solid state switches that control the power flow from the source to the load, and passive components like resistors, capacitors, and inductors, acting as sensing and energy storing elements. A semiconductor integrated circuit can incorporate on a single die a large number of small signal transistors as well as limited values of passive components (resistors, capacitors, and lately even inductors) and power transistors carrying a few Amperes. For larger levels of power, external discrete transistors built with specialized processes are utilized in conjunction with the IC. In this chapter we will see how ICs and discrete transistors require very different methods of fabrication. We will first discuss the integrated circuits typically incorporating the desired power management control algorithm and the process and package technologies utilized for their construction. Subsequently we will discuss the discrete power transistors, called to duty when the power levels cannot be handled monolithically by the integrated circuit, and the process and package technologies utilized for their construction.

2.2 Integrated Circuits Power Technology: Processing and Packaging

The power of the integrated circuit process lies in its ability to etch a high number of electrical components on a small silicon die and interconnect them to perform the desired actuation function. The main electrical components on board an IC are

Bipolar NPN transistors

Bipolar PNP transistors

Diodes

CMOS transistors

DMOS transistors

Resistors

Capacitors

The electrical properties of some of these components are discussed in Chapter 3. In this section we will illustrate the physical structure of these components as they are generated on the surface of a silicon die.

Diodes and Bipolar Transistors

Semiconductor crystals derive their amplification properties from bringing together materials of opposite electrical properties, namely N-type and P-type materials.

N-type materials are materials that, even if neutrally charged, have an excess of free electrons, or negative charges. In other words these electrons are very weakly tied to their nucleus and hence easy to move around in the form of an electric current.

In homogeneous materials atoms bond together by sharing their outer shell electrons: a kind of holding hands by sharing one electron with a neighbor atom. In the case of silicon (column IV of the Periodic Table of Elements) each atom shares its four outer shell electrons with four neighbor atoms. If we now introduce inside silicon one atom from column V of the Periodic Table of Elements, namely one having five outer shell electrons, this atom will bond with four neighboring silicon atoms but will have an excess of one electron un-bonded or free to move around. As this electron moves around, the foreign atom is left with a positively charged nucleus. Notice that the entire compound is still electrically balanced but the only difference now is that we have an electron that is much easier to move around. Column V elements like phosphorus (P), arsenic (As), and antimony (Sb) are called donor materials because they produce an excess of electrons inside column IV materials like silicon. Similarly if we introduce inside silicon one atom from column III of the Periodic Table of Elements, namely one having three outer shell electrons, this atom will bond with three neighboring silicon atoms but the fourth silicon neighbor will not get an electron. A positively charged 'hole' is created, namely an incomplete bond between two atoms made of one single electron instead of two. Eventually due to thermal agitation this hole will get filled by an electron. This means that the foreign atom has now an extra electron and is left negatively charged, while somewhere out there a silicon atom is missing an electron and is hence positively charged. In other words, the hole is moving freely around the silicon lattice. Column III elements like boron (B), gallium (Ga), indium (In), and aluminum (Al) are called acceptor materials because they readily accept an electron from a nearby silicon-silicon bond creating an excess of holes inside silicon.

A material *doped* with donors, meaning that it has an excess of negatively charged free electrons, is referred to as an *N-type material*, while one doped with acceptors, meaning that it has an excess of positively charged holes, is referred to as a *P-type material*. An N- and a P-type material brought together will form a junction. The simplest semiconductor element, the rectifying diode in Figure 2–1, is formed by such a junction between a P- and an N-type material. A positive potential applied to the P side will push the excess of holes toward the junction where they will recombine with excess electrons in the N-type material, sustaining a current flow in this "forward" direction. Most of the current in the P region is made by the movement of holes, while most of the current in the N region is created by moving electrons. This device is called bipolar, referring to a conduction mechanism based both on electrons and holes. If a negative potential is applied to the P-material, and a positive one is applied to the N-material, the charges are pushed away from the junction, resulting in zero conduction. The property of passing current only in one direction is the rectifying effect of a diode.

Figure 2–1 illustrates the diode conduction mode, in which a forward bias voltage V pushes a current I through the diode. Notice that the physical current in the wire is made of electrons (represented by negative circles) moving in the opposite direction of the conventionally positive current. Inside the diode the current is made of electrons in the N-material and holes (positive circles) inside the P-materials. The P-to-metal contact (anode) provides a mechanism for exchanging holes in the semiconductor for electrons which can travel in the external circuit.

A diode is a two terminal device, which, in conduction mode, yields from the cathode (N side) the same amount of current injected from the

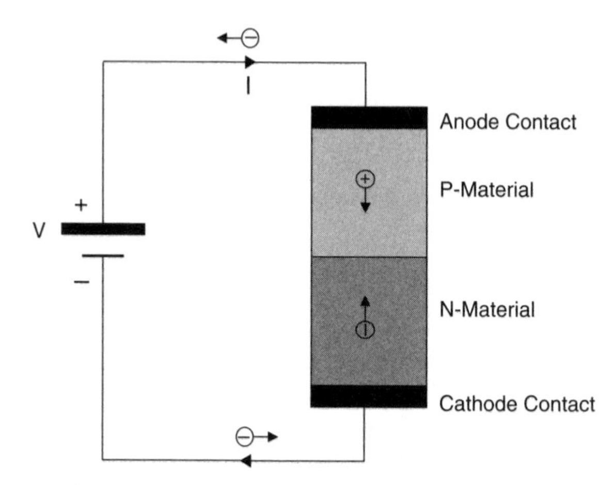

Figure 2–1 Diode in conduction mode.

anode (P side). A diode is a passive device lacking the ability to amplify, or modulate such flow of current.

Amplification requires a third terminal with the ability to modulate the current flow.

If we add a P to the N side of our PN junction, we create a PNP structure. The PNP structure is a three terminal device with two junctions, the PN junction, or emitter-base junction, normally positively biased, and the NP junction, or base collector junction, normally negatively biased. If the intermediate N layer (base) is thin enough and the base-emitter junction is forward biased, a positive charge injected from the emitter can reach the collector without significant recombination in the base. While the charge moves from one side (emitter) to the other (collector), its amount is determined by the magnitude of the positive potential V_{BE} applied to the forward-biased base-emitter junction (see Figure 2–2). A small voltage variation in this junction produces a large current variation in the collector. On the other hand, the thin base assures little charge recombination in the base, namely a small current flow in the base, need be supplied in order to sustain a large current flow from the emitter to the collector. Typically a 1 μA current in the base can sustain a 100 μA current flow from emitter to collector, resulting in a gain of 100 from input (base) to output. This is the amplifying effect in a PNP transistor. A PNP transistor moves charges from a positive potential to a grounded (zero potential) load; this is referred to as current sourcing. If the load is at a positive potential then the dual of the PNP, the NPN transistor (see Figure 2–3), will be able to move charges from the positively biased load to ground. As for the diode, the PNP transistor (or its dual, the NPN transistor) is a bipolar device because its conduction mechanism is based on both electrons and holes. For

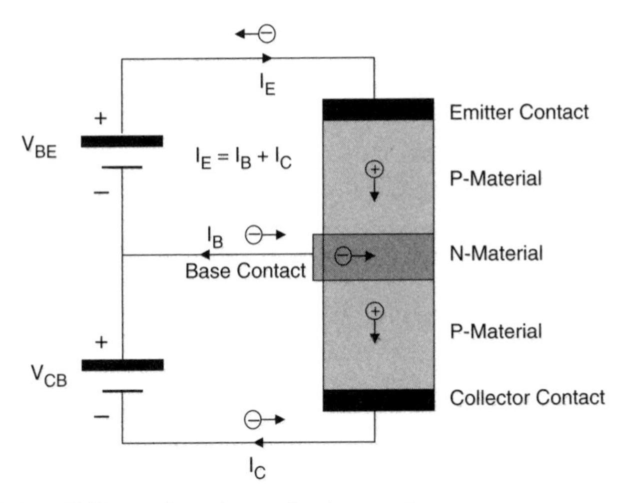

Figure 2–2 PNP transistor in conduction mode.

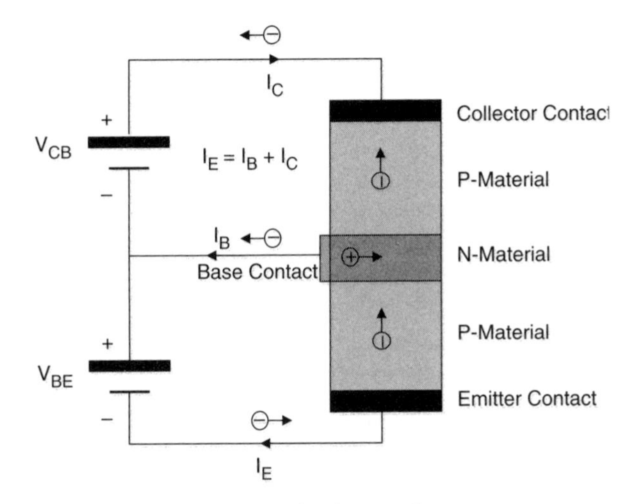

Figure 2–3 NPN transistor in conduction mode.

example in the PNP transistor, the bulk of the current flow is made of holes, the majority carriers in emitter and collector but minority carriers in the base. In the base a small percent (0.5%) of holes recombines with electrons, which are continuously supplied as base current. The base current also sustains a small current of electrons that flows from the base to the emitter (another 0.5% of the collector current). As explained earlier, a total base current—typically 1% of the collector current—is necessary to sustain the transistor conduction state.

Figure 2–3 shows the NPN transistor in principle. In reality semiconductor integrated circuits are built in a planar fashion, meaning all the components and their terminal mast will be etched via a lithographic process on the surface of a wafer. Figure 2–4 shows a realistic construction of an NPN transistor in a modern *Bipolar-CMOS-DMOS* (BCD) integrated circuit process. Starting with a substrate P+ material offering mechanical support, a layer of lightly doped silicon material is grown (P–EPI for epitaxial or superficial growth). This layer is then doped with donor and acceptor materials, according to the rules explained previously, to produce a device that is both electrically viable and topologically accessible. The emitter (Emitter) and base (P_{WELL}) diffusions are clearly marked in Figure 2–4. The current flow descends vertically from the emitter through the base into the collector N-material. The collector material is a composite of lightly doped N-material (HVN_{WELL}) that determines the voltage breakdown characteristics of the device, followed by a heavily doped N-material (N_{BL} for N buried layer) which offers a low resistance horizontal path to the collector current. Finally, the stack of N-materials SINK (for sinker), N_{WELL}, and N+ complete the path in the vertical direction, allowing the current to resurface at the collector (Collector) contact. Finally a protection layer (top layer) is deposed on top of the entire die to prevent contamination.

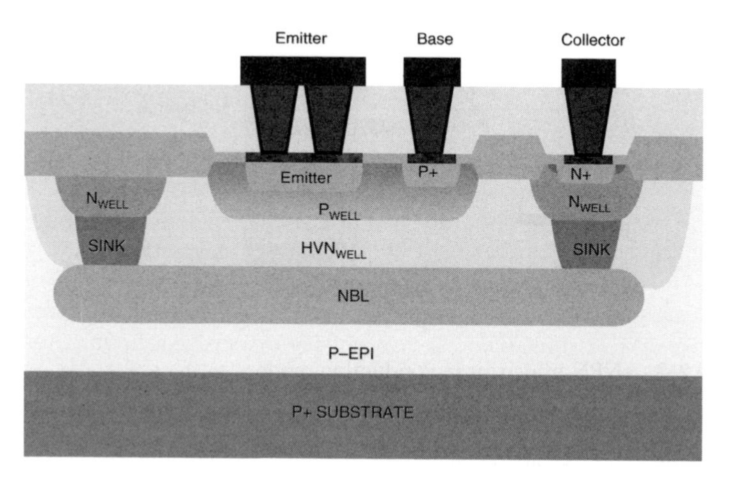

Figure 2–4 NPN transistor construction.

The PNP transistor is illustrated in Figure 2–5. The bulk of the current flow is horizontal from emitter to collector and the buried sequence of N-matierals here is utilized to provide a path for the base current to resurface back to base contact.

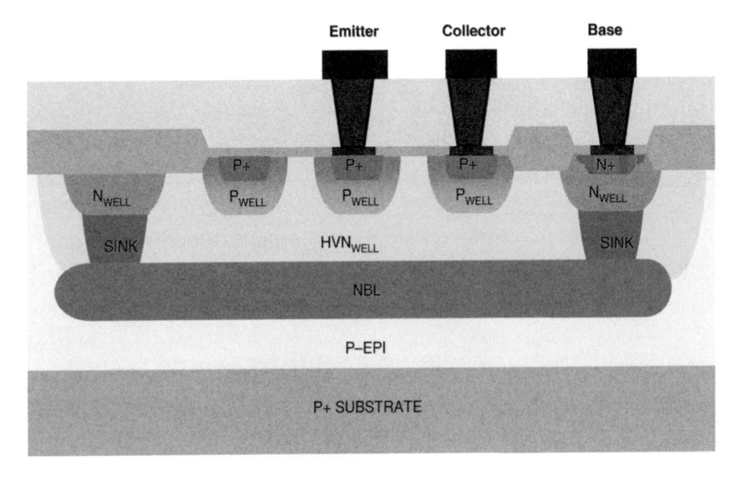

Figure 2–5 PNP transistor construction.

Metal-Oxide-Semiconductor (MOS) Transistors

N-MOS and P-MOS transistors are analogous respectively to NPN and PNP transistors but their conduction mechanism is based completely on one type of carrier: holes for the PMOS and electrons for the NMOS. For the PMOS (Figure 2–6) two P-type regions are separated by one N-type region. Such an N-type region is exposed to a "gate" or plate that can be polarized negatively, attracting the positive charges inside the N-material to the point of forming a conduction channel (enhancement of the channel). Hence the material is enhanced into a P-type for the duration of the applied gate voltage polarization and current can flow between what has become a simple sequence of three P-type materials from the source to the region under the gate to the drain. The gate plate—originally made of metal in older processes, now typically made of polysilicon—is isolated from the semiconductor by a thin layer of oxide material, which explains the name MOS (Meta-Oxide-Semiconductor) transistor. In this structure the gate voltage plays the role of the base current in the bipolar transistor, namely sustaining the transistor current flow. However since enhancement in the PMOS is produced electro-statically, meaning in absence of charge movement, this device has a perfect transfer of current from source (the dual of the emitter in the bipolar transistor) to drain (the dual of the collector in the bipolar transistor). The lack of base current, a net loss in the bipolar transistor, makes these devices valuable in many competing applications.

Figure 2–7 shows the construction of an N-channel MOS transistor with the two N diffusions (N) separated by a P-material (P_{WELL}) and the

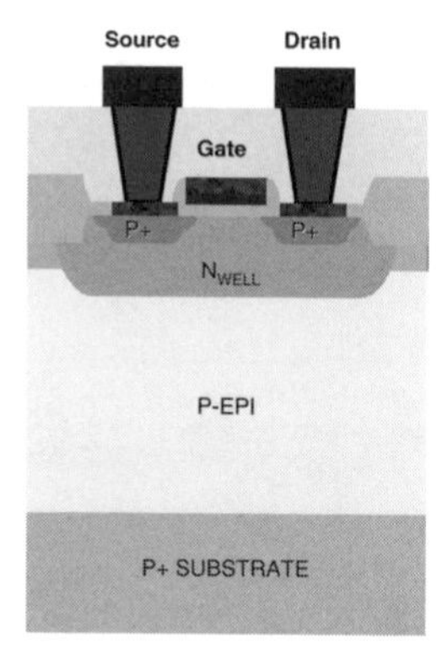

Figure 2–6 PMOS transistor.

gate (Gate) separated from the P_{WELL} by a thin oxide layer assuring the electrostatic action of depletion. Such a P-type region is exposed to a gate, or plate that can be polarized positively, attracting the negative charges inside the P-material to the point of forming a conduction channel (enhancement of the channel). Hence the material is enhanced into an N-type for the duration of the applied gate voltage polarization and current can flow between what has become a simple sequence of three N-type materials from the source to the region under the gate to the drain.

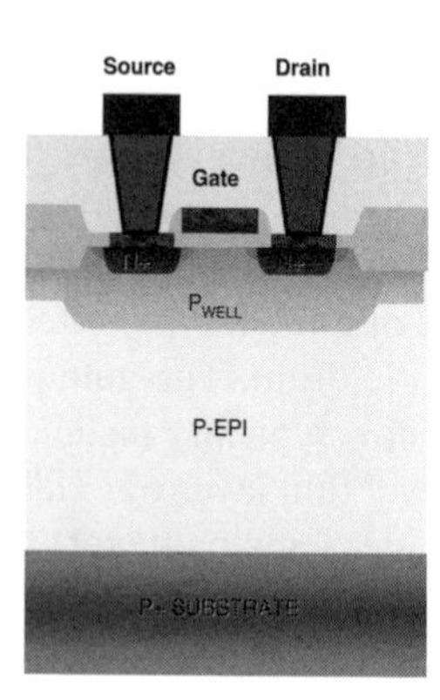

Figure 2–7 NMOS transistor.

DMOS Transistors

In MOS transistors conductivity rises as the gate length, or separation of source and drain, decreases. Making small openings in the oxide to depose a tiny gate requires expensive machinery and sophisticated lithographic processes. In some instances the problem can be circumvented by producing a small effective gate length by subsequent diffusion of two opposite materials in a wide opening. In the DMOS in Figure 2–8, (The D stands for double-diffused) a small gate length is obtained by following a deep P diffusion (Source P_{DIFF}) with a shallower N diffusion (N). The two materials penetrate with different lengths under the gate area, with the denser P+ traveling farther than the lighter N-matieral. Proper dosage and conditions will create an effective gate (the residual P_{DIFF} material not eaten up by the N diffusion) that is much smaller than the drawn gate length. In this structure the current flow proceeds from source, under the gate, and horizontally to the Drain contact. Sinker (SINK) and Buried Layer diffusions here have a protection function (anti-latch action).

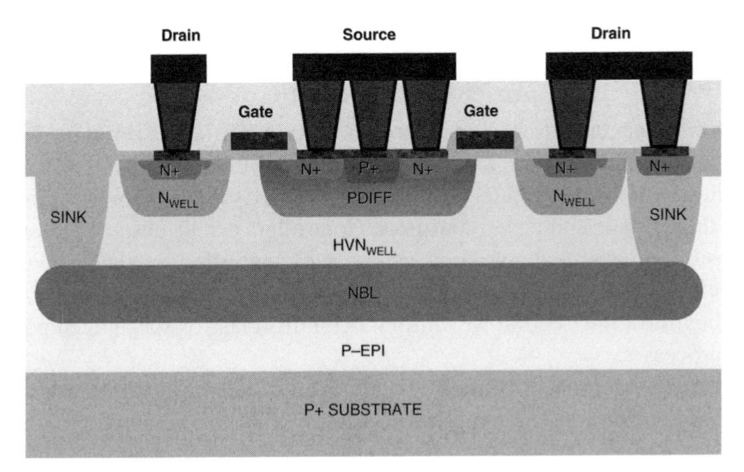

Figure 2–8 DMOS N-channel transistor.

CMOS Transistors

When we connect the source of a PMOS transistor to a positive supply, the source of an NMOS transistor to ground, and we short together the respective gates, we obtain the CMOS, or complementary MOS transistor, an inverting element that is at the foundation of logic design.

Passive Components

In addition to active components (components that can amplify a signal) like transistors, integrated circuits processes also provide a slew of passive components like resistors, capacitors, and lately even inductors. Figure 2–9 is an example of a resistor, a two-terminal device simply obtained by a long and narrow deposition of an N diffusion material.

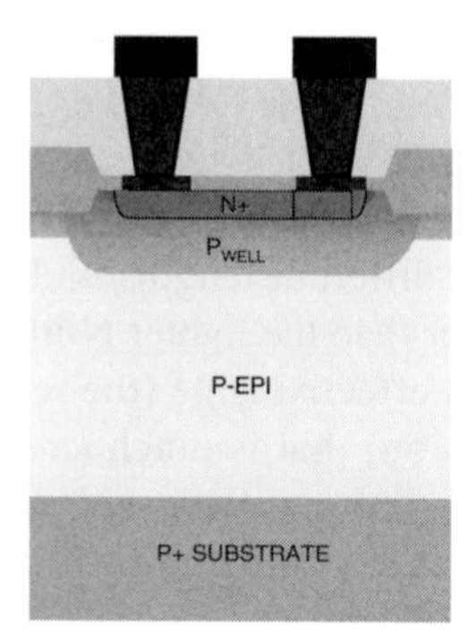

Figure 2–9 N+ resistor.

A Monolithic Process Example

Power management integrated circuits come in different varieties. If we narrow them down to voltage regulators, we still need to distinguish controllers from fully featured voltage regulator ICs that incorporate on die the driver stage and power transistor. Controllers can be designed in every possible process technology, however true monolithic regulators require specialized processes capable of integrating signal and power transistors on board. In this section we will focus on this class of specialized power IC processes.

One such process is the BCD process, capable of integrating bipolar transistors for precision applications, with CMOS for dense signal processing and DMOS for power handling.

Figure 2–10 shows a cross section of a generic low voltage BCD process. It illustrates the power of a monolithic planar process that is able to offer an impressive variety of devices all on the same surface of a die, all obtained at the same time, and with a single construction process. This process is suitable for many applications including motherboard DC-DC voltage regulator applications.

Packaging

Silicon dies must be enclosed in packages for protection and handling. IC packaging is a very important subject and can be more challenging than the IC design itself. For example, a package that lets moisture in will soon render the chip inside useless. In modern portable applications like cell-phone handsets the challenge is often to have a package that is no bigger than the die itself, hence the emerging popularity of chip-scale-package (CSP) like the one illustrated in the upper right corner of Figure 2–11. In high power applications heat dissipation is a crucial issue—the package is

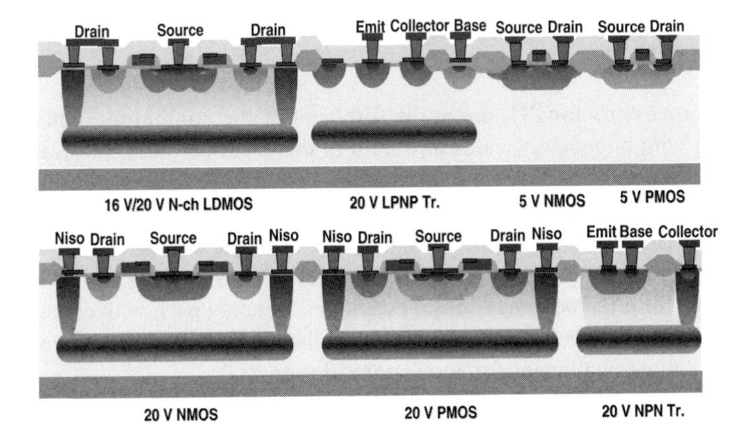

Figure 2–10 Cross section of low voltage BCD process.

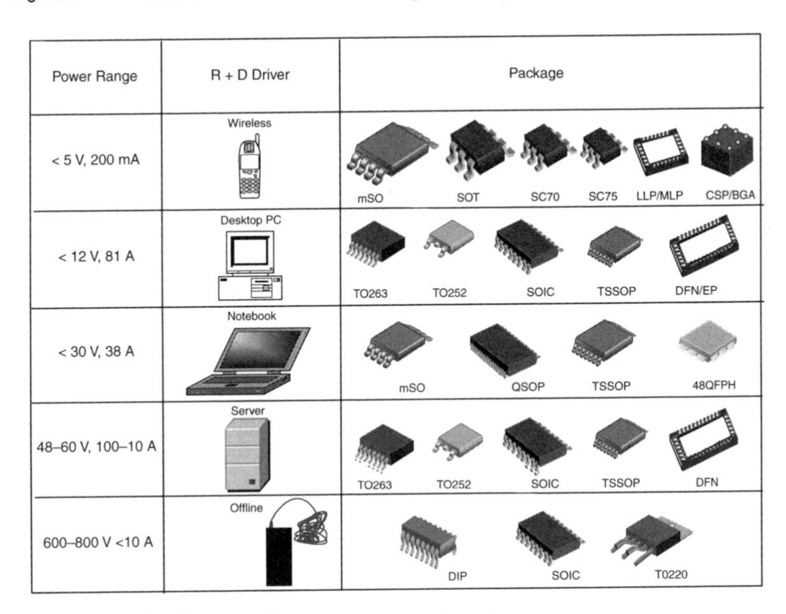

Figure 2–11 Package options versus target systems.

often the narrow bottleneck through which heat has to escape from the die and hence its thermal resistance has to be minimal like in the TO220 package illustrated in the lower right corner of Figure 2–11. In between we find a slew of package shapes and forms that fit the intended application, delivering proper power, voltage, current, or size characteristics.

2.3 Discrete Power Technology: Processing and Packaging

Microprocessors for PCs are at the forefront of the computing industry, leading with huge nano-scale chips built in multi billion dollar fabrication plants. So far, the success of the semiconductor industry has been assured by Moore's law—a concept that underscores the fast-paced dynamic of the industry. However, new chips in smaller footprints are upping the trend for increasing power densities to amazing levels. At every new technology juncture, the CPU becomes denser and hotter. Keeping pace with changing densities, compounded with the need for disposing the resulting heat, is creating more challenges for applications designers.

Providing power from the AC line is also becoming an issue for designers. The number and growth rate of electronic appliances is driving a huge demand for power, prompting concerns for power distribution and energy conservation and spurring a slew of protocols and initiatives aimed at minimizing the waste of power. These requirements are pushing technology advancements beyond the traditional cost-oriented model of minimizing the appliance's Bill of Materials (**BOM**) to look for new solutions.

At the core of all power management solutions, from the wall to the board, are power transistors. The evolution of discrete semiconductors is essential for supporting Moore's law, and thereby maintaining the industry's healthy growth. Not surprisingly, designing and mass producing cost-effective discrete transistors capable of efficiently handling power requires increasingly sophisticated semiconductor processes and packaging.

From Wall to Board

Electric power is transferred to the CPU in two crucial steps: from the high voltage AC line to an intermediate DC voltage and from there to the low voltage regulator (VR), which is needed to power the CPU. The high voltage "planar technology" transistor underlying this AC-DC conversion must sustain voltages in the 600–700 V range and few Amperes of current; meanwhile, the low voltage "trench" transistor powering the CPU has to handle a few volts with hundreds of Amperes. Both conversions have to be accomplished with the lowest possible power losses. It stands to reason, then, that such diverse performance requirements are satisfied by two quite different discrete MOSFET transistor technologies, "planar" for high voltage and "trench" for low voltage.

Power MOSFET Technology Basics

Conduction Losses

Power MOSFET technologies comprise a number of key elements that impact on-state, or conduction losses. These elements include a substrate to provide mechanical stability, a region used for blocking the drain potential in the off-state, and the conduction channel that provides gate control.

The greatest penalty of on-resistance for high voltage power MOSFETs is found in the epitaxial region. For conventional high voltage devices, the construction requires thick, highly resistive epitaxial material to support the 600 V blocking requirement. For devices below approximately 200 V, this region becomes less significant. Advanced high voltage devices utilize a technique called "charge balance" which is used to reduce conduction loss in the epitaxial region.

With power MOSFETs, the conduction channel resistance is determined by the channel length, the distance through which carriers must flow, and channel width, is the amount of transistor channel that is constructed in parallel. Lower resistance is achieved by increasing the channel width for a given silicon area. Due to the low conduction losses in the epitaxial region of low voltage devices, the channel density is critical for reducing conduction losses.

Switching Losses

The channel construction technique has a significant impact on the switching performance of a power MOSFET. The amount of polysilicon gate area that overlaps with the epitaxial region, the N+ source diffusions, and the source metal are key design parameters. This area, in conjunction with the thickness of the dielectric materials between these regions, sets parasitic capacitances that must be charged and discharged during each switching event.

Planar Power MOSFET Technology

The best choice for channel construction for high voltage power MOSFET devices is planar construction, as shown in Figure 2–12. In this type of construction, the polysilicon and the channel are displaced on the horizontal silicon surface of a planar device. Due to the conduction losses in the epitaxial region of high voltage devices, there would be a minimal benefit of a high channel density construction. In addition, low capacitances of the planar channel provide low switching losses. Planar construction, when combined with the charge balance epitaxial structure, provides optimized performance of a high voltage power MOSFET.

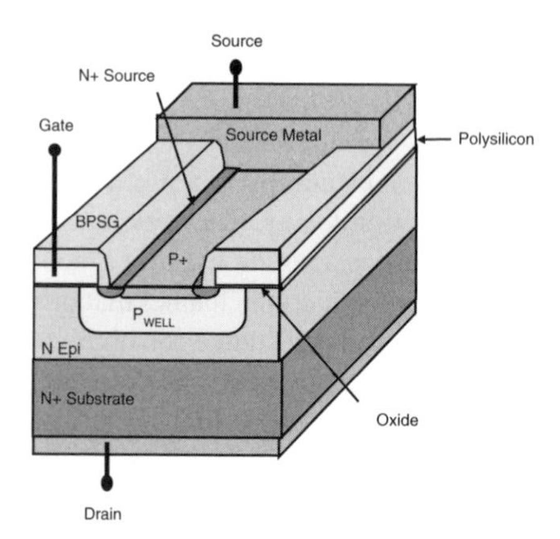

Figure 2–12 Planar DMOS transistor cross-section.

An example of this type of planar MOSFET technology is the FCP11N60 SuperFET™ from Fairchild Semiconductor. This product typifies a new generation of high voltage MOSFET that offers very low on-resistance and low gate charge performance. It does this using proprietary technology utilizing the advanced charge balance technique. Such advanced technology is tailored to minimize conduction loss, provide superior switching performance, and withstand extreme dv/dt rate and higher avalanche energy. Consequently, this kind of device is very well suited for various AC-DC power conversion designs using switching mode operation when system miniaturization and higher efficiency is needed. The future holds ongoing improvements in this type of technology for better conduction and switching loss performance.

Power Trench MOSFET Technology

For a low voltage power MOSFET device, channel conduction is best constructed utilizing a trench channel structure, which is illustrated in Figure 2–13. This construction technique places the polysilicon and channel vertically in the silicon epitaxial region. As a result, the channel density is maximized, providing a significant conduction reduction when compared to a planar device. In addition, low conduction losses per unit area allow the chip size to be reduced, improving switching losses. Also, capacitances are reduced through a careful tailoring of the capacitor dielectric thicknesses. This combination of low resistance and low switching losses of power trench MOSFETs provides the optimal solution for powering the CPU.

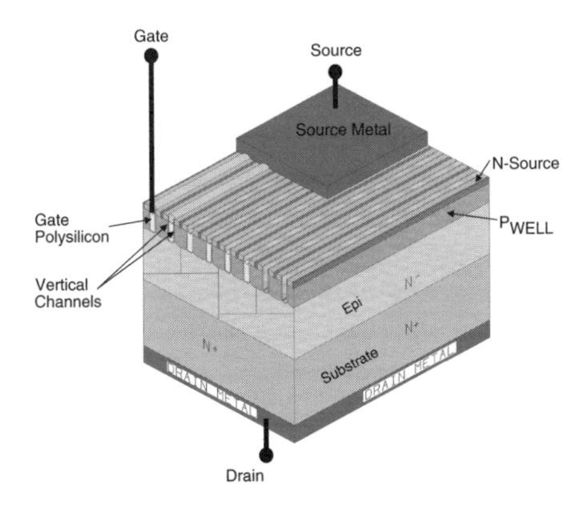

Figure 2–13 Trench MOSFET (channel structure).

An example of this technique is the delivery of 74 A continuous (93 A peak) without heatsink to Prescott class CPUs using a three-phase buck converter that utilizes planar DMOS discrete transistors in the power stage. In this example the *buck converter* utilizes devices such as Fairchild's FDD6296 high side MOSFET DPAK (one per phase) and a FDD8896 low side MOSFET DPAK (two per phase), in combination with a FAN5019 PWM controller (one) and a FAN5009 driver (one per phase).

Ongoing changes to these technologies will further enhance both the conduction and switching performance of the existing trench MOSFETs. As a result, the improvements will deliver increasingly better performance.

Package Technologies

Today, much work is being done to develop low parasitic (i.e., ohmic resistance, wire inductance) packages.

Figure 2–14 shows a power Ball Grid Array (BGA) package capable of delivering unprecedented levels of power thanks to the substitution of the wire bonds solder balls. A surrounding drain frame structure, which dramatically reduces the package resistance and inductance parasitics, is another important benefit of BGA packaging.

For example, in a server application, one BGA-packaged FDZ7064S device on the high side, and two FDZ5047N on the low side, can deliver 40 A/phase with a power density of 50 W/in^2. Hence, a four-phase implementation can easily deliver 200 A to the CPU.

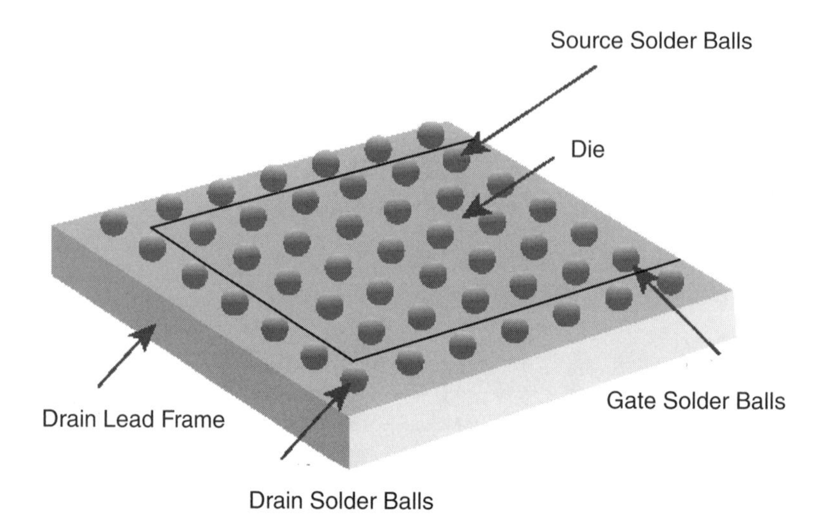

Figure 2–14 Illustration of a power BGA package.

2.4 Ongoing Trends

As wall-to-board power challenges will continue to escalate, MOSFET transistor processing and packaging solutions will continue evolving. A system approach to power distribution will assure the best mix of processes and package technologies for the powering of modern appliances. At the motherboard level (DC-DC conversion), the need to efficiently dispose of the heat in increasingly smaller spaces will continue to drive the need for trench and package technology that offers lower and lower parasitics. At the *silver box* level (AC-DC conversion), the need to draw efficient power from the AC line will drive future offline architectures toward the use of more planar discrete transistors of increased sophistication in order to support existing and new features like *Power Factor Correction* (PFC) with fewer overall power losses.

Chapter 3

Circuits

Modern circuit design is a "mixed signal" endeavor thanks to the availability of sophisticated process technologies that make available bipolar and CMOS, power and signal, and passive and active components on the same die. It is then up to the circuit designer's creativity and inclination to assemble these components into the analog and/or logic building blocks necessary to develop the intended system on a chip. While the digitalization of traditional analog blocks continues, new analog blocks are invented all the time. Examples of new analog functions are charge-pump voltage regulators, MOSFETs, and LED drivers. A contemporary example of digital technology cutting deep into analog core functions is the digitalization of the frequency compensation in the control loop of switching regulators. In this case while the feat has been accomplished—and it can indeed be exhilarating to move *poles* and *zeros* (see glossary) around with a mouse click—it is not clear that the feature of digital frequency compensation, and its associated cost in silicon, is always justified. So while digital technology—circuits and processes—continues to gain ground, analog keeps reinventing itself and rebuilding around a central analog core of functions that is tough to crack. We don't expect to see the digitalization of an analog circuit like the band-gap voltage reference—namely a digital circuit taking the place of the current analog one—happening any time soon. In this section we will discuss a number of analog, digital, bipolar, and CMOS circuits. It would be hopeless to try to report systematically all the building blocks for mixed-signal circuit design, or even just the main ones. Instead we will adopt the technique of "build as you go." With this in mind we will start from the single transistor and build up to some complex functions like linear and switching regulators that are at the core of power conversion and management.

Part I Analog Circuits

In this section we will discuss some fundamental analog building blocks for power management. We will review quickly the main properties of the elementary components, the transistors, so that we can use them to build elementary circuits like current mirrors and buffer stages. We will then use these elements and circuits to generate the analog building blocks like operational amplifiers and voltage references. Finally we will combine these analog building blocks into functional circuits. Given the subject of this book, not surprisingly, the functions we are interested in are voltage regulators, which are at the center of power distribution and management. The process of assembling elementary electrical components into a fully functional electronic product—namely the system design of an electronic product—can all be implemented on a single die, leading to a monolithic single integrated circuit, or can be spread over many chips, for example a discrete power transistor chip and a controller IC assembled in a module. Modern circuit design, both at the discrete and IC levels, relies on a mix of bipolar and CMOS elements. Power management integrated circuits can now be built on mixed bipolar CMOS and DMOS processes if the level of performance and complexity justifies it. System design will mix and match such ICs with external discrete components that will again range from bipolar to CMOS and DMOS with the selection generally being driven by cost first and performance second.

In the rest of this section we often draw bipolar circuits, but every circuit discussed has its counterpart in CMOS. By substituting the NPN with its CMOS dual, the N-channel MOS transistor, and the PNP with its dual, the P-channel MOS, all the functions discussed in bipolar can be replicated in CMOS.

3.1 Transistors

NPN

The NPN transistor (Figure 3–1) is the king of the traditional bipolar analog integrated circuits world. In fact in the most basic and most cost effective analog IC processes, the chip designer has at its disposal just that; a good NPN transistor. The rest, PNPs, resistors and capacitors are just by-products, a notch better than parasites. For intuitive, back-of-the-envelope type analysis, it is sufficient to model the transistor mostly in DC, keeping in mind that the bandwidth of such an element is finite. When complexity, like small-signal AC behavior, is added to the model, computing simula-

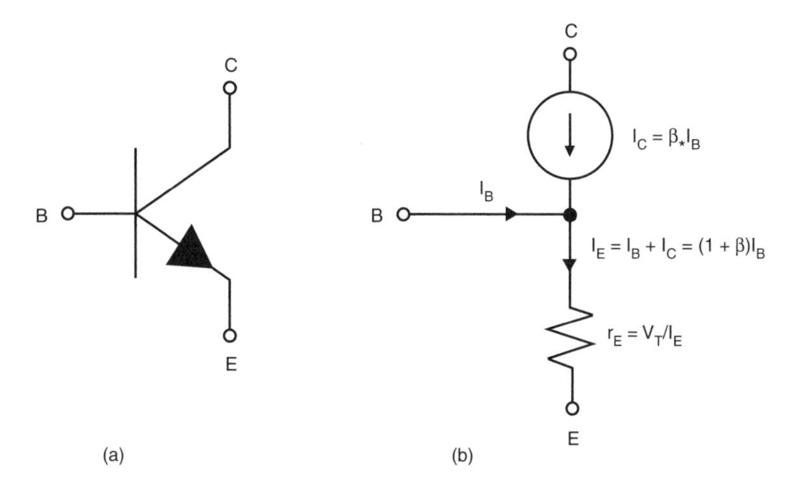

Figure 3–1 NPN Transistor (a) symbol and (b) model.

tions should be used since the math quickly becomes hopeless. In Figure 3–1 the NPN transistor is shown with its symbol (a) and its DC model (b). In this component, the current flow enters the collector and base and exits through the emitter. Simply stated, the transistor conducts a collector current I_C which is a copy of the base current I_B amplified by a factor of beta (β). It follows that the emitter current I_E is one plus beta times the base current. A typical value for the amplification factor is 100. NPNs have excellent dynamic performance, or bandwidth, measured by their cutoff frequency (f_T); easily above 1 GHz.

PNP

The PNP transistor (Figure 3–2) is complementary to the NPN, with the current flow entering the emitter and exiting the collector and base, the opposite of what happens in the NPN. Simplicity dictates that PNPs are a by-product of the NPN construction; hence they often have less beta current gain and are slower than NPNs. A typical value for their amplification factor is 50 and their cutoff frequency (f_T), is generally above 1 MHz.

Trans-Conductance

In addition to current gain, and bandwidth f_T, another important element of the transistor model is its trans-conductance gain G_M, namely the amount of current in the emitter as a result of a voltage input in the base-emitter junction. The small signal transistor model in Figures 3–1 and 3–2 shows

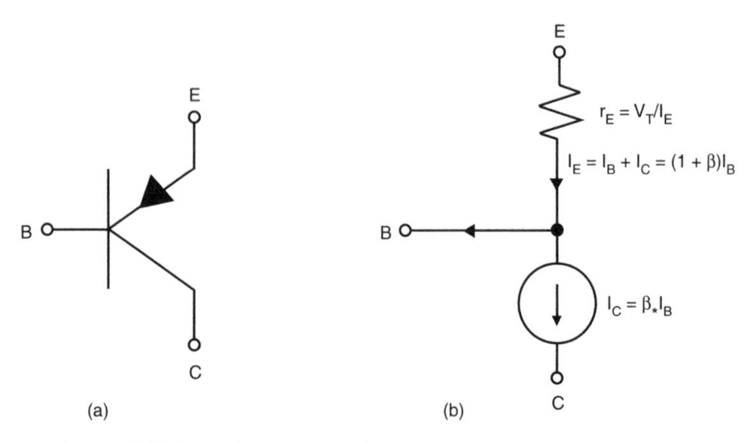

Figure 3–2 PNP Transistor (a) symbol and (b) model.

that the base-emitter voltage of a transistor—the infamous 0.7 V roughly constant voltage—is modulated by the resistance r_E where

$$r_E = V_T/I_E \qquad \text{Eq. 3–1}$$

$$V_T = KT/q = 26 \text{ mV at ambient temperature of } 25°\text{C} \qquad \text{Eq. 3–2}$$

where K is the Boltzman constant, T is the temperature in degrees Kelvin, and q is equal to the electron charge in Coulombs.

It follows then that a small signal voltage ΔV applied at the transistor base-emitter junction will act solely on the resistor r_E and develop a corresponding current dI.

$$dI = \Delta V/r_E \rightarrow dI/\Delta V = G_M = 1/r_E \qquad \text{Eq. 3–3}$$

Therefore, the trans-conductance gain G_M is the exact inverse of r_E. Since we deal more easily with resistors than trans-conductors, we will continue to represent the trans-conductance gain with the resistor r_E explicitly drawn in the model or simply implied in the transistor symbol.

Transistor as Transfer-Resistor

A transistor with 1 mA of emitter current will exhibit an emitter resistance of 26 mV/1 mA or 26 Ω according to Eq. 3–1. This, as any resistance in an emitter, produces an amplified resistance as seen from the base. In fact staying with this numeric example, an emitter current of 1 mA, in addition to a 26 mV drop in the emitter-base voltage, will produce a base current variation of approximately 10 μA (1 mA divided by an amplification of β + 1 or 101). From the base vantage point a 26 mV fluctuation in response to a base current fluctuation of 10 μA is interpreted as a resistance of 26 mV/10 μA = 2.6 kΩ. Naturally such transfer of resistance from low in the emitter to high in the base is the property that gives the name transistor or, transfer resistor to the electrical component.

Transistor Equations

The voltage to current relation in a bipolar transistor follows a logarithmic law given by

$$V_{BE} = V_T \times \ln(I/I_O) \qquad \text{Eq. 3–4}$$

where V_T is the thermal voltage and I_O is a characteristic current that depends on the specific process. This has some pretty interesting implications; for example, if the transistor from Eq. 3–4 carries a current x times higher, we can write

$$V_{BE}' = V_T \times \ln(x \times I/I_O) \qquad \text{Eq. 3–5}$$

The increase in voltage from the factor of x increase in current will be

$$\Delta V_{BE} = V_{BE}' - V_{BE} = V_T \times \ln(x) = (kT/q)\ln(x) \qquad \text{Eq. 3–6}$$

Given that $V_T = 26$ mV at ambient temperature, we see easily that doubling the current in a transistor ($x = 2$) will raise its V_{BE} by 18 mV (say from 700 mV to 718 mV) and a 10x increase in current will raise the V_{BE} by 60 mV. In gross approximation we can consider the V_{BE} of a transistor constant around 0.7 V, but to be more precise the V_{BE} shifts logarithmically with the current.

The relative insensitivity of the transistor V_{BE} to current variations is exploited in building current sources and voltage references.

Naturally the opposite is true for the current variation as a function of voltage. In fact if we invert the previous equation we have

$$I = I_O \times \exp(V_{BE}/V_T) \qquad \text{Eq. 3–7}$$

which shows that the current varies exponentially with the V_{BE}. We already know that a variation of 18 mV on the V_{BE} will double the current in the transistor. For a quick estimate of variations in current due to small voltage variations, we can linearize the exponential law and find that the current will vary at roughly two percent per millivolt. This strong dependence of current on the V_{BE} explains why the transistor is normally driven with current, not voltage.

This also explains how difficult it is to deal with offsets, or small voltage variations between identical transistors. Two identical transistors biased at the same identical voltage will have their current mismatched with a two percent error if their V_{BE} differs by just 1 mV.

MOS versus Bipolar Transistors

The dual of bipolar NPN and PNP transistors in CMOS technology are the P-channel and N-channel MOS transistors in Figure 3–3. The general function of the transistors are the same independently as their implementation but there are pros and cons to using both technologies. Generally speaking, the base, the emitter, and the collector in the bipolar transistor are analogous to the gate, source and drain in the MOS transistor, respectively. The bipolar transistors' main problem, which is not present in CMOS, is their need for a base current in order to function. Such current is a net transfer loss from emitter to collector. While the base current is small in small signal operation, in power applications, where the transistor is used as a switch, the base current necessary to keep the transistor on can be very high. This high base current can lead to implementations with very poor efficiency. With the popularity of portable electronics and the need to extend battery life, it is no wonder that CMOS often tends to have the upper hand over bipolar technologies. The advantage of bipolar over CMOS is that it has better trans-conductance gain and better matching, leading to better differential input gain stages and better voltage references. The best performance processes are mixed-mode Bipolar and CMOS (BiCMOS) or Bipolar, CMOS, and DMOS (BCD) processes in which the designer can use the best component for the task at hand.

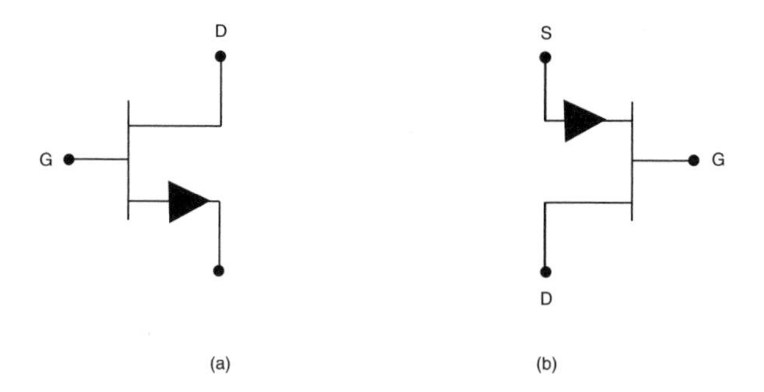

Figure 3–3 (a) N-channel MOS transistor and (b) P-channel MOS transistor.

The symbols in Figure 3–3 (a) and (b) are an easy-to-draw shorthand clearly mocking the bipolar counterparts of MOS transistors. In the technical literature there is a great proliferation of symbols for the MOS transistor. The most complete symbol is shown in Figure 3–4 (a) and (b) and exhibits a fourth terminal representing the "bulk" connection (typically ground for N-channel and positive supply for P-channel) and a more elaborate representation of the vertical segments representing the gate.

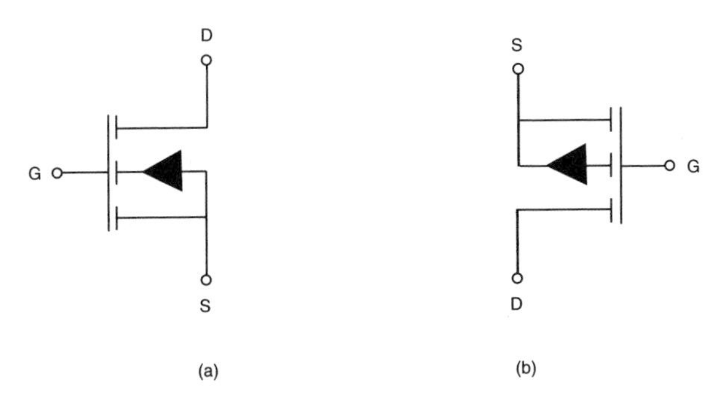

Figure 3–4 (a) N-channel MOS transistor and (b) P-channel MOS transistor complete of "bulk" terminal.

Another popular version is shown in Figure 3–5 (a) and (b): here the arrow is dispensed with and the gate is simplified to look like a capacitor (two parallel lines). In the rest of this book each representation is used at one point or another both because the corresponding material has been generated at different points in time and because variety is a true representation of the industry practice.

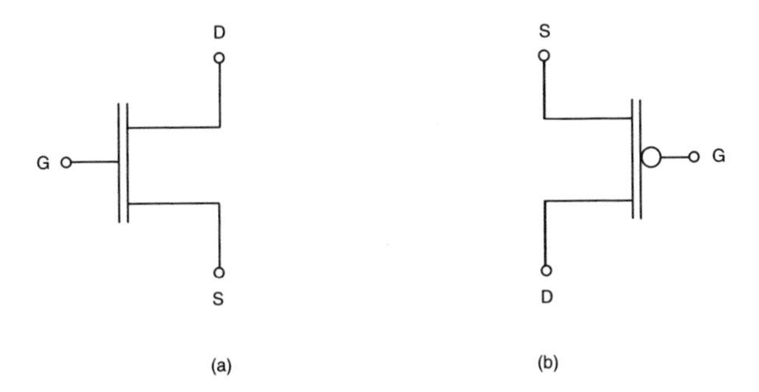

Figure 3–5 Alternative symbols for N-channel MOS (a) and P-channel MOS (b).

3.2 Elementary Circuits

In this section we will build increasingly complex and thus increasingly functional blocks, leading to some useful power management circuits.

Current Mirror

Current mirrors are a very common way to implement current sources or active loads. The foundation of a current mirror is the fact that two identical transistors driven by the same V_{BE} will carry identical currents. In Figure 3–6 the two transistors having a gain of β are connected in a mirror configuration; namely the same base and same emitter potentials. Such configuration yields a virtually perfect unity gain I_{OUT}/I_{IN} except for the base currents, which introduce a systematic error of β /2+. For example for β = 100 the error is roughly two percent.

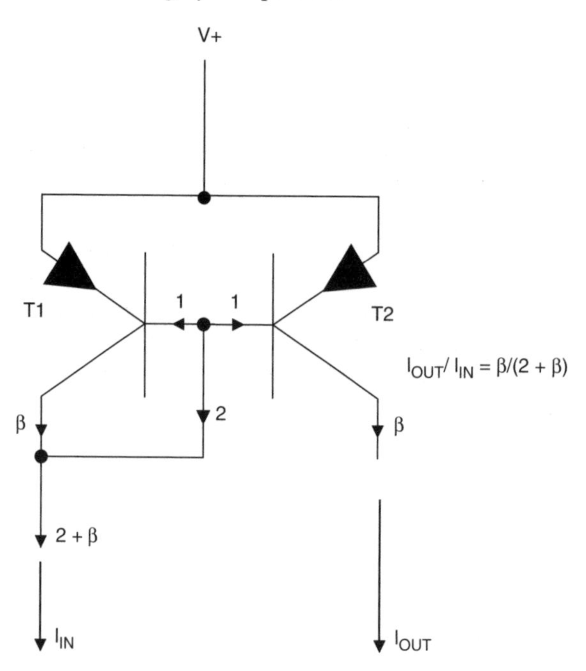

Figure 3–6 PNP current mirror.

Current Source

Current sources are a very popular means to set relatively constant bias currents.)

In Figure 3–7 the relatively constant voltage of the V_{BE} of T2 is forced across resistor R and the ensuing current is available at the collector

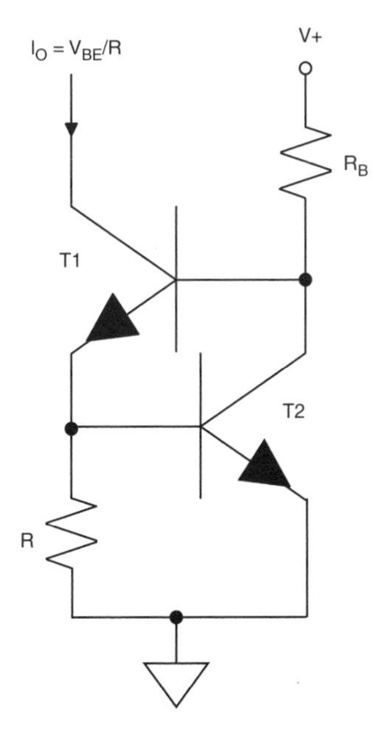

Figure 3–7 NPN current source.

of T1. Suppose that the supply V+ changes from 5 V up to 10 V, the current inside T2 will roughly double, but its V_{BE} will only increase by 18 mV, say from 0.7 V up to 0.718 V. Accordingly the current I_O will increase by 18 mV/R. In conclusion an initial voltage variation of 100 percent results in an error of only 18 mV/700 mV, or 2.6 percent.

Differential Input Stage

In Figure 3–8 an NPN differential stage is illustrated.

The following is a calculation of the trans-conductance gain dI/dV of this stage.

$$dI_1 = dV/2r_E \qquad \text{Eq. 3–8}$$

$$r_E = V_T/I_E \qquad \text{Eq. 3–9}$$

Substituting Eq. 3–9 into Eq. 3–10 we have

$$dI_1/dV = I_E/2V_T \qquad \text{Eq. 3–10}$$

For example with $I_E = 10\ \mu A$ we have a trans-conductance $dI/dV = 10\ \mu A/52\ mV = 1/5.2\ k\Omega$. Notice that the trans-conductance gain of this stage is a simple linear function of its bias current I_E.

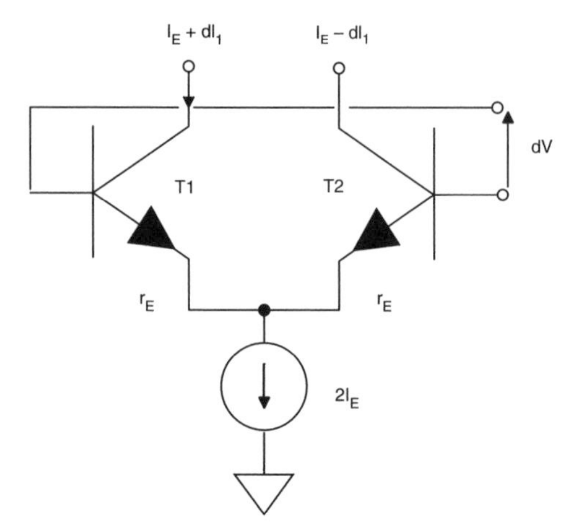

Figure 3–8 NPN differential stage.

Differential to Single Input Stage

In Figure 3–9 an NPN differential-to-single stage is illustrated.

The combination of a differential stage and a mirror allows the building of a differential input to single output stage, a fundamental input stage block in operational amplifiers. Thanks to the turn-around effect of the mirror, the gain of this stage is double the one calculated in the previous step.

$$2dI/dV = 1/r_E = I_E/V_T = 10\ \mu A/26\ mV = 1/2.6\ k\Omega\ I \qquad \text{Eq. 3–11}$$

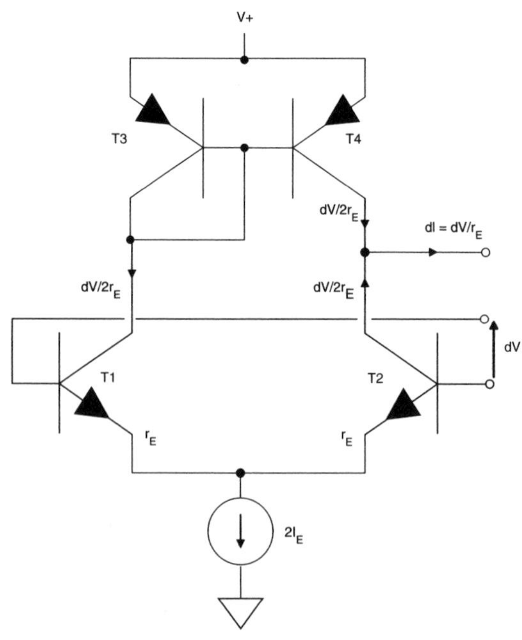

Figure 3–9 NPN differential-to-single stage.

Buffer

The function of a *buffer* is to transfer the voltage transparently from its input to its output while increasing dramatically the current drive. A voltage driven transistor, as discussed previously, is an ideal buffer thanks to its property of yielding a current that increases exponentially with the applied voltage. Since an NPN can only source current out of its emitter and a PNP can only sink current into its emitter, if we want to drive a bipolar (source or sink) load, we will have to use both types of transistors in the configuration of Figure 3–10. For example, if the current source I is 0.1 mA and the beta gain of each transistor is 100, then the buffer can drive a current of 0.1 mA × 100 = 10 mA.

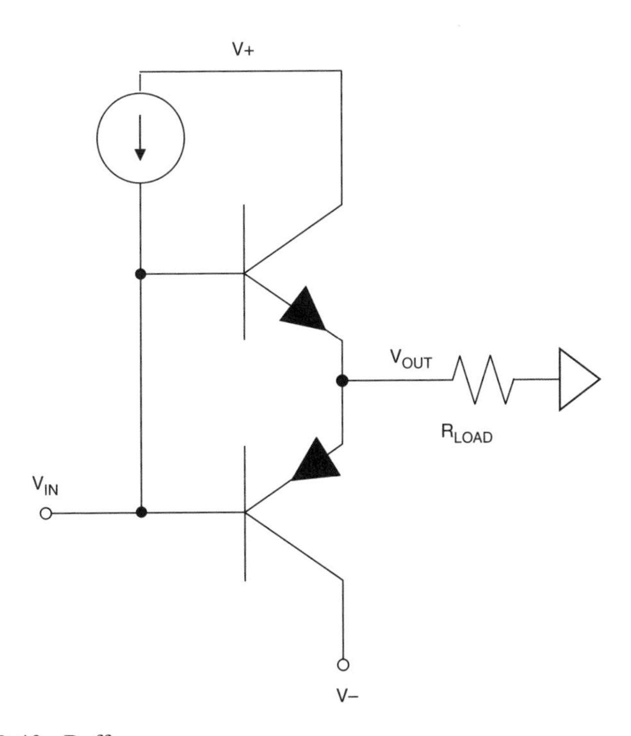

Figure 3–10 Buffer.

3.3 Operational Amplifier (Opamp)

As the name implies, if we finally put together all the elementary blocks above (transistors, current mirrors, current sources, differential stages, and buffers) we finally come to something usable, the *operational amplifier.*

Figure 3–11 shows a basic opamp essentially composed of three stages: the input differential-to-single stage, the gain stage, and the output buffer stage. The input stage shown here is inverted to the one in

Figure 3–9, namely with respect to the PNP differential pair and NPN mirror (also called active load). The intermediate stage is shown as a simple NPN transistor, and more often will be a full-fledged *Darlington stage* (two cascaded NPN transistors gaining beta squared, or β^2). The output stage is the buffer discussed in the previous section.

Inverting and Non-Inverting Inputs

The opamp in Figure 3–11 is shown as an open loop. Before closing the loop—connecting the inverting input to the output for negative feedback— it is a good idea to find out the inverting versus the non-inverting input.

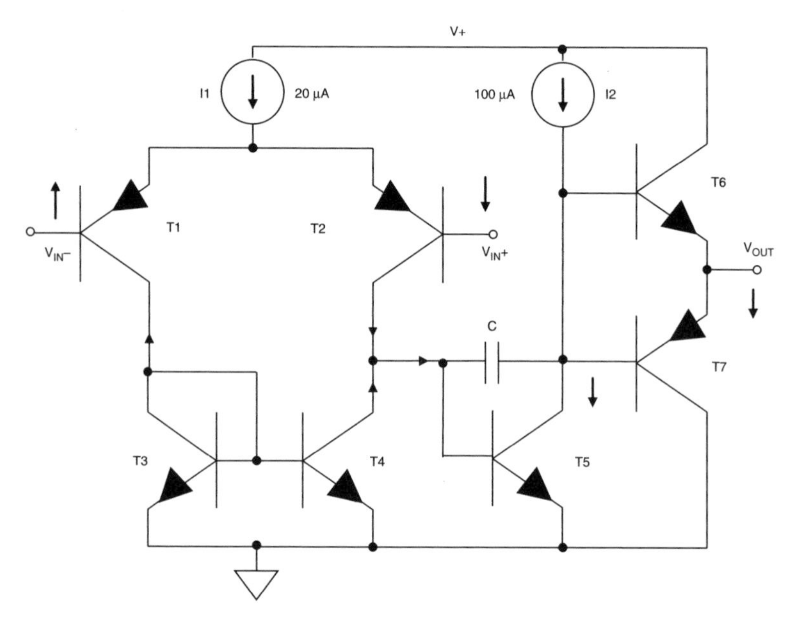

Figure 3–11 Bipolar opamp schematic.

The arrows in Figure 3–11 help in determining the input sign; note that an arrow on top of a wire indicates a small signal current flow in that wire while a floating arrow near a node indicates a small voltage signal acting on that node. Applying a positive voltage to the $V_{IN}-$ input (and correspondingly a negative one to the $V_{IN}+$) we cause more current flow in the base of T5. The collector of T5 will draw more current, pulling down the buffer input and thus the output. Since the output moves low when $V_{IN}-$ moves high, $V_{IN}-$ is indeed the inverting input, as its name seemed to imply at the start.

Rail to Rail Output Operation

In Figure 3–11 the output cannot get any closer to $V+$ than the sum of the V_{BE} of T6 and the V_{CESAT} of the current source (V_{CESAT} of T2 in the current mirror of Figure 3–6 when driven by a constant current sink I_{IN} is indeed a current source). Similarly, the output cannot get any closer to ground than the sum of the V_{BE} of T7 and the V_{CESAT} of T5.

In order to have low dropout operation (also referred to as rail to rail output operation) the shorter path between output and $V+$ or ground must be a V_{CESAT}.

In Figure 3–12 the principle of output rail-to-rail operation is illustrated. Current mirroring plays a heavy role here: mirrors T5:T7, T8:T9, and T6:T10 with ratios of 1:6, 1:8, and 1:8 respectively, provide a balanced current bias for the circuit.

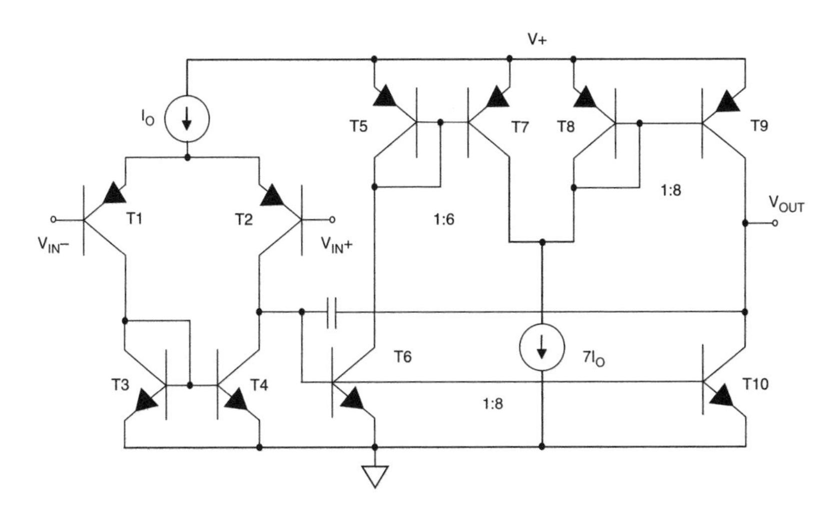

Figure 3–12 Low dropout opamp.

CMOS Opamp

As explained earlier, the bipolar opamp in Figure 3–12 can be easily replicated in CMOS by substituting NPN with N-channel MOS transistors and PNPs with P-channel MOS transistors. In Figure 3–13 transistors T1, T2, and T7 are P-channel and T4 through T6 are N-channel, resulting in a simple CMOS version of an opamp.

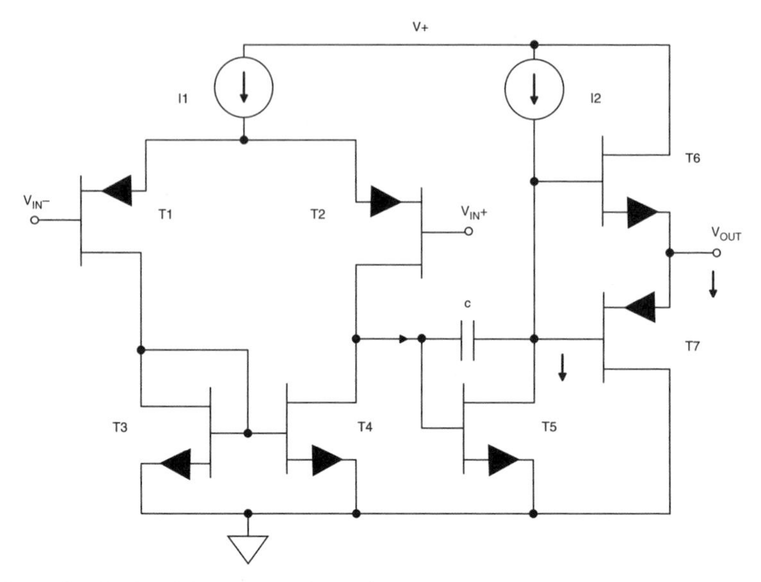

Figure 3–13 CMOS opamp schematic.

Opamp Symbol and Configurations

In Figure 3–14 we have the opamp in some common configurations. Notice how in closed loop configuration the feedback network (R1 and R2) sets the forward gain. The same feedback network returns to the input an amount of output signal that is inversely proportional to the gain. The max amount of feedback signal is returned in the case of the unity gain buffer configuration, where all the output signal is returned to the input. From a loop stability standpoint then, the unity gain buffer configuration appears to be the most critical.

DC Open Loop Gain

The DC gain of the bipolar opamp in Figure 3–11 is calculated as follows: if a small signal dV_{IN} is applied to the input differential ($V_{IN}+ - V_{IN}-$), the output of this first stage will produce a current equal to dV_{IN}/r_E. This current drives the base of T5, which develops a collector current β_5 times higher. This current is further amplified by T6 (or T7 depending on the polarity of the incoming current) by another factor of β_6. Finally this current is delivered to the load R_L. Mathematically

$$dV_{IN} \times (1/r_E) \times \beta_5 \times \beta_6 \times R_L = dV_{OUT} \qquad \text{Eq. 3–12}$$

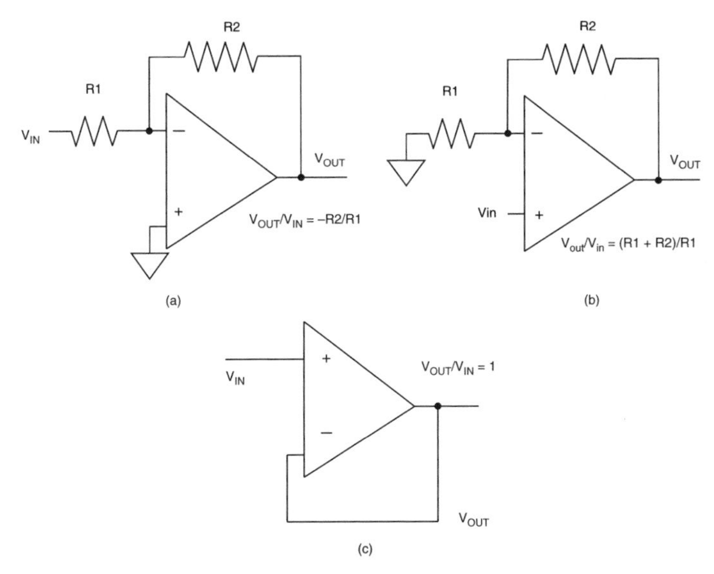

Figure 3–14 Opamp symbol and configurations: (a) inverting, (b) non-inverting, and (c) unity gain buffer.

from which, assuming for simplicity the two β gains are identical, the open loop DC gain is

$$G_{DCOL} = dV_{OUT}/dV_{IN} = \beta^2 \times R_L/r_E \qquad \text{Eq. 3–13}$$

For example, if r_E and R_L are both 2.6 kΩ, (r_E is 2.6 kΩ at I_E = 10 μA) and the β are both 100, the open loop gain is 10,000. This means that to move 1 V at the output only 1 V/10,000 (100 μV), of signal swing is needed at the input. Commercial products exhibit even higher gains. With differential input variations ($V_{IN}+ - V_{IN}-$) in the order of μV, no wonder an opamp may have volts swinging at its output with no appreciable voltage visible at its direct differential inputs. Accordingly, when a non-inverting input is connected to ground—as happens in many configurations—the inverting pin will appear to be grounded as well. The term "virtual ground" refers to such input.

AC Open Loop Gain

To be useful, the opamp will be ultimately connected in a closed loop configuration. A closed electrical loop is subject to oscillations or frequency instabilities due to parasitic reactive components (capacitors and inductors) present in each component in the loop and causing phase shifts.

Oscillation occurs in any regenerative closed loop system, especially those in which a signal injected in any point returns with equal or higher amplitude after a circulation (loop gain $\rightarrow 1$) and roughly equal phase (low phase margin). Such oscillations are eliminated if the open loop gain is made to be un-regenerative, meaning it assumes a value smaller than unity, at the critical frequency where the parasitic components become active. Intuitively, if an electric signal is cyclically multiplied (in a closed loop circuit) by a factor higher than one, (amplified) its amplitude will continually increase (regenerative loop) leading to self-sustained oscillations. Alternatively, the same signal multiplied cyclically by a factor lower than one (attenuated) will eventually be reduced down to zero (no oscillations). In traditional bipolar design, the most notorious source of phase shift is the PNP with its low f_T frequency around 1 MHz. Hence the AC open loop gain needs to be less than unity at that frequency. In that case the system will be stable with 45° of phase margin or better (stability criterion). In calculating the AC loop gain, we will assume that all the calculations are conducted at approximately the cutoff frequency of 1 MHz as this is the zone of interest for stability. This assumption allows the use of a simplified expression for the elements of the loop gain. At the basis of such circuit analysis simplification is the property that capacitors behave like short circuits (a piece of wire) and inductors behave like open circuits (a wire cut open) at sufficiently high frequencies. The same technique used for calculating the DC gain is applied here, the difference being that at the high frequency chosen for this analysis, the current out of the input stage will bypass the transistor T5 in Figure 3–11. Instead, the current will go through the capacitor C, developing at its output a voltage in proportion to its impedance of amplitude $1/\omega C$ where $\omega = 2\pi f$ is the pulsation frequency. The capacitor then presents this voltage to the output buffer which will pass it unchanged to the opamp output

$$dV_{IN} \times (1/r_E) \times 1/(\omega C) = dV_{OUT} \qquad \text{Eq. 3–14}$$

$$G_{ACOL} = dV_{IN}/dV_{OUT} = 1/(\omega C r_E) = 1/(2\pi f C r_E) \qquad \text{Eq. 3–15}$$

Such gain has to be less than or equal to one at $f = f_T$ hence by setting $G_{ACOL} = 1$ we have

$$1 = 1/(2\pi f_T C r_E) \qquad \text{Eq. 3–16}$$

from which we can calculate the compensation capacitor

$$C = 1/(2\pi f_T r_E) = 1/(2 \times 3.14 \times 1\text{ MHz} \times 2.6\text{ k}\Omega) = 61\text{ pF} \qquad \text{Eq. 3–17}$$

This value is in the right ballpark but integrating a 60 pF capacitor may take quite a lot of die space. Since f_T is a given parameter, depending on the process at hand, $r_E = V_T/I_e$ ends up being the only parameter to play with. For example if I_e is reduced from 10 to 5 μA, r_E will double and C can then be reduced to 30 pF.

3.4 Voltage Reference

The voltage reference is the last ingredient necessary to build a voltage regulator, otherwise known as the king of power management and power conversion. The most popular voltage references are based on active circuits, like the Widlar circuit which will be the focus of the following section.

Positive TC of ΔV_{BE}

From Eq. 3–6

$$\Delta V_{BE} = V_T \times \ln(x) = (k \times T/q)\ln(x) \qquad \text{Eq. 3–18}$$

Taking the derivative with respect to temperature we have

$$d/dT(\Delta V_{BE}) = k \times q \times \ln(x) = [(k \times T/q)/T]\ln(x) = \Delta V_{BE}/T \qquad \text{Eq. 3–19}$$

Normalizing to the amplitude of ΔV_{BE} we have the expression for the incremental temperature variation of ΔV_{BE}

$$(1/\Delta V_{BE}) \times d/dT(\Delta V_{BE}) = 1/T = 1/300°\text{C}^{-1} \qquad \text{Eq. 3–20}$$

Namely, the ΔV_{BE} variation in temperature, normalized to its amplitude, is 0.33%/°C positive. For example if we apply Eq. 3–20 rewritten as $d/dT(\Delta V_{BE}) = \Delta V_{BE}/T$, a ΔV_{BE} of 18 mV will have a temperature variation of 18 m/300 = +0.6 mV/°C and a ΔV_{BE} of 600 mV will have a temperature variation of 600 m/300 = +2 mV/°C.

Negative TC of V_{BE}

The V_{BE} as a well known negative Temperature Coefficient (TC)

$$d/dT(V_{BE}) = (V_{BE} - V_{BG})/T + 3V_T/T = -2 \text{ mV/°C} \qquad \text{Eq. 3–21}$$

or in relative terms for $V_{BE} = 0.6$ V

$$(1/V_{BE}) \times d/dT(V_{BE}) = -2 \text{ mV}/0.6 \text{ V} = -1/300 \qquad \text{Eq. 3–22}$$

Comparing Eq. 3–20 with Eq. 3–22 we have

$$(1/V_{BE}) \times d/dT(V_{BE}) = (1/\Delta V_{BE}) \times d/dT(\Delta V_{BE}) \qquad \text{Eq. 3–23}$$

namely the relative variations of V_{BE} and ΔV_{BE} are identical in value and opposite in sign. This property is at the basis of the design of temperature independent circuits.

Table 3–1(a) and (b) formulas describe the equal in amplitude and opposite in sign temp behavior of the V_{BE} and ΔV_{BE}. Table 3–1(c) combines the two formulas into one.

Table 3–1 V_{BE} and ΔV_{BE} Temperature Dependency

(a) $\frac{1}{\Delta V_{BE}} \frac{d(\Delta V_{BE})}{dT} = \frac{1}{T} = \frac{1}{300} {}^\circ C^{-1}$
(b) $\frac{1}{V_{BE}} \frac{d(V_{BE})}{dT} = -\frac{1}{T} = -\frac{1}{300} {}^\circ C^{-1}$
(c) $\frac{1}{\Delta V_{BE}} \frac{d(\Delta V_{BE})}{dT} = -\frac{1}{V_{BE}} \frac{d(V_{BE})}{dT}$

In conclusion, since ΔV_{BE} and the V_{BE} have opposing behavior in temperature, equal amplitudes of each summed up will always lead to a resulting voltage with null temp coefficient.

Build a ΔV_{BE}

Now let's see how we can build practical circuits that can mix up ΔV_{BE} and V_{BE}. As a first step, we will build a circuit that behaves like a ΔV_{BE}. To this end, let us repeat for convenience the expression of the V_{BE}.

$$V_{BE} = V_T \times \ln(I/I_O) \qquad \text{Eq. 3–24}$$

I_O is proportional to the emitter area such that

$$I_O = kA \qquad \text{Eq. 3–25}$$

Hence two transistors of different areas, carrying different currents, will have different values for V_{BE} as follows:

$$V_{BE} = V_T \times \ln(I/kA) \qquad \text{Eq. 3–26}$$

$$V_{BE}' = V_T \times \ln(I'/kA') \qquad \text{Eq. 3–27}$$

And differentiating,

$$\Delta V_{BE} = V_{BE}' - V_{BE} = V_T \times \ln[(I/I')(A'/A)] \qquad \text{Eq. 3–28}$$

Setting $I/I' = x$ and substituting into Eq. 3–28 we have

$$\Delta V_{BE} = V_T \times \ln(x \times A'/A) \qquad \text{Eq. 3–29}$$

For example if $A'/A = 10$ and the two transistors carry the same current ($x = 1$) then $\Delta V_{BE} = 26$ mV $\times \ln 10 = 60$ mV.

In Figure 3–15 the two transistors T1 and T2 have the same current $I1 = I2 = 100$ µA, where $I1$ in T1 is set by the current source I and $I2$ is set by the V_{BE} coupling of the two transistors in conjunction with their area ratio $I2 = \Delta V_{BE}/R2 = 60$ mV/600 Ω = 100 µA. The voltage across $R3$ will be $(R3/R2) \times \Delta V_{BE}$ and since $R3/R2$ is 6 kΩ/600 Ω = 10, the drop across $R3$ is 10 × 60 mV = 600 mV. This $\Delta V_{BE}'$ voltage is actually an amplified ΔV_{BE} and thus has all the properties of the ΔV_{BE} including its positive TC.

In conclusion, Figure 3–15 shows a circuit that produces a 600 mV voltage with positive temperature coefficient of ΔV_{BE}.

Building a Voltage Reference

Adding the ΔV_{BE} voltage in Figure 3–15 to a proper V_{BE} value—as described later in more detail—should produce a voltage that is invariant to temperature, a *reference voltage*, fundamental to any servo control mechanism. The result is the circuit in Figure 3–16. It should be intuitive that matching of T1 and T2 is critical and best obtained if the two transistors see (are biased to) the same collector voltage. Since the collector voltage of T1 is equal to its base voltage (base and collector of T1 are shorted) it follows that the best voltage for collector of T2 is analogous to V_{BE1}. Since the collector of T2 is connected to the base of T3 we will need to make the base-emitter voltage of T3, V_{BE3}, identical to V_{BE1}. By constructing T3 identical to T1 and biasing it to the same current value (100 µA), its V_{BE} will indeed be virtually identical to $V_{BE}1$, fulfilling the above matching criteria.

In Figure 3–16 the V_{BE} (600 mV) of T3 is summed up to the $\Delta V_{BE}'$ (600 mV) of resistor R3 to add up to a temperature invariant voltage of 1.2 V at the V_{REF} node, namely $V_{REF} = V_{BE} + \Delta V_{BE}' = 1.2$ V This V_{OUT} is

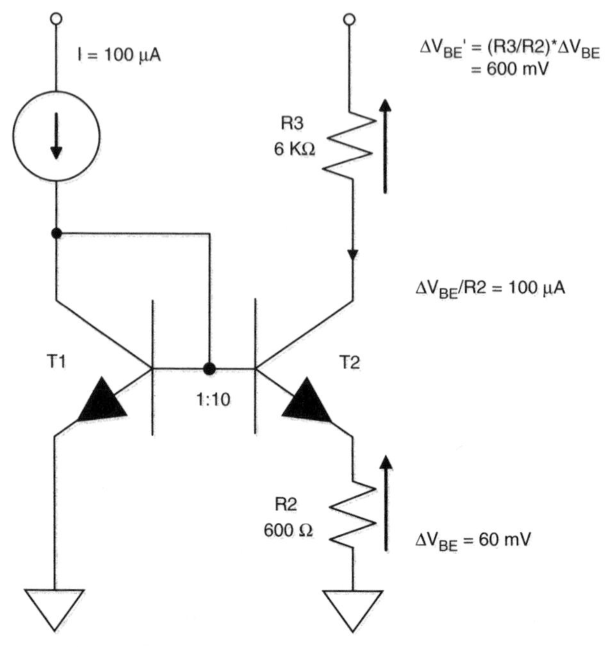

Figure 3–15 $\Delta V_{BE}'$ circuit.

temperature invariant and its value is equal to the band-gap of the silicon. We can then write that

$$V_{REF} = V_{BG} = V_{BE} + \Delta V_{BE} = 1.2 \text{ V} \qquad \text{Eq. 3–30}$$

This analysis is correct and a good lead to design voltage references. However it is a bit of an oversimplification. In reality any voltage reference circuit will have some slight dependence on temperature. A plot of V_{REF} over temperature is slightly curved and that curve will generally exhibit a true $dV_{OUT}/dT = 0$ at only one temperature point, typically at ambient temperature for a well done design. The circuit in Figure 3–16, yielding a voltage equal to the silicon band-gap is referred to as a band-gap voltage reference. This particular implementation is also called a Widlar voltage reference, after its inventor. A band-gap voltage reference can yield easily TC flatness in the order of 50 ppm/°C.

Fractional Band-Gap Voltage Reference

Naturally all the terms in Eq. 3–30 can be divided by any number higher or lower than one, leading to voltage references that are correspondingly lower than (fractional) and higher than (multiple) the V_{BG}. If k is the dividing factor we can write

$$V'_{REF} = V_{BG}/k = V_{BE}/k + V_{BE}/k$$

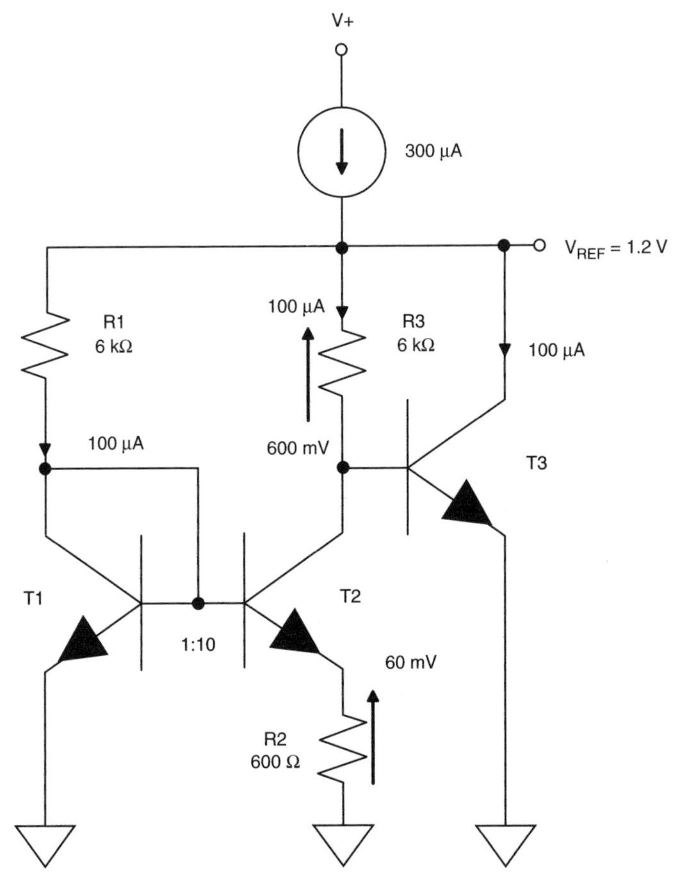

Figure 3–16 Voltage reference.

For example, for $k = 4$ we have

$$V'_{REF} = V_{BG}/4 = 300 \text{ mV} = 150 \text{ mV} + 150 \text{ mV}$$

The circuit in Figure 3–16 can be modified easily, like the one in Figure 2–14 to produce a V_{BE} drop of 150 mV ($\Delta V_{BE3} \times$ R2/R4) and a ΔV_{BE} drop of 150 mV ($V_{BE1}-V_{BE2} \times$ R2/R3). Sum the two drops (300 mV drop across R2) and we first come up with a 300 mV fractional band-gap voltage drop floating across R2. Then this drop is shifted down to the output by T4 (the drop across R5 is identical to the drop across R2 if T4 and T3 are identically biased). Notice also that here V_{CC} can be as low as 1 V, leaving room for the 300 mV reference, 660 mV V_{BE4}, and some minimum headroom for the current source. For more implementation details see, for example, patent number 4,628,247 (go to *www.uspto.gov* and search by the patent number or author name) by this author.

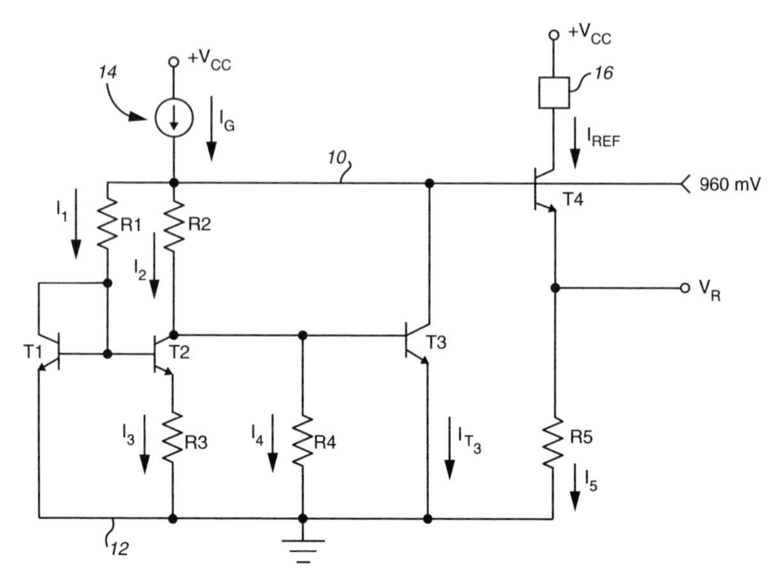

Figure 3–17 Fractional band-gap voltage reference.

3.5 Voltage Regulator

The voltage regulator is a circuit that keeps a constant set voltage across its load despite changes in load, temperature, and power supply, among other things. It follows that such a device has to have the temperature properties of a voltage reference, a good driving capability, and flexibility in output setting.

Intuitively, an opamp plus a reference should suffice in building a basic voltage regulator. In Figure 3–18 such a device is shown. However, an opamp is limited in driving capability, generally to a few tens of milli-Amps. In order to boost the driving capability, a power pass element like transistor T1 in Figure 3–19 is introduced.

The device in Figure 3–19 is a positive linear voltage regulator. Positive because V_{OUT} is positive (negative voltage regulators are available but are less common) and linear because the pass element transferring current to the load is biased in the linear region, operating in the presence of both high voltage and high current at the same time. When the pass element is operated in *switch mode*, meaning it is full on during part of the operating cycle and full off for the remainder of the time, we have a switching voltage regulator. As a reminder, full on means that the transistor is in the saturation region with high current conduction and low voltage between the emitter

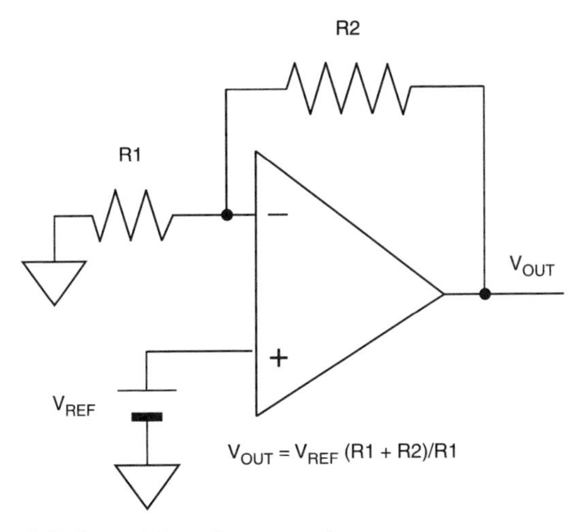

Figure 3–18 A light-weight voltage regulator.

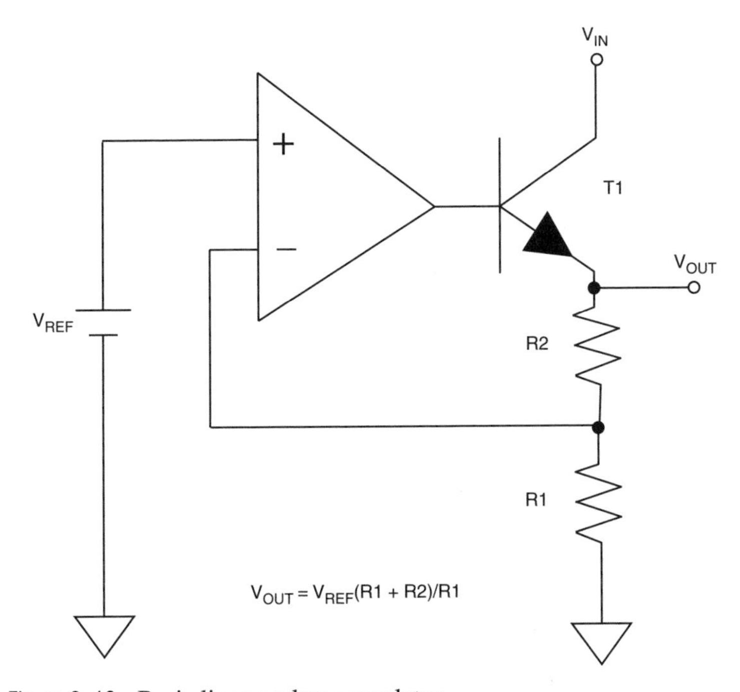

Figure 3–19 Basic linear voltage regulator.

and collector (for a bipolar pass element). Full off means that the transistor sustains the entire supply voltage without any conduction of current.

There are many variations of the basic configuration in Figure 3–19; including a PNP pass element as opposed to an NPN; in which case the PNP implements a low dropout operation. Other variations include N-channel DMOS instead of NPN and P-channel PMOS instead of PNP.

Linear regulators are very popular but they consume a lot of power, as the current transferred to the load also dissipates power inside the pass element biased in the linear region.

The power delivered to the load is

$$P_{OUT} = V_{OUT} \times I \qquad \text{Eq. 3–31}$$

while the power wasted in the pass transistor is

$$P_D = (V_{IN} - V_{OUT}) \times I \qquad \text{Eq. 3–32}$$

Hence the efficiency η is

$$\eta = P_{OUT}/(P_{OUT} + P_D) = V_{OUT} \times I/(V_{IN} \times I) = V_{OUT}/V_{IN} \qquad \text{Eq. 3–33}$$

For example, if the output is half the input, half of the power goes to waste. On the other hand if the output is close to the input, this configuration becomes increasingly effective thanks to its simplicity and low noise operation.

3.6 Linear versus Switching

Linear circuits are simple and elegant and remain the foundation of analog and digital circuit design. In fact, no matter how digital or logic intensive a circuit is, certain fundamental blocks, like voltage references, remain analog; not just analog but bipolar. So it happens that every time a CMOS designer needs to build a precise band-gap voltage reference he has to dig deep down in the scraps of his arsenal, and pull out the parasitic bipolar transistor as the only means to build a decent voltage reference. However as complexity builds, linear circuits become impractical, consequently at a system level switching circuits tend to prevail. With switching regulators we trade noise for efficiency as these circuits are inherently efficient but also inherently noisy due to their switching nature. In the next section we cover the foundations of switching regulators, the other crucial elements of power conversion and power management.

Generally speaking linear circuits, like opamps, are still hugely popular as general purpose devices as well as building blocks for power management. There is a great variety of opamps, from high speed to low power to input rail-to-rail and/or output rail-to-rail operation. Another class of linear circuits that is very popular is the *Low Drop Out* (LDO) linear regulator, which finds broad acceptance in low noise environments like the cell phone handset. As power goes up linear regulators become impractical and end up yielding to switching regulators.

3.7 Switching Regulators

There are many types of switching regulators. Here we will cover two main types, the inductor based step down, or buck converter, widely used in low voltage DC to DC conversion and the transformer based flyback converter used in offline AC to DC conversion. The techniques illustrated to analyze these two cases can be utilized to explore any other architecture.

3.8 Buck Converters

Switching regulators operate in switch mode, meaning the energy is transferred cyclically to the output—where it is stored in a capacitor—in finite increments (Figure 3–20). In this mode of operation the pass transistor is

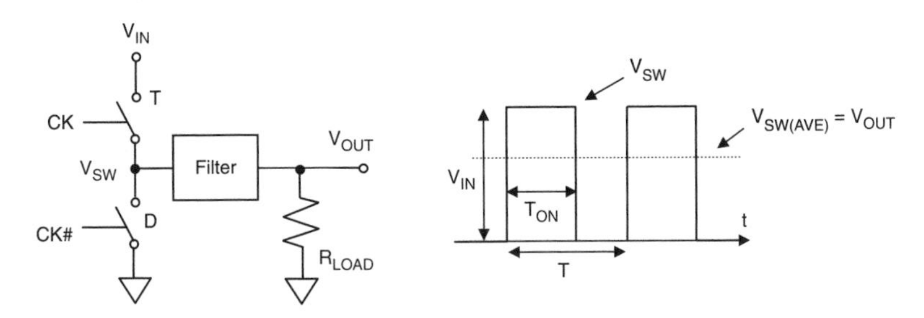

Figure 3–20 Switching regulator principle.

full on during the charge transferring part of the operating cycle and full off for the reminder of the cycle time.

Since the pass element is switching between V_{IN} and ground at a clock frequency f with an on time equal to T_{ON}, then the waveform at its output is a square wave and the time average over the period T of that voltage waveform ($V_{IN} \times T_{ON}/T$) will become the output DC voltage (V_{OUT}) when filtered

$$V_{IN} \times T_{ON}/T = V_{OUT} \qquad \text{Eq. 3–34}$$

being

$$T_{ON}/T = DC \qquad \text{Eq. 3–35}$$

where DC is the duty cycle. Finally we have that in this type of switching regulator

$$V_{OUT} = DC \times V_{IN} \qquad \text{Eq. 3–36}$$

From Eq. 3–35, Eq. 3–36 and being that $T = 1/f$ we have

$$T_{ON} = DC \times T = V_{OUT}/(V_{IN} \times f) \qquad \text{Eq. 3–37}$$

We will use this expression later.

Naturally this switching regulator requires a clock for cycling at frequency f, and a filter to average the switching waveform at the output; this is normally a second order LC filter with the inductor holding on to the circulating current and the capacitor holding on to the output voltage during the off time. During on-time the partially depleted capacitor's charge is restored. It is clear that even with heavy filtering the output of a switching regulator is inherently rippled, as opposed to the flat output of a linear regulator.

Switching Regulator Power Train

Figure 3–21 shows a higher level of detail of the output stage, the power train, including the pass transistor T, the diode D, and the LC filter.

In a steady state the average inductor current is equal to the load current. The inductor feeds the load as well as the output storage capacitor through the transistor T during its on time. During this "charge" time, the inductor current I_{RCH} ripples in proportion to

$$I_{RCH} = (V_{IN} - V_O) \times T_{ON}/L \qquad \text{Eq. 3–38}$$

I_{RCH} is the positively sloped portion of the triangular ripple waveform, corresponding to the charge time T_{CH}, also equal to T_{ON} in Figure 3–21. During the off time the inductor current recirculates through the diode D, decaying an amount of

$$I_{RDISCH} = V_{OUT} \times (T - T_{ON})/L \qquad \text{Eq. 3–39}$$

I_{RDISCH} is the negatively sloped portion of the triangular ripple waveform, corresponding to the discharge time $T_{DIS} = T - T_{ON}$. Since the

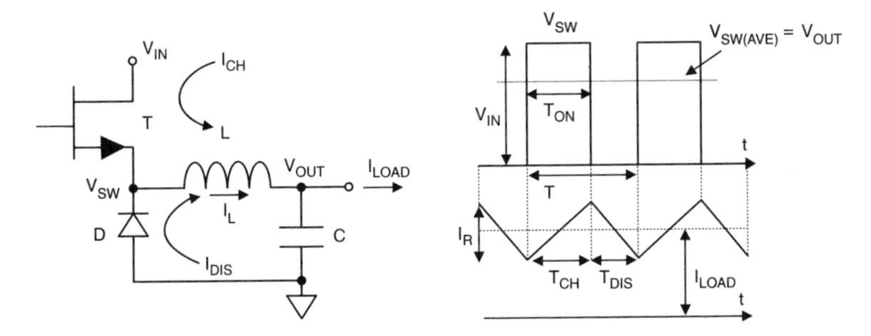

Figure 3–21 Power train.

current ripple waveform is continuous, the amplitude of the two portions of the current ripple must be identical.

$$I_R = (V_{IN} - V_O) \times T_{ON}/L = V_{OUT} \times (T - T_{ON})/L \qquad \text{Eq. 3–40}$$

where I_R is simply the symbol for the peak-to-peak ripple current. From Eq. 3–40 we find again that

$$V_{OUT}/V_{IN} = T_{ON}/T = DC \qquad \text{Eq. 3–41}$$

as we had already found with Eq. 3–36.

Actually, if we repeat the analysis accounting for the transistor on dropout V_{SAT}, the forward diode drop V_{DR}, and the series resistance R_{ESL} of the inductor L we have

$$DC = (V_{OUT} + V_{DIODE} + V_{ESL})/(V_{IN} - V_{SAT} + V_{DIODE}) \qquad \text{Eq. 3–42}$$

where

$$V_{ESL} = R_{ESL} \times I_{LOAD} \qquad \text{Eq. 3–43}$$

If we substitute Eq. 3–37 in Eq. 3–40 we have

$$I_R = (V_{IN} - V_O) \times (V_{OUT}/V_{IN})[1/(f \times L)] \qquad \text{Eq. 3–44}$$

Inverting, we have

$$L = (V_{IN} - V_O) \times (V_{OUT}/V_{IN})[1/(f \times I_R)] \qquad \text{Eq. 3–45}$$

For example if $V_{IN} = 12$ V, $V_{OUT} = 1.2$ V, $F = 300$ kHz, and $I_R = 1$ A, we need an inductor of value

$$L = 10.8 \times 0.1/300 \text{ kHz} \times 1 = 3.6 \text{ µH} \qquad \text{Eq. 3–46}$$

Output Capacitor

In first approximation V_{OUT} can be assumed to be constant. In second approximation the output will exhibit small voltage variations. Since the most significant variation of the output happens during load transients from light to full load or vice versa, the output capacitor is generally sized on the basis of such load transients.

Electrolytic Capacitors and Transient Response

Electrolytic capacitors are very popular in computing applications due to their low cost. These capacitors have very low cutoff frequency f_C, a typical value being 1 kHz.

Since f_C is, by definition, the frequency at which the capacitor reactance $X_C = 1/(2\pi fC) = ESR$ (Equivalent Series Resistor, the parasitic resistance in series with a capacitor), it follows that at frequency f_C we have

$$X_C/ESR = (1/2\pi f_C C)/ESR = 1 \qquad \text{Eq. 3–47}$$

from which

$$f_C = 1/(2\pi \times ESR \times C) = 1 \text{ kHz} \qquad \text{Eq. 3–48}$$

At clock frequencies around 300 kHz, typical of the switching operation, X_C will be 300 times smaller than ESR for an electrolytic capacitor. Hence, with electrolytic capacitors the output ripple is mostly due to the product of the capacitor's ESR resistance and the current ripple I_R.

If we have a 20 A load ($I_{LOAD} = 20$ A) transient and want a 60 mV ripple, we will need an ESR

$$ESR = V_R/I_{LOAD} \qquad \text{Eq. 3–49}$$

or

$$ESR = 60 \text{ mV}/ 20 \text{ A} = 3 \text{ m}\Omega \qquad \text{Eq. 3–50}$$

From Eq. 3–48 we can then calculate the capacitor C:

$$C = 1/(2\pi \times ESR \times f_C) = 1/(6.28 \times 3 \text{ m} \times 1 \text{ k}) = 53 \text{ mF} \qquad \text{Eq. 3–51}$$

Figure 3–22 illustrates a typical output perturbation in response to a load current step function for a 5 V in, 2 V out system. The drawing illustrates the effects of both the capacitor series resistance *ESR* and series inductance *ESL*.

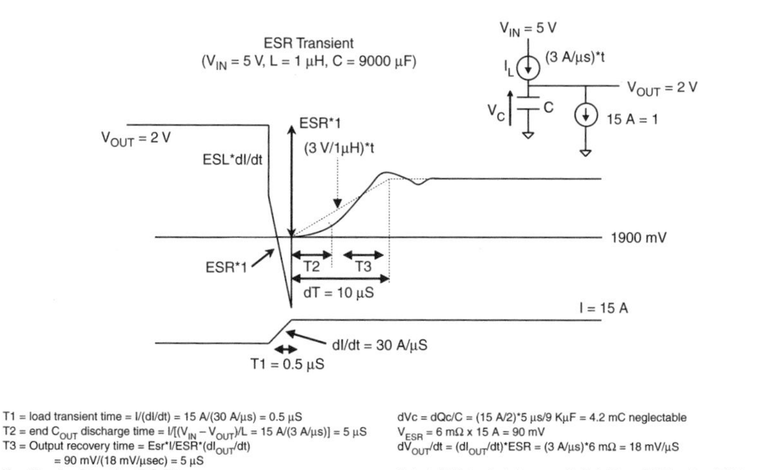

Figure 3–22 Loading transient response.

Ceramic Capacitors

At the opposite end of the capacitor spectrum are ceramic capacitors, characterized by a very low *ESR*, leading to much higher cutoff frequencies, in the order of 200 kHz. The high cost of these capacitors keeps their use to a minimum for fast response until the slower electrolytic capacitors come to bear the transient load. With such low *ESR*, the voltage ripple in a ceramic capacitor V_{RC} is generally due to its capacitive droop.

In a zero to full load transient, while the capacitor starts to supply the load, the current in the inductor builds up at a rate

$$dI_L = [(V_{IN} - V_{OUT})/L] \times dt \qquad \text{Eq. 3–52}$$

until the current equals the load, at which time the capacitor *C* ceases the depletion and initiates recharge. Hence the depletion time is

$$T_{DPL} = I_{LOAD} \times L/(V_{IN} - V_{OUT}) \qquad \text{Eq. 3–53}$$

During this time the depletion charge Q_{DC} is produced by half of the load current since the other half is supplied by the inductor current linear ramp. Hence

$$Q_{DC} = (1/2)I_{LOAD} \times T_{DPL} = (1/2)I^2{}_{LOAD}L/(V_{IN} - V_{OUT}) \qquad \text{Eq. 3–54}$$

it is also

$$Q_{DC} = C \times V_{DROOP} \qquad \text{Eq. 3–55}$$

where V_{DROOP} is the voltage droop across the capacitor corresponding to a Q_{DC} depletion charge. Substituting Eq. 3–54 into Eq. 3–55 we have

$$CV_{DROOP} = 1/2I^2{}_{LOAD}L/(V_{IN} - V_{OUT}) \qquad \text{Eq. 3–56}$$

From which

$$C = (1/2)I^2{}_{LOAD}L/[(V_{IN} - V_{OUT})V_{DROOP}] \qquad \text{Eq. 3–57}$$

For example, if $I_{LOAD} = 20$ A, $L = 3.6$ μH, $V_{IN} = 12$ V, $V_{OUT} = 1.2$ V, and $V_{DROOP} = 60$ mV, we have

$$C = 0.5 \times 400 \times 3.6 \text{ uH}/10.8 \times 60 \text{ mV} = 1.1 \text{ mF} \qquad \text{Eq. 3–58}$$

We can now calculate the corresponding *ESR* for a 1,100 μF ceramic capacitor

$$1/(2\pi \times 200 \text{ kHz} \times C) = ESR \qquad \text{Eq. 3–59}$$

$$ESR = 1/6.28 \times 200 \text{ kHz} \times 1.1 \text{ mF} = 0.72 \text{ m}\Omega \qquad \text{Eq. 3–60}$$

In conclusion the same 20 A load requires 53 mF electrolytic capacitors or just 1.1 mF ceramics. Based on cost, electrolytic is the way to go in this case as 1.1 mF ceramics would definitely be more expensive than 53 mF electrolytics.

It is more likely that using ceramics would raise the clock frequency *f* considerably, leading to a smaller inductor and hence to a smaller value of *C*. The price for raising the frequency is an increase in switching losses.

Losses in the Power Train

Ohmic Losses

MOSFET R_{DSON} high side	$I^2R_{DSON} \times DC$	Eq. 3–61
MOSFET R_{DSON} low side	$I^2R_{DSON} \times (1 - DC)$	Eq. 3–62
Sense resistor (in series with L)	I^2R_{SENSE}	Eq. 3–63
Output inductor	I^2R_{OUTIND}	Eq. 3–64
Input inductor	$I^2_{RMS}R_{ININD}$	Eq. 3–65
Input capacitor	$I^2_{RMS}R_{INESR}$	Eq. 3–66

Dynamic Losses

Output switching	$(V_{IN} \times I/2)(t_R + t_F) \times f_{CK}$	Eq. 3–67
Gate drive	$Q_G \times V_G \times f_{CK}$	Eq. 3–68

IC Losses

Controller	$V_{IN} \times I_{BIASCT}$	Eq. 3–69
Driver	$V_{IN} \times I_{BISDR}$	Eq. 3–70

The previous formulas give us an idea of the main sources of loss and help estimate their relative weight; accordingly they should help the designer pick the right passive and active components depending on his or her cost versus performance trade-off objectives. Even with all these losses the switching regulator is an inherently efficient device. When it comes to discrete transistor elements the right piece of silicon must also be selected in conjunction with the right package, as the same power dissipation in a smaller package can lead to unacceptable case and or junction temperatures. Figure 3–23 shows the detail of the switching losses. Notice how the low side driver operates in quasi-zero voltage switching. This transistor carries the current load for most of the time in applications with low duty cycle (very common in computing). For these reasons this transistor is optimized for ohmic losses, as opposed to switching behavior. Likewise, the high side transistor *Q1* carries current for a small time in application with low duty cycle, while it is subject to full switching swing.

Not surprisingly then, this transistor is optimized for switching performance, as opposed to ohmic.

Due to their different duties, high side and low side MOSFETs are substantially different technologically, so much so that a semiconductor company may often be leading in high side technology while lagging in low side technology or vice-versa.

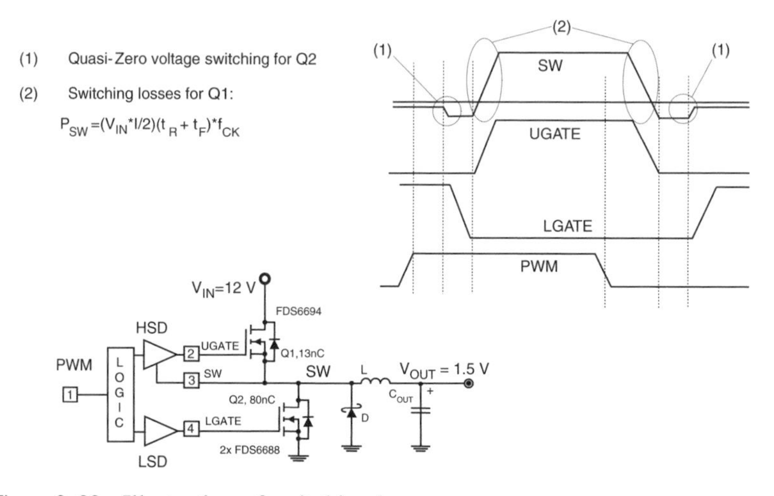

Figure 3–23 Illustration of switching losses.

The Analog Modulator

At the heart of the control scheme of a classic switching regulator is the analog modulator, of which one typical implementation is shown in Figure 3–24. The analog modulator is essentially a comparator, having a linear waveform V_T at its positive input.

From Figure 3–24 we can see graphically that the error voltage V explores the periodic piecewise-linear (saw-tooth) modulation waveform V_T in such a way that when V is entirely below V_T, the output V_{OC} is all positive (100 percent duty cycle or, $V_{OC} = V_{IN}$) and vice versa, when V is entirely at the top of V_T, the output V_{OC} is low (0 percent duty cycle or, $V_{OC} = 0$). The picture illustrates an intermediate case with an intermediate duty cycle, yielding a square wave V_{OC} switching between zero and V_{IN}. Hence the average output voltage varies linearly from 0 to V_{IN} while the error voltage V_ε covers the entire V_{TPP} excursion. Mathematically the input to output transfer function of this block will be

$$A_{VM} = V_{OC}/V_\varepsilon = V_{IN}/V_{TPP} \qquad \text{Eq. 3–71}$$

where A_{VM} represents the linear gain of the modulator, V_{OC}/V_{ε} is the expression of such gain as a function of the input signal V_{ε} and the output signal V_{OC}, and V_{IN}/V_{TPP} is its value. For example, if the amplitude of the triangular waveform is $V_{TPP} = 1$ V and the power supply voltage is $V_{IN} = 12$ V, then the modulator gain $A_{VM} = 12$. Notice how such gain is a function of the power supply, hence, the same block used in a 5 V application will gain less than in a 12 V application (5/12 times less in fact or $A_{VM} = 5$).

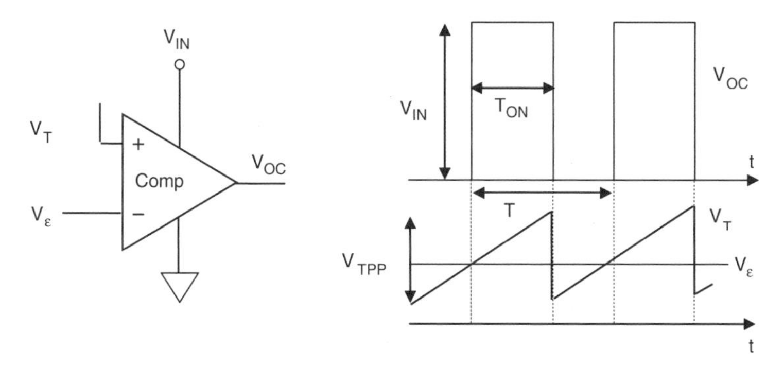

Figure 3–24 Analog modulator.

Driver

The type of driver described here has a function similar to the buffer stage shown in Figure 3–10, but it is suitable to use in a switching application where the signal to the buffer is large, swinging from ground to V_{IN} and the output stage works in the saturation region.

In Figure 3–25 the clock *CK* signal drives through the *I* and 2*I* current sources the buffer T1/T2. In the illustration of Figure 3–25 the clock opens or closes the switch *CK*. When the switch is closed the net current at the shorted gates of T1 and T2 is a sinking current of value *I* ($2I_{SINK} – I_{SOURCE} = I_{SINK}$). Such sink current I_{SINK}, will drive off T1 and on T2, producing a low output at the node V_{DRO}, which will be at a potential just above V_{SW}. Alternatively, with *CK* off, the net current at the shorted gates of T1 and T2 is a source current which will drive off T2 and on T1, producing a high output (node V_{DRO}), which will be at a potential just below V_{CHP}. In turn the T1/T2 buffer output (node V_{DRO}) turns the high side transistor T_{HS} on and off. When V_{SW} is low (say zero voltage), the capacitor *C* is charged to V_{IN} by the diode D. In the next phase, when V_{SW} goes high (approximately V_{IN}), the node V_{CHP} goes to roughly $2 \times V_{IN}$, providing the necessary headroom for transistor T_{HS} to turn on. The action of biasing the node V_{CHP} above V_{IN} via the capacitor/diode C/D action is referred to as *charge pumping*.

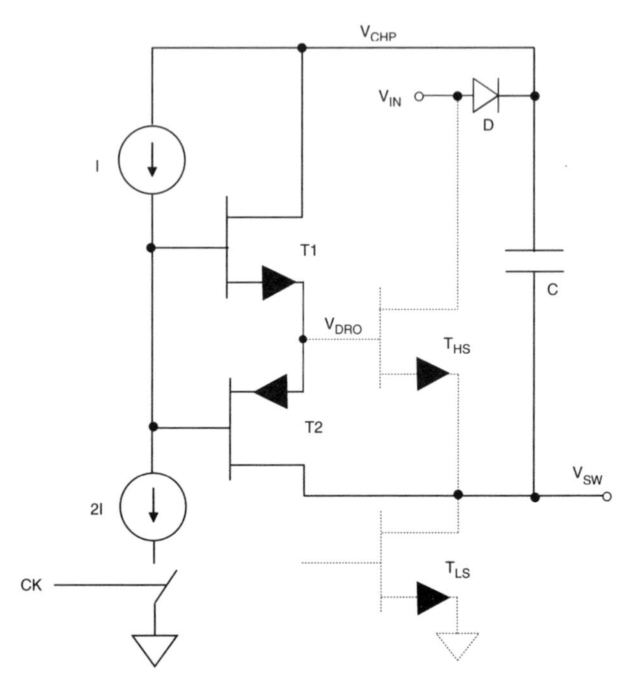

Figure 3–25 Driver stage.

Switching Regulator Block Diagram

In Figure 3–26 we have the entire switching regulator complete with power train, driver and control loop. The output voltage V_{OUT} is sensed via the divider R1/R2 and compared to the reference voltage V_{REF}. The G_M amplified difference of these two signals (V) drives the modulator, which has a square wave output that is buffered and reproduced to the intermediate output node V_{SW}. The LC filter yields the average of this square wave to the final output node V_{OUT}.

Switching Regulator Control Loop

In this section we add some math to the concepts discussed intuitively in the previous sections, arriving at a formula for stability of the control loop. In order to calculate the loop gain, we open the loop as indicated in Figure 3–27, inject a small signal dV_I at one end of the loop and follow it to the other end; multiplying the input signal by the gain of the block in front of it moves the signal forward. Using this technique at the other end of the loop we obtain the returning signal V_R.

We have

$$dV_I \times (G_M) \times Z_{COMP} \times A_{VM} \times A_{LC} \times \alpha = dV_R \qquad \text{Eq. 3–72}$$

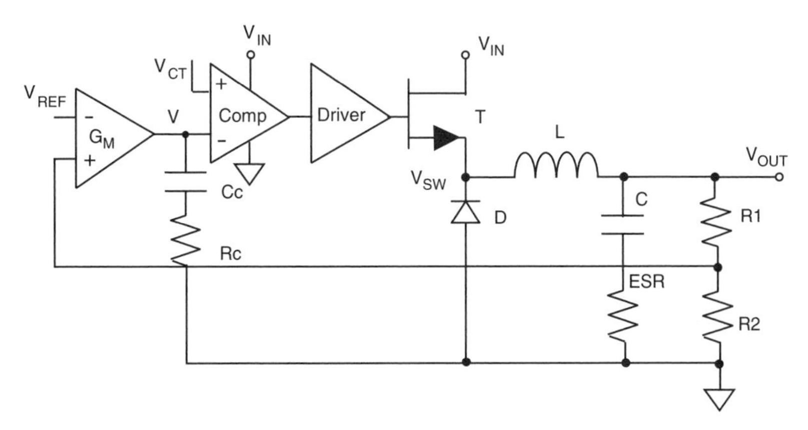

Figure 3–26 Switching regulator block diagram.

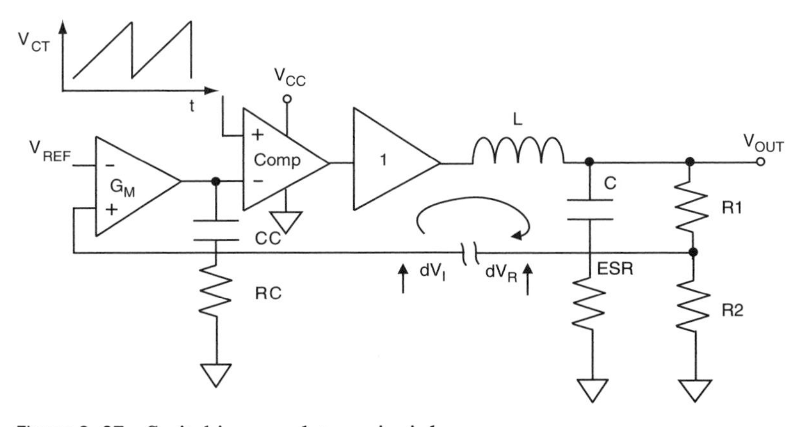

Figure 3–27 Switching regulator principle.

where

$$Z_{COMP} = (1 + pR_C C_C)/pC_C \quad \text{Eq. 3–73}$$

$$A_{VM} = V_{CT}/V_{IN} \quad \text{Eq. 3–74}$$

$$A_{LC} = (1 + pC \times ESR)/(1 + p^2LC) \text{ (double pole approximation)} \quad \text{Eq. 3–75}$$

$$\alpha = R2/(R1 + R2) \quad \text{Eq. 3–76}$$

Hence

$$AOL=dV_R/dV_I=A_V\times[(1+pR_CC_C)/pC_CR_C]\times[(1+pC\times ESR)/(1+p^2LC)] \quad \text{Eq. 3–77}$$

where the expression of A_V is

$$A_V = R_C G_M (V_{IN}/V_{CT}) \times \alpha \qquad \text{Eq. 3–78}$$

is the *mesa* gain (the Bode plot in Figure 3–28 explains the name of this parameter).

Notice how we have a system with one pole in the origin $1/pC$, two coincident poles $1/(1 + p^2LC)$ and two zeros $(1 + pR_CC_C)$ and $(1 + pC \times ESR)$.

The graph in Figure 3–28 shows the placement of all the poles and zeros. The crossover of the Bode plot with the horizontal axis is the system bandwidth f_{BW}. The double pole position in the Bode plot space is uniquely defined by the coordinates (A_V, f_{pLC}). The move from this coordinate to the coordinate of the next zero $[f_{ZESR}, A_V/(f_{ZESR}/f_{pLC})^2]$ is simply dictated by the double slope decay associated with the double pole. Similarly, the crossover coordinate $(1, f_{BW})$ is determined by the single final pole slope.

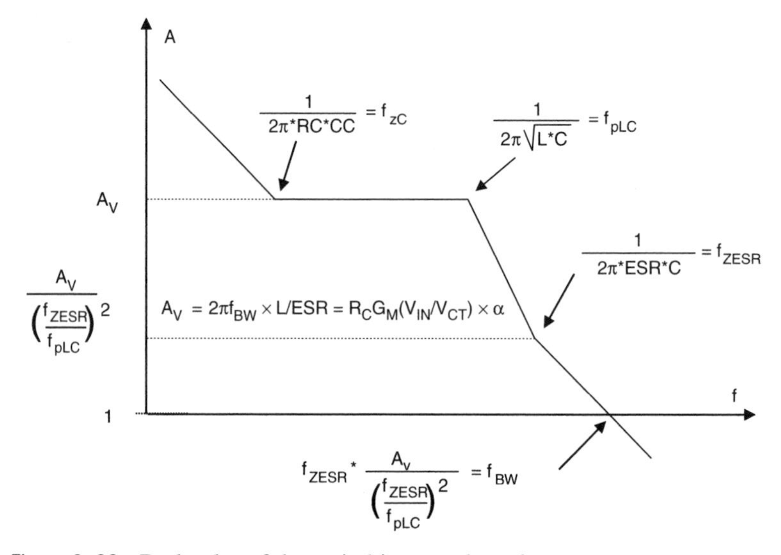

Figure 3–28 Bode plot of the switching regulator loop.

Now we simplify Eq. 3–77 as at the crossover frequency we can neglect 1 in the $(1 + pR_XC_X)$ expressions and enter $AOL = 1$ at $p = j\omega = j2\pi f_{BW}$ obtaining the value for A_V

$$A_V = 2\pi f_{BW} \times L/ESR \qquad \text{Eq. 3–79}$$

For example if $C = 53$ mF, $ESR = 3$ mΩ (from voltage ripple constraints), $L = 3.6$ μH (from current ripple constraints), and $f_{BW} = 30$ kHz from dynamic constraints, from Eq. 3–79 we will need a mesa gain of

$$A_V = 6.28 \times 30 \text{ kHz} \times 3.6 \text{ µH}/3 \text{ m}\Omega = 226 \qquad \text{Eq. 3–80}$$

We know the *ESR* zero to be at 1 kHz

$$f_{ZESR} = 1 \text{ kHz} \qquad \text{Eq. 3–81}$$

and

$$f_{pLC} = 1/2\pi(LC)^{1/2} = 1/6.28 \times 4.3 \times 10^{-4} = 10000/27 = 370 \text{ Hz} \qquad \text{Eq. 3–82}$$

Now from Eq. 3–78

$$A_V = R_C G_M (V_{IN}/V_{CT}) \times \alpha = 226 \qquad \text{Eq. 3–83}$$

and given $V_{IN} = 12$ V, $VCT = 1$ V, $\alpha = 1$, and $G_M = 26$ µA/26 mV = 1/1 kΩ, we have

$$R_C = 226 \times 1 \text{ k}\Omega/12 = 19 \text{ k}\Omega \qquad \text{Eq. 3–84}$$

Now from

$$f_{ZC} = 1/2\pi R_C C_C \qquad \text{Eq. 3–85}$$

setting f_{ZC} at 37 Hz (10x below the *LC* double pole) we have

$$C_C = 1/6.28 \times 19 \text{ k}\Omega \times 37 = 0.22 \text{ µF} = 220 \text{ nF} \qquad \text{Eq. 3–86}$$

And so all the main parameters of the control loop are set.

Input Filter

The input current is chopped as indicated in Figure 3–29 so it will need some input filtering.

Input Inductor L_{IN}

Assuming that at the input we need a current smoothed down to 0.1 A/µs with an input voltage ripple of 0.5 V, we have

$$dV = 0.5 \text{ V} \qquad \text{Eq. 3–87}$$

$$dI/dt = 0.1 \text{ A/µs} \qquad \text{Eq. 3–88}$$

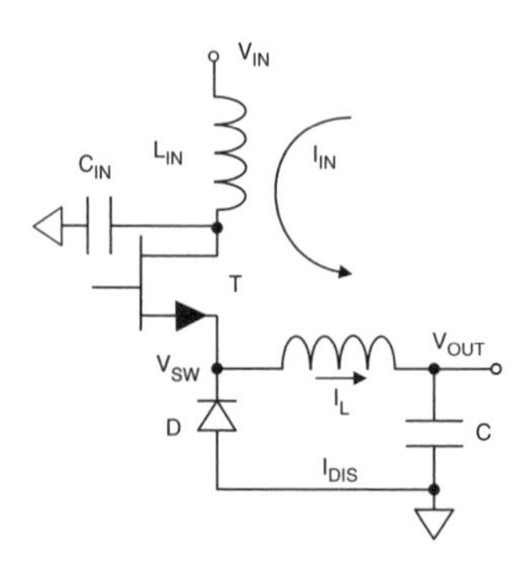

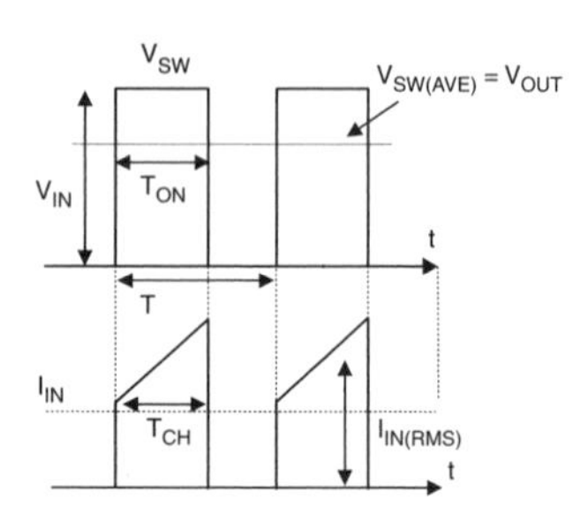

Figure 3–29 Input filter.

and from

$$L_{IN} \times dI/dt = dV \quad \text{Eq. 3–89}$$

$$L_{IN} = 0.5\ \text{V}/(0.1\ \text{A}/\mu\text{s}) = 5\ \mu\text{H} \quad \text{Eq. 3–90}$$

Input Capacitor

$$I = 20\ \text{A} \quad \text{Eq. 3–91}$$

$$dI/dt = 0.1\ \text{A}/\mu\text{s} \quad \text{Eq. 3–92}$$

hence the time to build up 20 A in the inductor is, from Eq. 3–89

$$dT = 20\ \text{A}/(0.1\ \text{A}/\mu\text{s}) = 200\ \mu\text{s} \quad \text{Eq. 3–93}$$

From the formula

$$C_{IN} \times dV/dt = I \quad \text{Eq. 3–94}$$

Knowing that

$$dV/dt = 0.5\ \text{V}/200\ \mu\text{s} = 2.5\ \text{V/ms} \quad \text{Eq. 3–95}$$

$$C = 20\ \text{A}/(2.5\ \text{V/ms}) = 8\ \text{mF} \quad \text{Eq. 3–96}$$

This capacitor has to sustain an RMS current defined as a function of the DC and peak current as follows:

$$I_{RMS} = I(DC - DC^2)^{1/2} \quad \text{Eq. 3–97}$$

from which

$$I_{RMS} = 20(0.1 - 0.01)^{1/2} = 20 \times 0.09^{1/2} = 20 \times 0.3 = 6\ \text{A} \quad \text{Eq. 3–98}$$

It is important to select the input capacitor capable of carrying the calculated RMS current.

Current Mode

So far we have analyzed control schemes based on a single control loop, the voltage control loop setting the output voltage. In any regulator when the output is low—say at start-up—the pass transistor will keep charging the output capacitor via the inductor until the output reaches final value. During this phase the voltage across the inductor is $V_{IN} - V_{OUT}$ and the current is building in the inductor at a rate $[(V_{IN} - V_{OUT})/L] \times t$. If this phase lasts too long, the current build up inside the inductor can be excessive. One way to control such build up is cycle-by-cycle current control using a secondary current control loop nested inside the primary voltage control loop. In the current control loop illustration in Figure 3–30 the current in the inductor is limited to V/R_{DSON}.

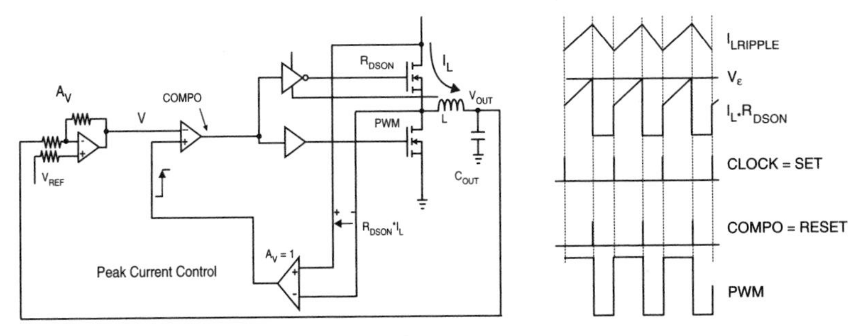

Figure 3–30 Current mode illustration.

Another interesting outcome is that now the entire block from the V voltage node to the I_L current node (inductor current) becomes a simple trans-conductance block with a transfer function that is simply $1/R_{DSON}$

$$I_L/V = 1/R_{DSON} \qquad \text{Eq. 3–99}$$

It follows that from a small signal analysis stand point, the inductor effect in the loop is effectively bypassed; the open loop gain loses the LC double pole and is left with only the C_{OUT} single pole. In this case the expression of the open loop gain becomes

$$AOL = A_V/(pC \times R_{DSON}) \qquad \text{Eq. 3–100}$$

This is a very simple expression compared to Eq. 3–77. A more complicated circuit yields a simpler transfer function! It follows that in principle a current mode regulator should be easier to compensate compared to a plain voltage mode control loop.

In this section we have covered some fundamental aspects of switching regulators and some general techniques for their analysis. With the tools provided we should be able to pick a PWM controller and match it to the power train and compensation elements. With this foundation the reader can venture into more complex aspects of circuital architecture including

- leading and trailing edge modulation
- valley and peak current control
- PWM versus PFM versus hysteretic control

Some of these aspects are discussed in the following chapters. For other aspects not covered here the reader should refer to the references in the further reading section at the end of this book.

3.9 Flyback Converters

Figure 3–31 shows a simplified block diagram of a flyback converter power train. In this voltage mode flyback architecture the energy is stored in the transformer when the switch SW is on and transferred to the load when the switch is off.

The use of a transformer with a turns ration of n:1 allows a lot of freedom as far as input versus output value setting. In a flyback converter the transformer stores energy during the on time of the SW1 transistor. The inductor windings are coupled in such a way (opposite windings as indicated by the dots on each transformer winding) that voltage on the two windings are of the opposite sign. This arrangement, coupled with the placement of diode D (we will approximate the forward drop of the diode to zero), is such that when current flows in the primary winding, it cannot flow in the secondary. Accordingly the energy associated with the primary current cannot be transferred to the secondary and it is stored in the transformer air gap. When the switch is open, the current ceases to circulate in the primary and the energy stored in the transformer gap releases via a current in the secondary. If the voltage on the secondary is V_{OUT} (assured by the control loop not shown here) then this voltage will reflect back on the primary via the turns ration, hence the voltage across the transformer primary will be $-nV_O$. This voltage subtracts to V_{IN} so that the final voltage across the open switch SW during the off phase is

$$V_{SW} = V_{IN} - (-nV_O) = V_{IN} + nV_O \qquad \text{Eq. 3–101}$$

This observation is important because the switch SW is most likely going to be a DMOS transistor and its voltage rating will have to be

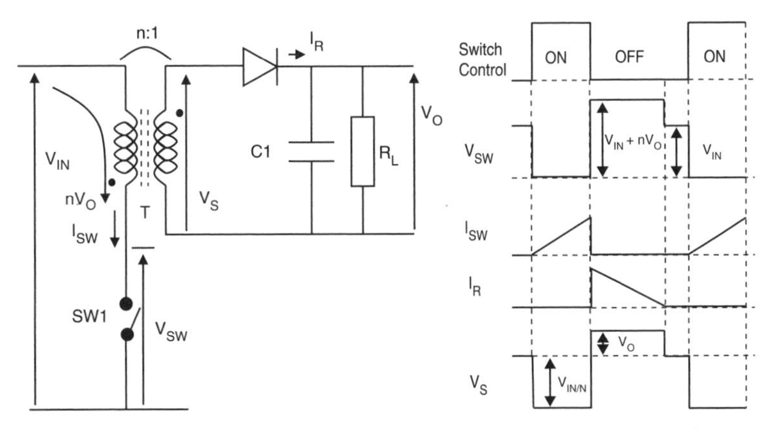

Figure 3–31 Flyback converter simplified block diagram and waveforms.

selected to be safely above $V_{IN} + nV_O$. On the secondary side, the average of the secondary current waveform I_R is the load current. The picture shows the case of light load, with secondary current reaching zero when the primary switch SW is still off. In the absence of current on the secondary there is no voltage on the secondary and no reflected voltage on the primary side, hence during this time interval the voltage across the primary winding is zero and the voltage across the switch SW is simply V_{IN}.

The control loop and its analysis techniques are similar to the one discussed for the buck converter and will not be repeated here.

The other advantage of the transformer, besides input-to-output voltage ratioing, is isolation. In high voltage applications isolation is mandatory not only in the forward path, but also in the feedback path. For this reason transformers in the forward path are a must in offline applications, while in the feedback path often opto-couplers (Figure 3–32) are utilized for signal isolation. In an opto-coupler the photo-diode emits light proportionally to its bias current. A portion of this light hits the corresponding phototransistor which in turn produces a current variation proportional to the incoming light. Since the coupling mechanism is based on light, the opto-coupler works with AC as well as DC feedback signals. In the following chapters we will encounter a few examples of such isolated architectures.

A conventional transformer is called to transfer energy, not store it, so it does not normally have an *air gap*, which is the place where energy is stored. In the flyback configuration, the transformer is hybridized to have an air gap and store energy as discussed earlier. For this reason this "transformer" is also referred to as a "coupled inductor" since the two windings, due to the energy storage twist, act essentially like inductors. Figure 3–33 is a nice illustration of the transformer ferrite core and its energy storage air gap.

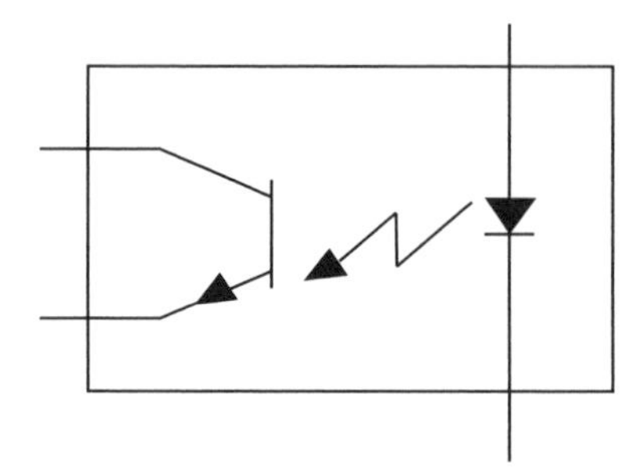

Figure 3–32 Symbol of opto-coupler.

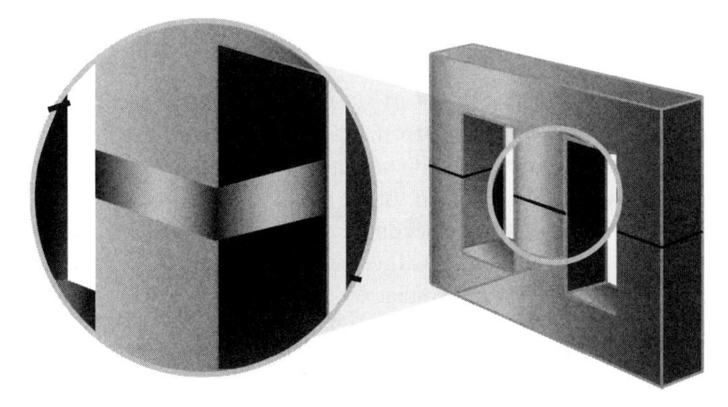

Figure 3–33 Gapped transformer illustration.

As with non-isolated converters, there is a long list of isolated converter architectures as well. We will encounter some of these architectures in the next chapters. For a more systematic treatment of these architectures the reader can refer to the provided references.

Part II Digital Circuits

In this section we will discuss some fundamental digital building blocks for power management. We will quickly review the main properties of the elementary components, the logic gates, so that we can use them to build higher level functions like flip-flops, shift registers, and communications input and output functions. There are many good reasons to mix analog and digital circuits. Soon we will see an example where adding a flip-flop to an analog regulation loop improves the noise insensitivity of the circuit.

Today's power management devices are often externally driven by a central processing unit. In order to interface with such CPUs, power management chips may include on board some or all of the logic elements mentioned above in the form of input-output communications cells. Finally digitalization of power, as will be discussed in detail later, is another reason for a mixed analog and digital approach to power management.

3.10 Logic Functions

NAND Gate

In Figure 3–34 we have a fundamental logic block, the *NAND gate* with its symbol, CMOS implementation, and truth table, the equivalent of the input to output transfer function we have for an analog block. The truth table can be easily proven by exercising it on the CMOS implementation schematic.

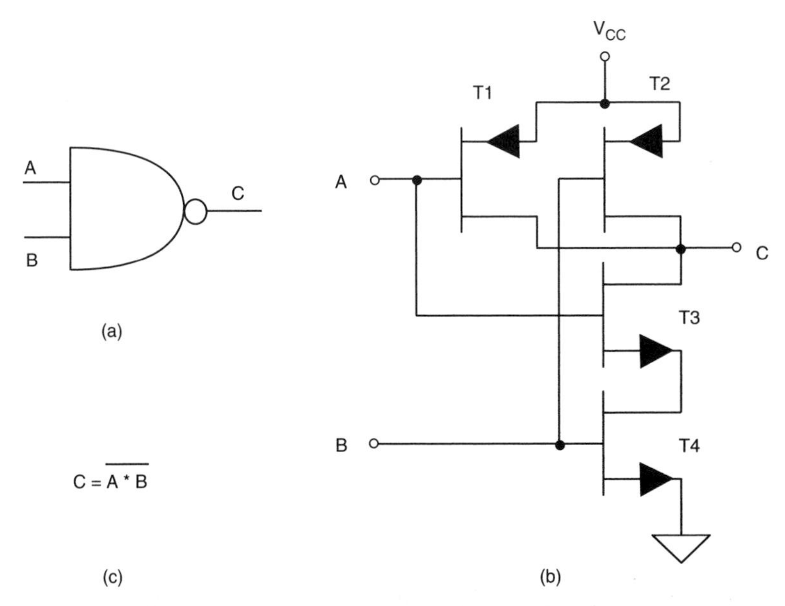

Figure 3–34 Logic NAND gate (a) symbol, (b) CMOS implementation, and (c) truth table.

Set-Reset R Flip-Flop

In Figure 3–35 we have put to use the NAND gates to build a *Set-Reset Flip-Flop*, or to be more precise, a Set#-Reset# one (# stands for the negation bar), the most elementary memory cell. In the truth table M stands for the memory state; when Set# = Reset# = 1 the output stays in the previous state. Naturally one inverter in front of each input will produce a Set-Reset Flip-Flop with the table shown in Figure 3–36.

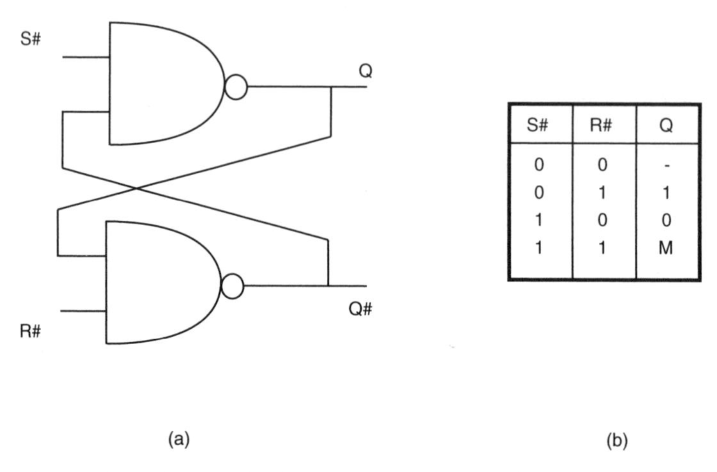

S#	R#	Q
0	0	-
0	1	1
1	0	0
1	1	M

Figure 3–35 Set#-Reset# Flip-Flop (a) logic schematic and (b) truth table.

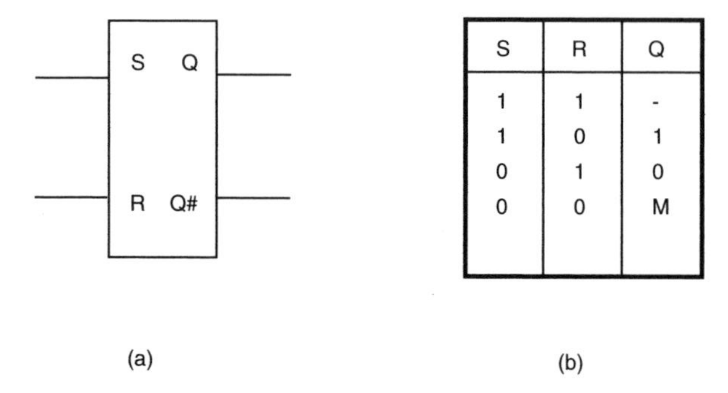

S	R	Q
1	1	-
1	0	1
0	1	0
0	0	M

Figure 3–36 Set-Reset Flip-Flop (a) symbol and (b) truth table.

Current Mode with Anti-Bouncing Flip-Flop

In Figure 3–37 we have put to use the Set-Reset Flip-Flop by inserting it into the current mode voltage control loop from Figure 3–30. The circuit in Figure 3–30 is subject to noise as the comparator can be triggered by any noise spike at any time. By inserting the flip-flop in the loop we create a synchronous system that is insensitive to noise. In fact, from Figure 3–37 and the table in Figure 3–36(b) we see that once reset is triggered (a spike to one and back to zero) the flip-flop is in a memory state until the next set spike. Hence a new charging cycle cannot be initiated by false triggering of the comparator.

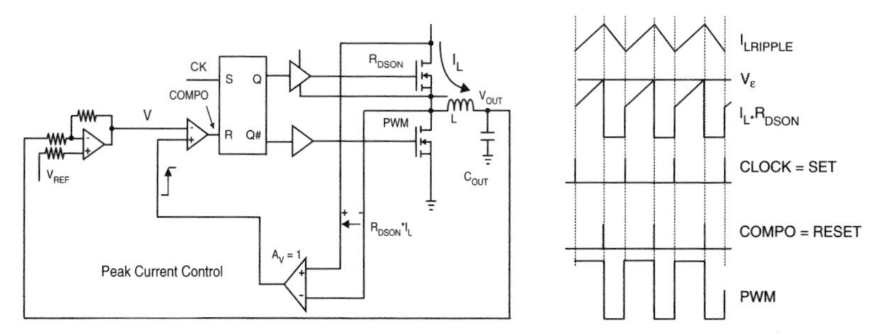

Figure 3–37 Current control with anti-bounce Set-Reset Flip-Flop.

Chapter 4

DC-DC Conversion Architectures

In the first two sections of this chapter, we will discuss in detail two buck converter cases. The first case is a high current buck converter for desktop, handling high current and thus requiring external power MOSFET transistors. The emphasis here will be on the advantages of a specific architecture for this application, called *valley control*. The second case is a low current buck converter for ultraportable applications. For such low power applications, the power transistors are integrated on board. In this case, the emphasis is on the design methodology and fast time to market. In the third section we will discuss the active clamp, a method to deliver instantaneous power to the load bypassing the output filter. This method is advantageous because the filter slows down the response of a regular buck converter regardless of the speed of the front end silicon. In the fourth section we will discuss battery charger system architecture for notebooks. Finally in the fifth section we will cover the subject of digital power, a new trend of implementing power with digital techniques in place of traditional analog ones.

4.1 Valley Control Architecture

Modern CPUs require very low voltage of operation (1.5 V and below) and very high currents (up to 100 A). Such power comes more and more frequently from the *silver box*, a power supply device typically used inside a desktop PC box that provides all the necessary offline power to the PC electronics. With a buck converter, this application results in very low duty cycle, on the order of 0.1 V/V, which stretches the limit of performance of the conventional **peak current-mode control** architecture. The proposed valley control technique brings new life to the buck converter application, allowing it to meet present day specifications more easily as well as remain a viable solution in the future.

Peak and Valley Control Architectures

This section describes the two different architectures illustrated in Figure 4–1.

Peak Current-Mode Control Based on Trailing Edge Modulation

In normal closed-loop operation, the error amplifier forces V_{OUT} to equal V_{REF} at its input, while at its output the voltage V_{ε} is compared to the high side MOSFET current (I_L) multiplied by R_{DSON} (on resistance of the DMOS). When $I_L \times R_{DSON}$ exceeds the error voltage, the PWM comparator flips high, resetting the flip-flop and consequently terminating the charge phase by turning off the high side driver and initiating the discharge phase by turning on the low side driver. The discharge phase continues until the next clock pulse sets the flip-flop, initiating a new charging phase.

Valley Current-Mode Control Based on Leading Edge Modulation

Valley current-mode control operation mirrors that of peak current-mode control, but it has significant advantages. In normal closed-loop operation, the error amplifier forces V_{OUT} to equal V_{REF} at its input, while at its output its voltage V_{ε} is compared to the low side MOSFET current (I_L) times R_{DSON} (notice that in the previous case V_{ε} is compared to the high side MOSFET current). When $I_L \times R_{DSON}$ falls below the error voltage, the PWM comparator flips high, setting the flip-flop and consequently initiating the charge phase by turning on the high side driver and terminating the discharge phase by turning off the low side driver. The charge phase continues until the next clock pulse resets the flip-flop, initiating a new discharging phase.

Current Sensing

The fact that valley current-mode control relies on sensing of the decaying current (the current in the low side MOSFET) has one useful implication for current sensing. If lossless current is implemented, the sensing is done across the low side MOSFET, which is normally on for 90 percent of the time in this type of application. Since the on-time of the low side

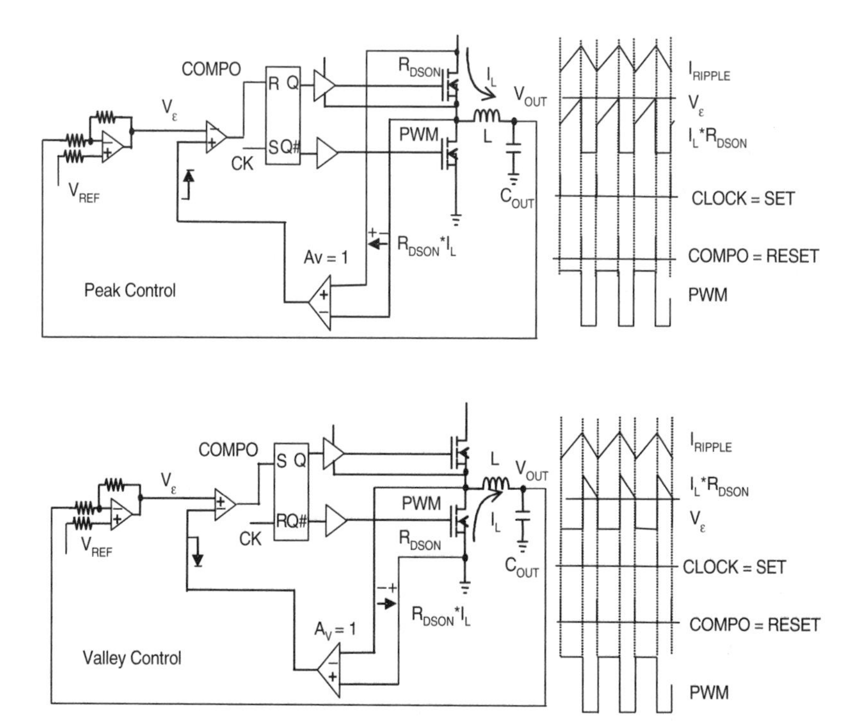

Figure 4–1 Peak and valley control.

MOSFET is almost ten times wider than that of the high side MOSFET, sampling and processing of the low side device current are much easier to accomplish in comparison to high side sensing. Sensing of the high side current at low duty cycles is so undesirable that some solutions in the market have been based on sensing low side current *and* a trailing edge current control strategy. However, the current information comes after the fact—namely after the current has peaked, has started the decaying phase, and can be utilized for cycle-by-cycle peak current control only at the next cycle. In addition to sampling the current, a mechanism must be provided to hold the sampled information until the next cycle. The sample-and-hold mechanism adds complexity to the circuitry, and more importantly adds a delay or phase shift, which tends to compromise the frequency stability of the control loop.

Maximum Frequency of Operation

In the case of very low duty cycle operation with either valley or peak current-mode control, the maximum frequency of operation is limited by the minimum possible on-time of the high side driver. While in both cases the same set of initial physical limitations determines the high side driver minimum pulse width, the peak current-mode control has in addition a limiting settling time requirement, namely the pulse must be wide enough to allow the current to be measured. This additional limitation applies to the cases of lossless high side sensing and to sensing with a discrete high side sense resistor.

Frequency of Operation for Peak Current-Mode Control

Assuming that the settling time for sensing the high side current is $T_{ONP\text{-}MIN} = 100$ ns, then with $DC = 0.1$ V/V, we have a minimum period of operation T_{MINP},

$$T_{MINP} = T_{ONP-MIN}/DC = 100 \text{ ns}/0.1 = 1\ \mu\text{s}$$

which corresponds to a maximum frequency of operation

$$f_{MAXP} = 1/T_{MINP} = 1 \text{ MHz}$$

Frequency of Operation for Valley Current-Mode Control

In valley current-mode control where we sample the low side current, the limitation discussed above is far less strict. Assuming an analogous minimum pulse width for the low side device,

$$T_{MINV} = T_{ONV-MIN}/(1 - DC) = 100 \text{ ns}/0.9 = 110 \text{ ns}$$

which corresponds to a maximum frequency of operation

$$f_{MAXV} = 1/110 \text{ ns} = 9 \text{ MHz}$$

The converter still has to meet the constraint of minimum on-time of the high side driver. Transition times of 10 ns and below are obtainable today. Assuming a minimum on-time of the high side device of two transition times, we have

$$T_{ONHV-MIN} = T_R + T_F = 10 + 10 = 20 \text{ ns}$$

$$T_{MINHV} = T_{ONHV-MIN}/DC = 20 \text{ ns}/0.1 = 200 \text{ ns}$$

This yields a maximum frequency of operation,

$$f_{MAXHV} = 1/T_{MINHV} = 5 \text{ MHz}$$

While today's conventional monolithic and discrete technologies do not permit practical operation at such a high clock rate, it appears that as these technologies improve, only valley current mode control will be able to easily track the speed curve and operate at such high frequencies.

Transient Response of Each System

In this section we discuss the transient response of the two systems. The advantages of valley control are obvious from Figure 4–2. This system is inherently fast and able to turn on immediately in response to a step current, as opposed to peak control, where a delay (T_{DELAY}) as high as a full clock period is to be expected. In both cases the current ramps up (builds up linearly inside the inductor) with a slope that is determined by the inductance and saturated voltage appearing across the inductance and limited by the maximum duty cycle DC_{MAX}.

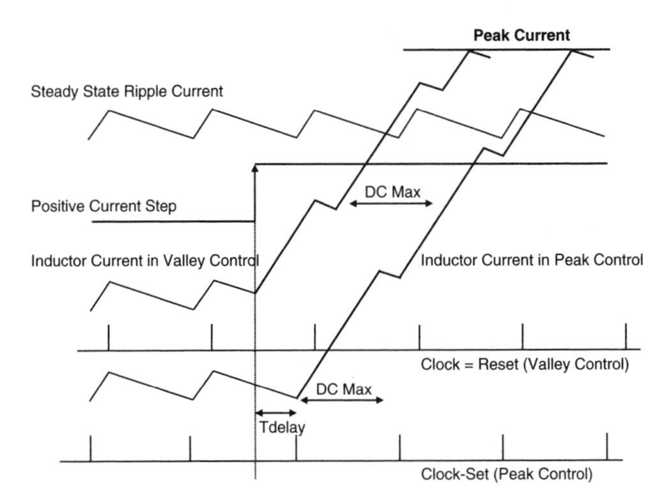

Figure 4–2 Positive current step.

As an example, if the clock is 700 kHz per phase, a full period delay corresponds to 1.5 μs.

Traditional peak current-mode control architecture will need enough output capacity to hold up for one extra 1.5 μs in comparison to valley current-mode control. Consider for this example that an output capacitor of 1 mF will discharge an extra 100 mV with a 65 A load in 1.5 μs.

The comparative responses to a negative load current step are illustrated in Figure 4–3. Here again the advantage of valley control architecture is evident. During a negative load current step, the valley control scheme is able to respond with zero duty cycle. On the other hand, after

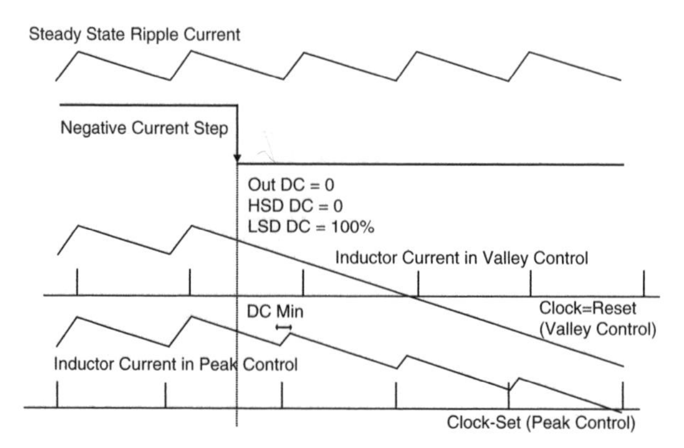

Figure 4–3 Negative current step.

each clock pulse with peak current-mode control, the controller forces a minimum width high side on-time. This minimal on-time is determined by the speed of the current control loop. Thus, it is seen that the valley control scheme offers superior transient response with a negative load step as well.

Valley Control with FAN5093

The FAN5093 is a two-phase interleaved buck controller IC that implements the valley control architecture based on leading edge modulation. The current normally is sensed across the low side MOSFET R_{DSON} (for lossless current sensing); although for precision applications a physical sense resistor can be placed in series with the source of the low side MOSFET.

Figure 4–4 shows the two PWM switching nodes of the two-phase buck converter, with the FAN5093 clocking each phase at a frequency of 700 kHz.

In Figure 4–5 we show the response of the voltage regulator to a 25 A per phase positive current step.

In Figure 4–6 we show the response of the voltage regulator to a 25 A per phase negative current step.

Figure 4–7 shows the FAN5093 application and highlights the two-phase interleaved architecture of this buck converter. Multiphase is discussed in more detail in Chapter 7. As evidenced in Figure 4–4, *interleaving* consists of phasing the two channels 180 degrees apart so the load current is provided in a more time-distributed fashion, leading to lower input and output ripple currents. In other words, if the load is too high to be handled by a single phase, there are two ways to solve the problem. The more traditional way is brute force: to beef up the circuit by paralleling as many MOSFETs as necessary. The new concept introduced by multiphase is interleaving, to take the same numbers of transistors that we

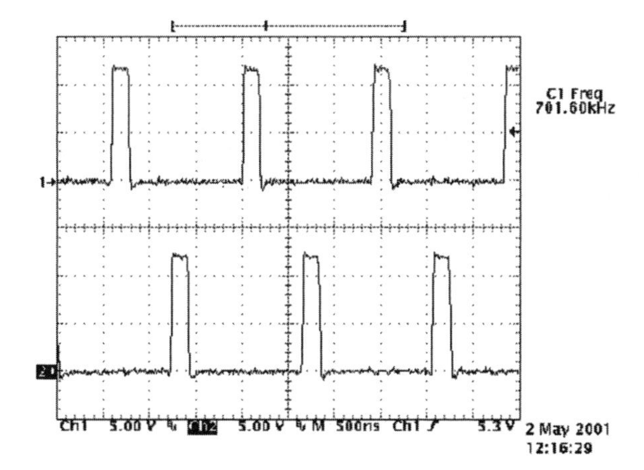

Figure 4–4 Interleaved buck converter: $V_{IN} = 12$ V, $V_{OUT} = 1.5$ V, $f_{CK} =$ 700 kHz per phase. Top waveform: switching node of Phase 1. Bottom waveform: switching node of Phase 2.

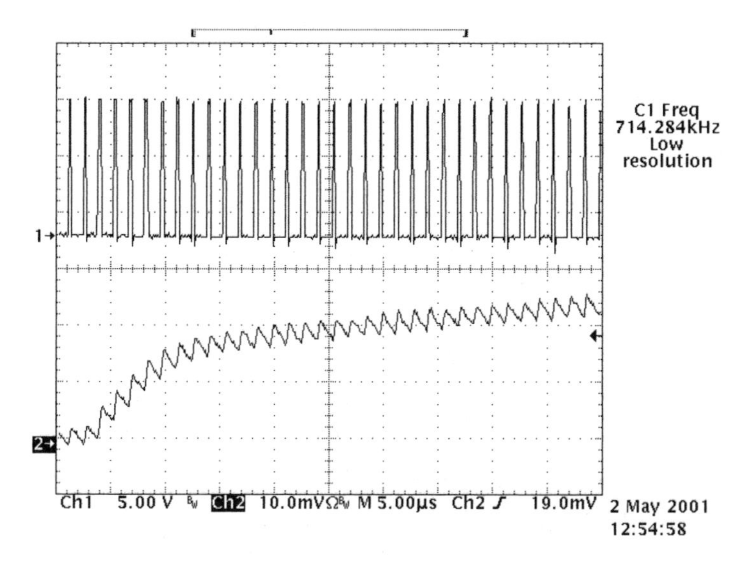

Figure 4–5 Regulator response to a positive current step. Top waveform: switching node of Phase 1. Bottom waveform: Phase 1 current.

wanted to parallel and operate them out of phase. Now we have reduced input and output ripple, and hence we can get by with smaller input and output passives.

The IC whose die layout is shown in Figure 4–8 incorporates the controller and the drivers and works in conjunction with an external DMOS transistor

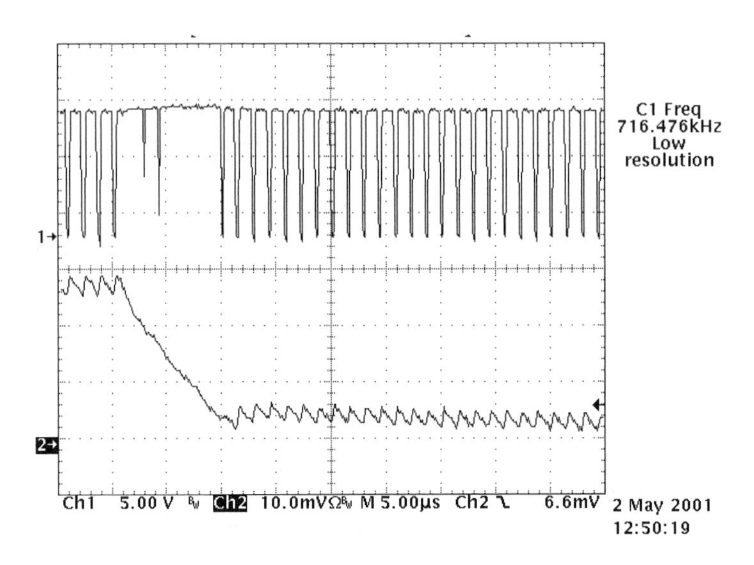

Figure 4–6 Regulator response to a negative current step. Top waveform: switching node of Phase 1. Bottom waveform: Phase 1 current.

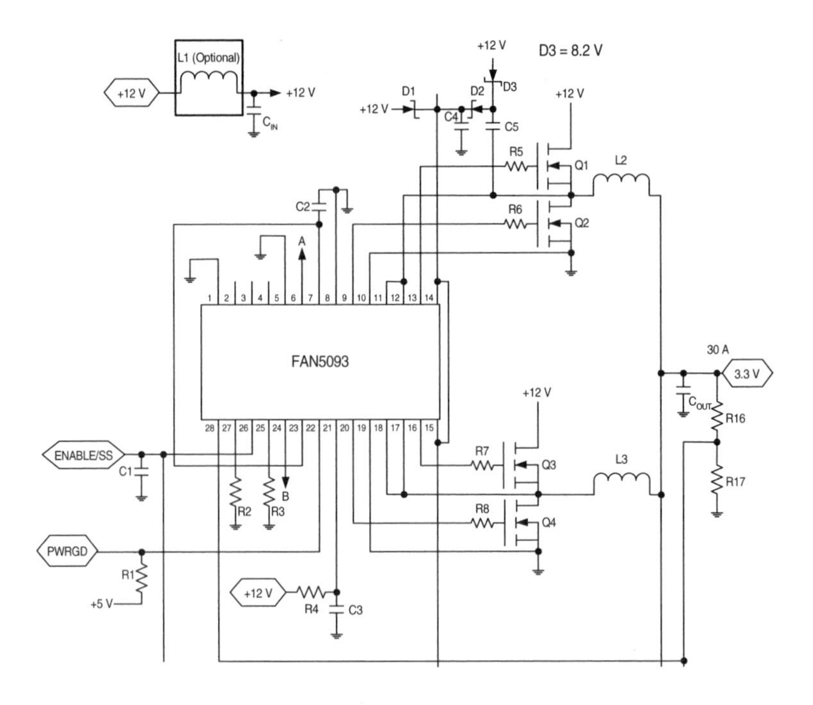

Figure 4–7 FAN5093 application diagram.

to handle 30 A with 3.3 V. For further details, a full data sheet of the FAN5093 is provided in Appendix A. The IC is built in a 30 V, 0.8 μm BiCMOS mixed signal process with excellent Bipolar and CMOS performance.

Figure 4–8 FAN5093 die picture.

Conclusion

We have shown that the valley current-mode control buck architecture based on leading edge modulation has superior transient response characteristics when compared to the traditional peak current-mode control buck architecture based on trailing edge modulation. These transient response characteristics translate directly into a reduced number of output capacitors and consequently lead to a more cost-effective solution in a smaller board space. This advantage, while already measurable today, will become more marked in the future when progress in discrete and controller technologies will enable multi-MHz frequencies of operation at reasonable efficiencies.

4.2 Monolithic Buck Converter

A New Design Methodology for Faster Time to Market

Until recently, the prototype of a new power management subsystem would be built only after its various components were physically available for prototype construction. However, a new trend is emerging, where a *virtual prototype* is built by the subsystem manufacturer far ahead of the availability of physical components. From a power chip designer's perspective, the benefit is that a good behavioral model of the voltage regulator can be utilized prior to the transistor-level design, reducing time-consuming full-chip simulations to a minimum. From the system designer/customer's perspective, the benefit is that behavioral models will be available far ahead of final silicon. Therefore, the system designer can quickly test his virtual subsystem using behavioral simulations to provide timely feedback to the chip designer before the chip is frozen into silicon. When the physical subsystem prototype is finally built, testing and debugging will be much faster and easier to finalize thanks to the previous virtual iterations.

In the model (Figure 4–9), the platform designer launching the system Px at time zero will wait six months for delivery of silicon Sx+1 for his next platform Px+1. But the designer could immediately obtain behavioral models of the silicon Sx+1 from the silicon vendor, who is already twelve months into the development cycle of that silicon.

Since Moore's law seems to hold well no matter what, the end result should be an improvement in productivity rather than a reduction in the development cycle. This results in a higher number of platform varieties launched in a unit of time.

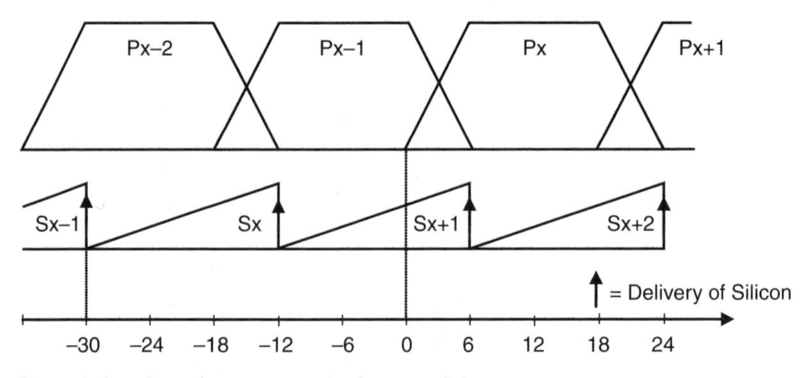

Figure 4–9 Development cycle time model.

The Design Cycle

This section explores the various steps of designing the controller for a buck converter, from the construction of a simple behavioral macro model and the subsequent transistor-level Simulation Program with Integrated Circuit Emphasis (SPICE) simulation to the silicon implementation. The time duration for each phase is also discussed. Finally, we will compare the waveforms obtained with behavioral simulation versus SPICE simulations and pictures taken at the oscilloscope from the physical prototype. We will see that the three different methods produce quite similar results.

The FAN5301

Figure 4–10 shows the block diagram of the FAN5301, a high-efficiency DC to DC buck converter, while Figure 4–11 shows the application. The architecture provides for high efficiency under light loads and at low input voltages, as well as optimum performance at full load. Further detail is provided in a later discussion of the behavioral block diagram.

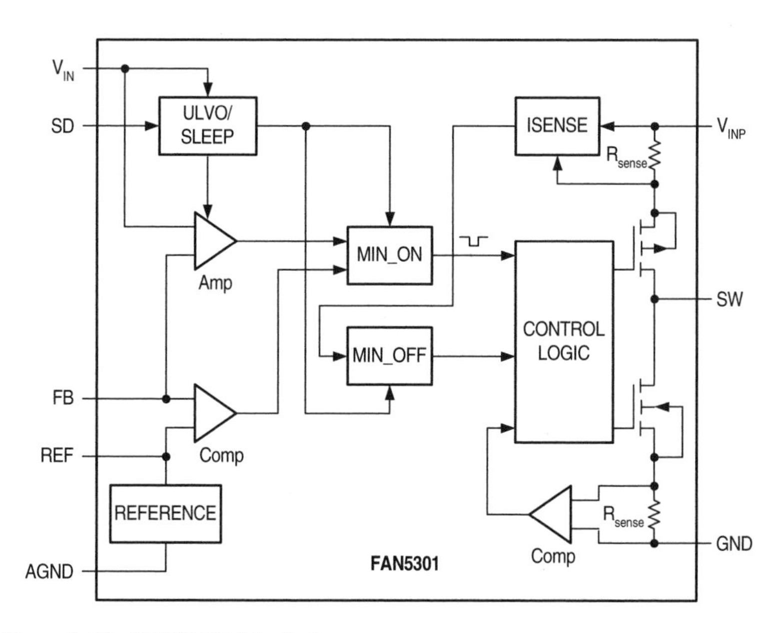

Figure 4–10 FAN5301 block diagram.

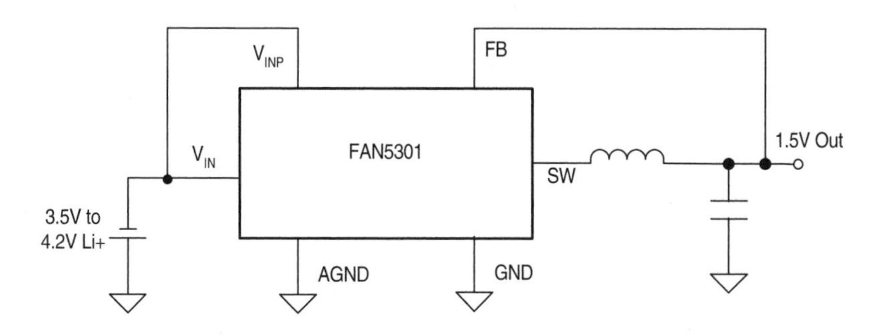

Figure 4–11 FAN5301 application.

The Behavioral Model

Figure 4–12 shows the behavioral model of the entire power supply, complete with the controller as well as the external components. The controller is based on a minimum on-time, minimum off-time architecture.

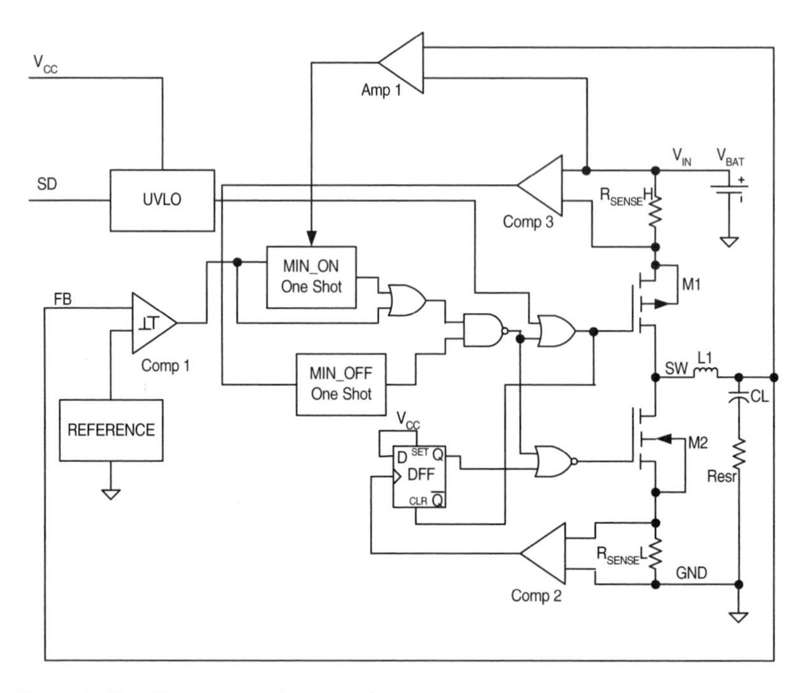

Figure 4–12 Voltage regulator model.

Light Load Operation

The main control loop in light load operation is the minimum on-time section in Figure 4–12, consisting of a *hysteretic comparator* (Comp1) that controls a "one shot" circuit (MIN_ON One Shot) and driving the high side PMOS switch M1. The one shot circuit fires on for a duration of time that remains steady at constant input voltage and increases as the voltage across the pass transistor ($V_{IN} - V_{OUT}$) decreases.

During on-time, the high side driver transistor M1 is turned on for a duration equal to the one shot (MIN_ON One Shot) on-time, and then turned off. When M1 turns off, M2 is turned on until the inductor current goes to zero, at which point both transistors are turned off until the output voltage falls below a set threshold. At this point the one shot fires again, initiating another cycle.

Full Load Operation

In full load, the minimum on-time block is bypassed by the hysteretic comparator (Comp1), which forces the output to swing with a ripple equal to the comparator hysteresis. At full load, the current in the inductor is continuous and the operation is synchronous.

Over-Current

The minimum off-time one shot (MIN_OFF One Shot) in Figure 4–12 is controlled with a cycle-by-cycle current limit comparator (Comp3) that samples the current flowing through M1 via R_{SENSE}. In over-current, the high side driver is turned off for the time that is set by the MIN_OFF One Shot. The high side driver is then turned back on. If the over-current persists, M1 is turned off again after a short time set by the Comp3 and MIN_OFF One Shot loop delay.

Completing the circuit shown in Figure 4–12 is an under voltage lockout circuit (UVLO block) and a precision-trimmed band-gap reference (V_{REF} block).

One Shot

To illustrate the level of complexity of the regulator behavioral model, Figure 4–13 dives into a representation of the MIN_ON One Shot device shown in Figure 4–12. The one shot consists of a current source I_{RAMP} that charges a capacitor C1, which can be reset via the switch S1. The comparator "looks" at the ramp level with respect to the control reference voltage CONT. The ramp time provides the duration of the output pulse.

Comparator

Delving further down into the nesting of the controller, Figure 4–14 shows the block diagram of the comparator Comp in Figure 4–13. The PSPICE (a popular brand and flavor of SPICE) behavioral model of the comparator uses "primitive" SPICE-level blocks (like the one in Figure 4–15). A summing block follows the inverting stage into the GLIMIT (or gain/voltage limiter), into the inverter, and then into the output. The GLIMIT function provides the comparator gain while the limit function allows the designer to restrict the voltage output to a reasonable range, like 0–5 V. The resistor provides some convergence help.

A *SPICE deck* like the one in Figure 4–15 describes each primitive functional block in Figure 4–14.

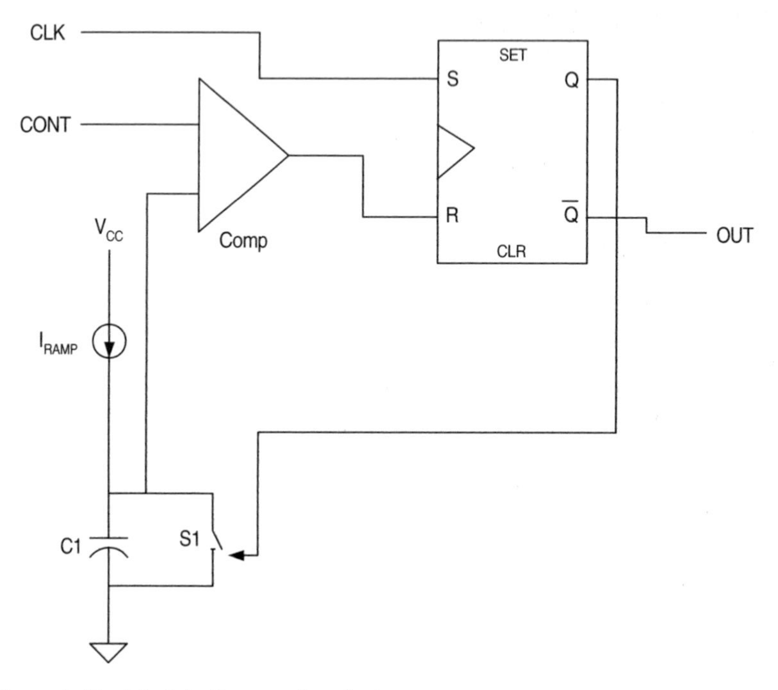

Figure 4–13 Model of a one shot device.

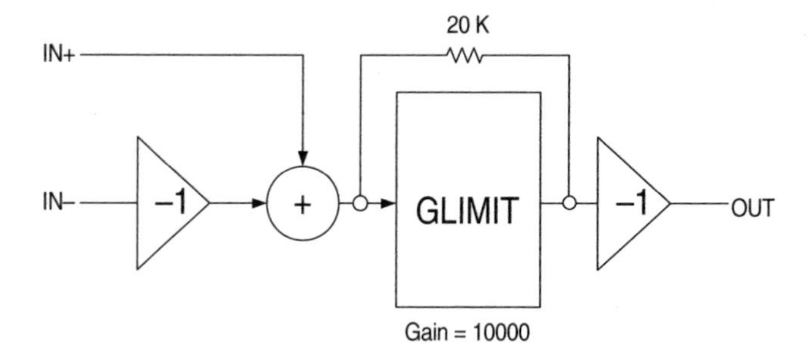

Figure 4–14 Comparator behavioral model.

Results

The waveforms in Figure 4–16 show the transient response of the regulator to a step function load from 0 mA to 100 mA as produced by the behavioral simulation. Input voltage is 3.3 V and output voltage is 1.25 V. The graph in Figure 4–17 shows the same transient response from a transistor-level SPICE simulation (Spectre on Cadence platform). Finally, Figure 4–18 illustrates the same transient from the physical prototype.

```
E_HS62_GAIN2  $N_0101 0 VALUE {-1*V($N_0100)}
R_HS62_R1  $N_0102 $N_0101 20k
E_HS62_GLIMIT1  $N_0100 0 VALUE {LIMIT(V($N_0102)*10000,0,-5)}
E_HS62_SUM1  $N_0102 0 VALUE {V($N_0103)+V($N_0104)}
E_HS62_GAIN1  $N_0104 0 VALUE {-1*V($N_0105)}
R_HS62_R4  0 $N_0103 10e6
R_HS62_R3  0 $N_0104 10e6
R_HS62_R5  0 $N_0102 10e6
```

Figure 4–15 GLIMIT SPICE deck.

Although the corresponding waveforms in Figures 4–16, 4–17, and 4–18 are not identical, they are sufficiently similar to infer consistent functional behavior from each. Some of the differences can be attributed to the unavoidable variation on the external components such as inductor parasitics, capacitor ESR/ESL, and noise (in the case of Figure 4–18). Also complicating the comparison is the fact that the laboratory equipment cannot duplicate the instantaneous current load change the way that simulations can. On the simulation side, differences in SPICE operating parameters (Spectre versus PSPICE) and sampling can affect the output wave shapes due to interpolation and sampling errors.

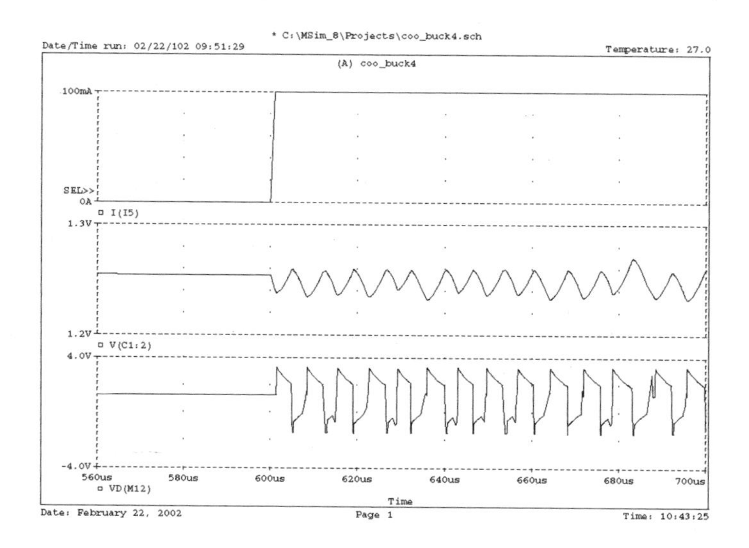

Figure 4–16 Transient response from behavioral model.
Top Trace: I_{LOAD} 0–100 mA
Middle Trace: V_{OUT}
Bottom Trace: *SW pin*

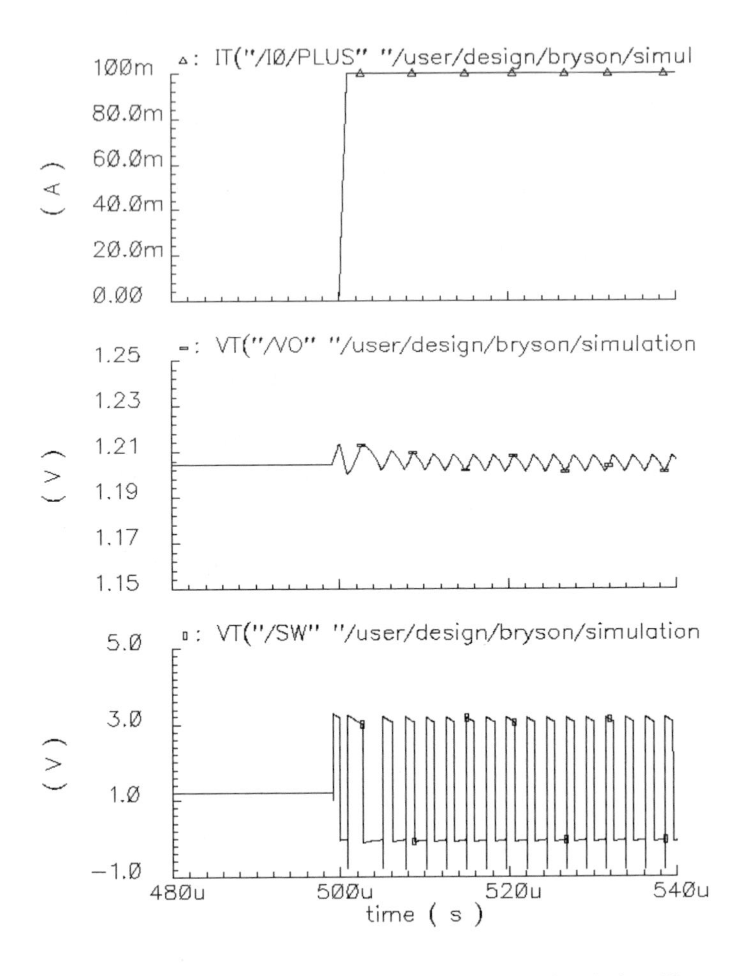

Figure 4–17 Transient response from transistor-level simulation (Spectre).
Top Trace: I_{LOAD} 0–100 mA
Middle Trace: V_{OUT}
Bottom Trace: *SW pin*

Timing

The behavioral simulation took three and a half minutes running on a Pentium II, 366 MHz platform providing data for an 800 μs simulation. The full chip SPICE simulation, operating on more than 500 individual components (transistors and passives), lasted nearly four hours running on a SUN Ultra10. The behavioral model was built during the initial architectural

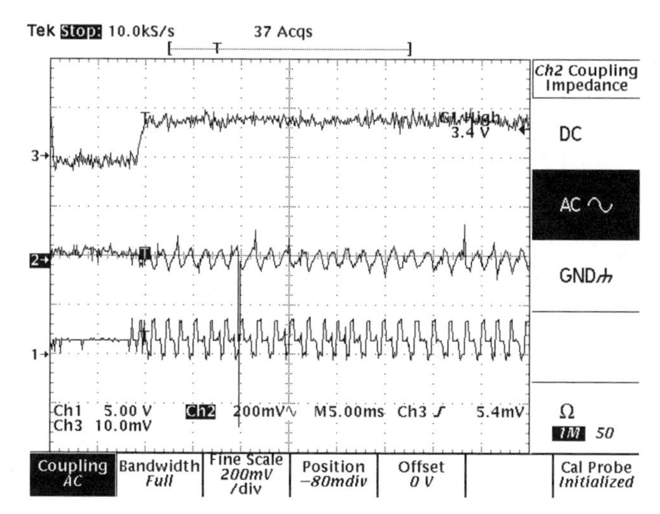

Figure 4–18 Transient response from silicon prototype.
Top Trace: I_{LOAD} 0–100 mA
Middle Trace: V_{OUT}
Bottom Trace: *SW pin*

development after the preliminary data sheet was available and took two weeks to build. This model was able to eliminate many simulation iterations at the full chip level and gave the customer preliminary results six months ahead of functional silicon.

Conclusion

Behavioral models of voltage regulators and power management subsystems already are a reality. These models increase productivity, reduce the number of simulations and silicon and system iterations, and require less time to design. Finally, the speed and simplicity inherent in the behavioral models allow for the extension of the simulation approach to new levels of circuit complexity, in line with the growing complexity of modern systems.

4.3 Active Clamp

Introduction

Present day microprocessors require over 100 A of supply current at voltages near 1 V. Future processor generations are projected to require greater current at supply voltages even lower. Furthermore, the load demand can exhibit abrupt changes from light load (<0.5 A) to full load. These processors can impose load steps with *di/dt* (rate of increase of a current step *di* in a given time interval *dt*) on the order of a few A/ns when the processor switches between inactive and active modes, or vice versa. We will discuss some of the ramifications for the design of the power supply required to supply these microprocessor loads. We focus on the use of a paralleled active circuit, which can be thought of as an active clamp, the details of which are explained in the next paragraph. Breadboards of these circuits have been built and tested, and a prototype IC design has been tested.

The discussion of active clamps is organized as follows. The first section outlines the application, while the second section outlines the design of an active clamp device that may be used in parallel with a switchmode converter to alleviate the demand on the passive filter components. The third section reports the test results for the breadboards of these circuits, and the final section wraps up additional issues.

Application

Figure 4–19 shows the interconnection of the active clamp circuit with a standard DC-DC converter. The active clamp is designed to work in parallel with the output of a conventional switching regulator. Its function is to do nothing—i.e., appear as a high impedance terminal—during normal steady state switching regulator operation. In the event that a load transient drives the output of the switching regulator beyond a specified tolerance band, the clamp activates to hold the output voltage within this specified band. As such, the active clamp must be designed to handle the full load current, but only for short intervals on the order of tens of microseconds. Hence, the active clamp circuit must be designed for high peak current and high peak power capability but need not handle significant continuous steady state power. In order to provide a useful function, the device must be able to sink or source rated current within a hundred nanoseconds.

In order to test easily the potential benefits of a CPU power application by using the active clamp, we shall discuss a scaled down application requiring about 10 A from a switching regulator operating at about 300 kHz. For the 5–3.3 V buck application, a filter inductor of $L = 1.5\ \mu H$ will yield a peak-peak current ripple of about 2.5 A, a typical design. In order to achieve a peak-peak output ripple of 50 mV, a capacitor of only about 20 μF is needed if ESR is neglected. With a 20 μF ceramic chip capacitor with ESR typically less than 10 m, the ESR contribution to the output ripple voltage will only be on the order of 20 mV peak-peak. It is likely that some designers would select a larger capacitor when considering ripple, in order to further reduce the voltage ripple and to avoid the possibility of a capacitor reliability problem because of the large ripple

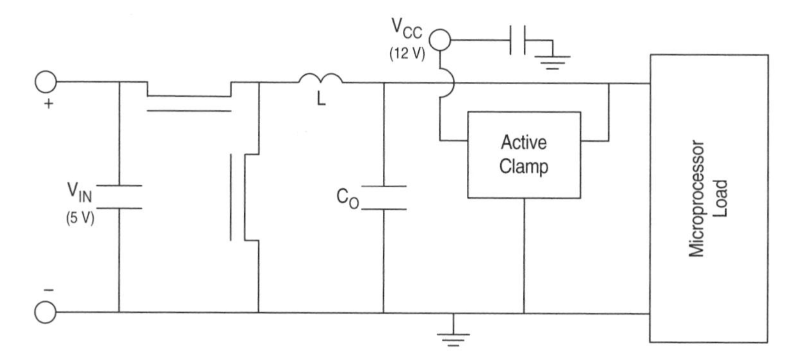

Figure 4–19 Interconnection of active clamp circuit with switchmode regulator.

current. Nevertheless, designs in the range of 20 μF to 50 μF are feasible with ceramic chip capacitors, or with a total capacitance approaching 200 μF due to higher ESR for solid tantalum chip capacitors.

When considering transient response to step load changes, the requirement on the output capacitor may be far more stringent. A very simple analysis based on an average model (as seen in Figure 4–20) to a load current of a voltage feedback scheme predicts a step response of the form of Eq. 4–1

$$v(t) = \frac{I_O}{\omega_c C_O}\sin(\omega_c t) \qquad \text{Eq. 4–1}$$

to a load current step of magnitude I_O amps, where C_O is the output capacitance and c is the crossover frequency of the feedback loop. Here we assume that all forms of damping including output capacitor ESR are negligible, current feedback is not used, the input impedance is stiff, and the duty cycle control does not saturate. The latter is the most significantly unrealistic assumption, as we will discuss below. We use the approximation of Eq. 4–1 to estimate the first peak of the voltage in the transient response, for a scenario in which duty cycle saturation does not play a role. With an aggressive crossover frequency of 100 kHz, C_O = 20 μF, and a load step of 10 A, the peak voltage transient is 800 mV. With a +5 percent tolerance band of about 3.3 V, this peak voltage is unacceptable since the tolerance band amounts to only +165 mV. With the same conditions, except for an output capacitance C_O = 200 μF, we see a voltage peak of 80 mV with this simple model. For a design with C_O = 2,000 μF, with all other conditions the same, we expect to see only 8 mV of voltage disturbance.

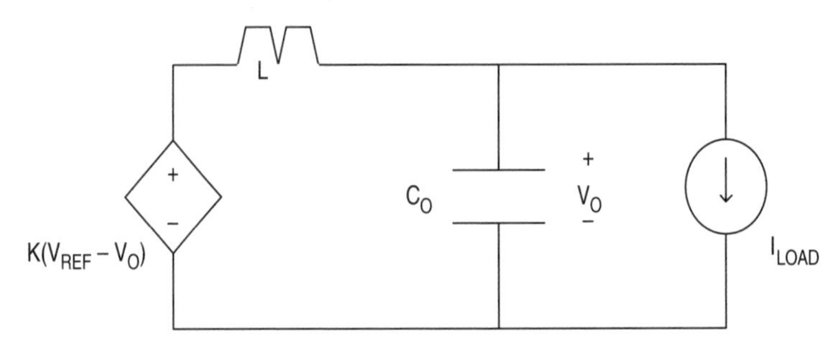

Figure 4–20 Simple average circuit model for voltage-mode controller buck converter.

The above analysis is unrealistic due to the fact that the duty ration control will nearly always saturate (i.e., reach its maximum or minimum) under a large load transient. A simple analysis incorporating saturation is depicted in Figure 4–21. Assume the converter output initially is 3.3 V,

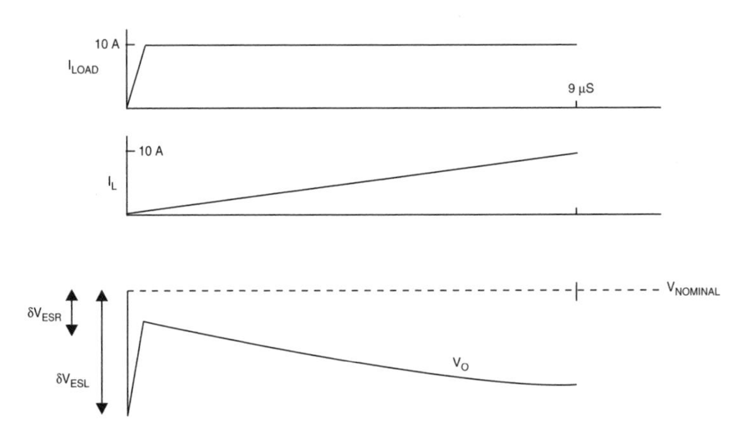

Figure 4–21 Circuit waveforms under duty cycle saturation.

supplying zero-load current. The load then steps to 10 A. At the end of the 9 µs transient, the output voltage can begin its recovery; therefore this is the approximate time point when the voltage bottoms out. The total charge removed from the output capacitor, assuming a ramp current waveform, is 45 µC. The analysis is independent of the output capacitor value and assumes that both a stiff (i.e., large input capacitance) input voltage and a 100 percent duty cycle can be applied. Now with a 20 µF output capacitance, the voltage transient will be about 2.25 V peak. With 200 µF or 2,000 µF, the transient voltage peaks at 225 mV or 22.5 mV, respectively. Since these estimates are optimistic (assuming stiff input voltage and 100 percent duty cycle), a design choice of 2,000 µF does indeed make sense.

The advantage of the active clamp is that the output capacitor will only need to be designed to handle the ripple current and not to contain the output voltage during load transients. As such, for the application discussed here, ceramic chip capacitors as small as 20 μF may be feasible when incorporating an active clamp circuit, whereas output capacitances of at least 400 μF would otherwise be needed. Therefore, an active clamp makes it possible to save 380 μF. Let's now consider an application in which the load is specified at 20 A maximum at 3.3 V, but the tolerance is considerably tighter at +2 percent. The output capacitor bank required to support a full step load transient in this application includes 20,000 μF of bulk hold-up capacitance in conjunction with a network of paralleled high frequency (lower inductance, lower ESR) tantalum and ceramic capacitors. In this scenario and in those involving the specification of future microprocessors, the active clamp will yield far greater benefits.

Note that during the activation of the clamp circuit following a load increase, charge is supplied from a bypass capacitor on the V_{CC} supply. This supply level would likely be set at a higher voltage (e.g., 12 V) than the 5 V supply, which probably would be used to supply the main DC-DC converter. As such, the total capacitance required to bypass the V_{CC} supply would be much smaller than would be required to hold up the output voltage directly at the output pin, in the absence of the clamp circuit. The use of a smaller capacitor on the V_{CC} supply is possible because the V_{CC} supply rail can be allowed to sag, perhaps by a few volts during a transient.

The active clamp circuit actually is built from two independent half circuits: the lower clamp and the upper clamp. We will begin by discussing the overall scheme of one of the half circuits. A functional schematic and control loop diagram for the upper clamp function is shown in Figure 4–22. The upper clamp function is activated whenever the output voltage goes above the high reference V_H.

The current sensing feedback around the opamp constitutes a minor feedback loop, forcing the clamp output transistor current to be approximately

$$I_{OUT} = -\frac{(V_O - V_H)R_F}{(R_I R_S)} \qquad \text{Eq. 4–2}$$

at low frequency. Dynamically, this minor loop has the response of the opamp connected in a standard differential gain of twenty-five connections. For an opamp such as the LM6171, the corner frequency for the minor loop occurs at about 3 MHz. The outer loop determines the overall speed of response. With an output capacitance of 20 μF, the outer loop exhibits a corner frequency at about 4 MHz. As such, one would expect to see a second order system, i.e., a ringing, transient response, characterized

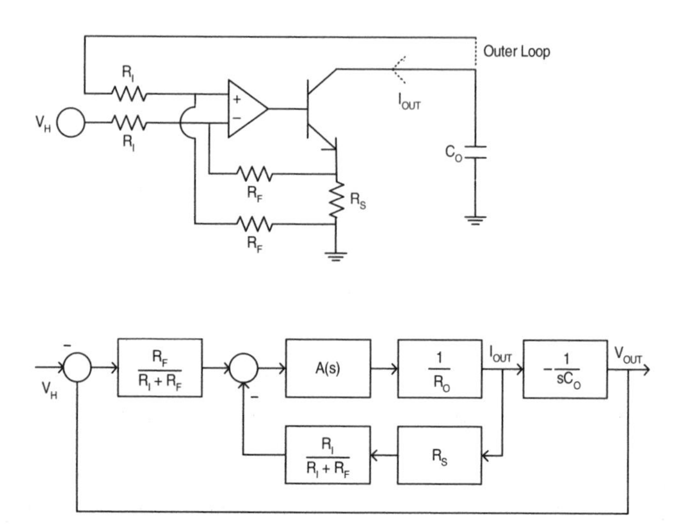

Figure 4–22 (a) functional schematic of clamp and (b) corresponding loop dynamics model.

by about a 45 percent phase margin. With larger output capacitances, the outer loop response slows down further, yielding larger phase markings and more damped transient responses. However, this will in no way impede the function of the clamp circuit, since the minor current loop responds equally fast with any output capacitance.

Eq. 4–2 is also of interest in determining the DC load regulation characteristic of the active clamp. This characteristic is governed by an effective impedance of 2 mΩ for our prototype design, which at 10 A yields a voltage regulation of 20 mV. Note that it is essential to include disparate sensing and forcing pins in an IC package, since the IC package and board traces leading to the output capacitor may very well impose more than 2 mΩ of resistive and/or inductive impedance.

Figure 4–23 shows a schematic for the complete active clamp circuit. This is essentially the circuit from Figure 4–22, but it uses compound transistor connections to realize the large output stages. A potential process for fabrication of the circuit should be a state-of-the-art high speed bipolar process, with minimum device feature size and availability of high performance opamp cells. A test board was built using IC kit parts with such a process.

Note that, given the novelty of this active clamp architecture, there is not much characterization data available for the type of operation that will be of interest, namely high-current density operation that is either saturated or unsaturated. However, some data on the kit parts was obtained in curve tracer evaluations and was used to guide the design of the active

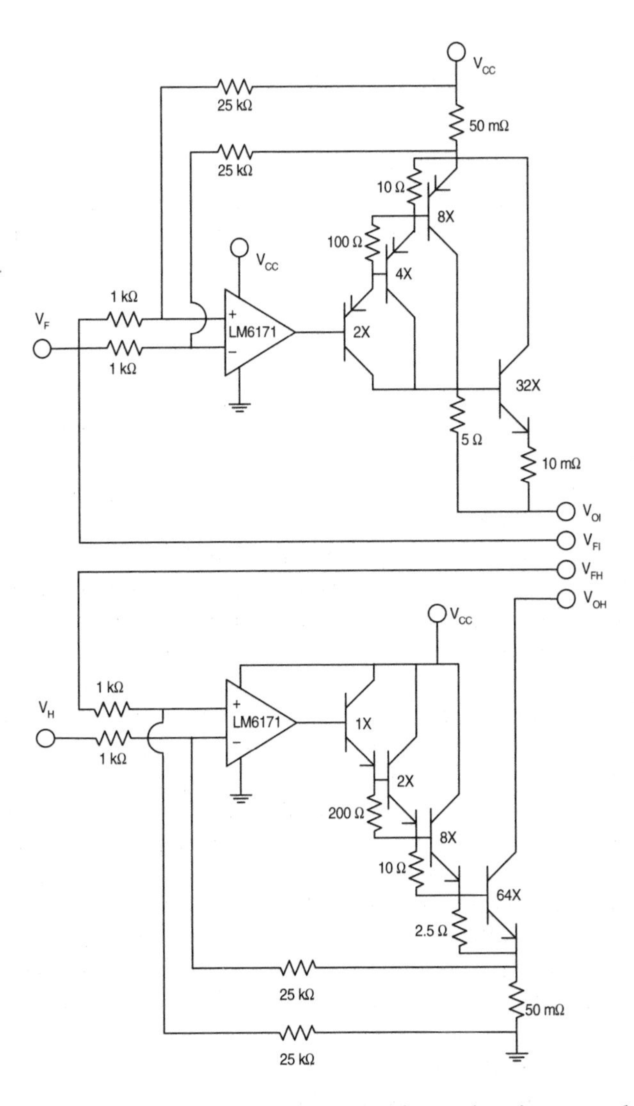

Figure 4–23 Schematic of active clamp circuit. Device sizes are relative to a 960 μm^2 emitter area cell.

clamp. The reason for operating with very high current densities is to conserve die area, but one of the consequences of operation at very high current densities (on the order of ten to twenty times the normal design current density) is degradation in device speeds. This will be discussed briefly in the "Test Results" section.

An overall strategy for the design of each of the two large compound transistor connections was to obtain a minimum current gain of five in each stage and to use four stages to avoid overloading the opamp output. A significant base-emitter "leak" resistor was designed into each of the stages of the compound transistor in order to provide a turn-off mechanism for each stage. Typical sizing of these base-emitter resistors was made to allow for a leak current on the order of one-fifth of the base current.

Test Results

A prototype of the active clamp circuit scaled for a 2.5 A load has been built using discrete opamps and IC kit parts all constructed in a high speed complementary bipolar process. The prototype active clamp circuit test circuit was interconnected with a DC-DC converter, operating at a switching frequency of 300 kHz, as shown in Figure 4–19. For this interconnection, an oscilloscope photo of the response to a step load change from 0 A to 3 A is shown in Figure 4–24(a). Note that the DC-DC converter is running at about its nominal output of 3.3 V with an output ripple of about 50 mV peak-peak and an approximate frequency of 300 kHz. A time-scale blow-up of this transient is shown in Figure 4–24(b), where the scale is 200 ns/division. In this figure, we can see the voltage spike due to the parasitic inductance (ESL) associated with the output capacitance and layout. As evidenced by the waveform for the opamp output that drives the compound pull-up device, the recovery of the clamp circuit occurs in about 300 ns. Note that the recovery is considerably slower than would be computed for the nominal rated base transit time TF and rated transition frequency $f_t \approx 3$ GHz for the process. This is likely due to high level injection effects, especially the *Kirk effect* (a dramatic increase in the transit time of a bipolar transistor caused by high current densities), which effectively broadens the base region of a bipolar device. In order to conserve die area, the output devices are driven at emitter current densities as high as ten to twenty times the level at which the onset of the Kirk effect occurs.

Waveforms complementary to those in Figure 4–24(a) and (b) are shown in Figure 4–24(c) and(d). For the latter two photos, the DC-DC converter supports a 3 A steady state load, which is interrupted. The upper clamp function behaves analogously to the lower clamp function.

Figure 4–24(a) illustrates the clamp function under 3 A step load. The upper waveform is the output node. The dashed horizontal trace depicts the upper clamp reference level of 3.45 V, and the solid horizontal trace depicts the lower clamp reference level of 3.15 V. Although divisions are not visible, the scale is 1 μs per division. The total time depicted is thus about 10 μs. The lower waveform is the gate signal for the pulse load circuit. Figure 4–24(b) shows the time-scale expansion to 200 ns/division of (a), except the second waveform, shown at 500 mV/division, is the opamp output waveform that drives the compound PNP pull-up device. Part (c) of Figure 4–24 illustrates the clamp function when the 3 A load is interrupted. Again, the upper waveform is the output node and the lower waveform is the gate signal for the pulse load. The scales are the same as in Figure 3-24(a). Figure 4–24(d) shows the time-scale expansion to 200 ns/division of (c), except the second waveform, shown at 500 mV/division, is the opamp output waveform that drives the compound NPN pull-down device.

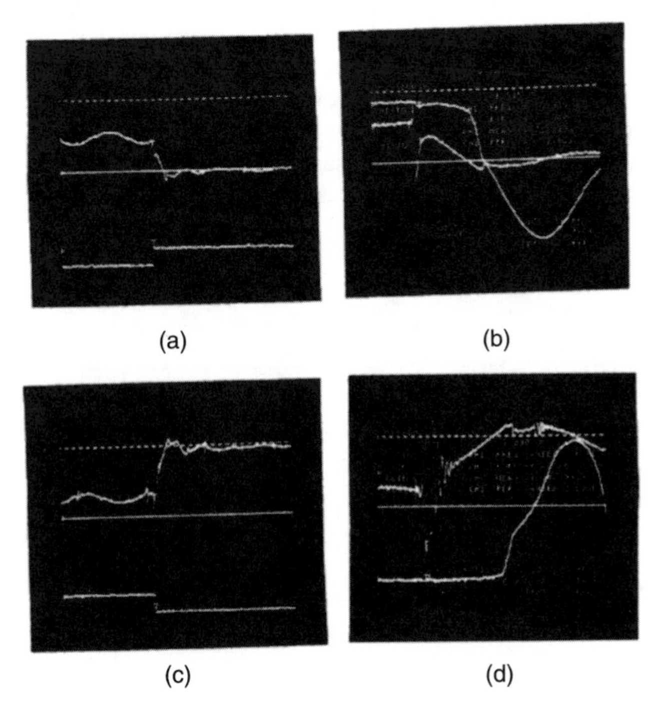

Figure 4–24 (a) Clamp function under 3 A step load. (b) Time scale expansion to 200 ns/division of (a). (c) Clamp function when 3 A load is interrupted. (d) Time-scale expansion to 200 ns/division of (c).

Comments

The design described in previous sections is a possible route to ease the constraints that microprocessor loads impose on power supply systems. However, one potential complication arises since the design requires two reference levels that are application-dependent and that are coordinated with the internal reference voltage of the main DC-DC converter. This complication is related to the difficulty of matching reference voltages residing on different circuits. One possible coordination scheme that has proven effective is to develop the two reference levels for the clamp circuit by real time low-pass filtering of the DC-DC converter output, as illustrated in Figure 4–25. With this approach, the reference levels are directly tied to the DC-DC converter output voltage level. Thus, the tolerance bands may be tightened substantially and are limited mainly by the output ripple voltage. Another approach for the design would be to implement the clamping function as part of an active filtering scheme. This approach was not taken in the design discussed in this section to allow for maximum efficiency. Nevertheless, such an approach may be warranted by higher current, lower voltage applications where the basic ripple filtering function may become more difficult.

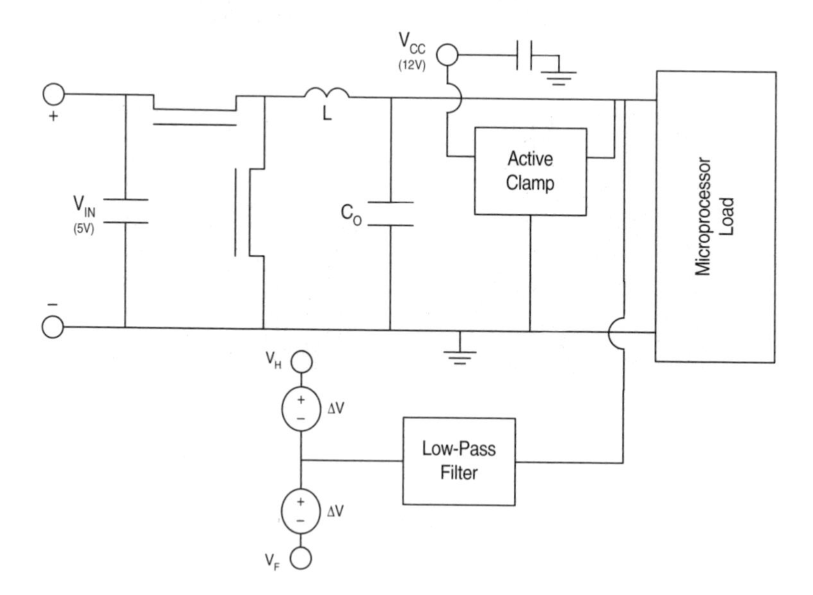

Figure 4–25 Possible scheme for supplying reference levels to clamp circuit.

4.4 Battery Charging Techniques: New Solutions for Notebook Battery Chargers

New solutions are necessary to satisfy the requirements of modern mobile charger applications. A modern charger for mobile computing today has to

- have high efficiency,
- communicate on a serial bus with other elements like the host microcontroller and smart battery,
- provide a wide range of charging currents and voltages,
- provide a very precise charge termination voltage, and
- enable fast charge.

The available solutions on the market thus far are few and fall far short of meeting the *OEM's* (Original Equipment Manufacturer's) expectations. The listed requirements, if not evaluated carefully, may lead to the design of "brute force" products that are too complex, too difficult to produce, and too expensive.

We will now briefly address all these issues.

High Efficiency

With some notebook models incorporating the charger inside the notebook body, there is no escape. The charger, like any other heat-producing element inside the notebook, must exhibit an average efficiency of 90 percent or better. This is not an easy task and can be accomplished only by means of a composite approach. Efficiency can be improved by optimizing the blocks discussed in the following sections.

The Controller

The controller IC should be designed in a fast, relatively small geometry (1.2 micron minimum feature) BiCMOS process for minimum standby power dissipation. The right process alone would not suffice. The entire low power design repertoire needs to be used in the design of the controller, including latched, or clock edge driven commutations as opposed to state driven commutations and choice of the *sweet point* for frequency operation. Operation at such a sweet point would produce the best compromise between size of the passive components and efficiency. The controller drives external discrete DMOS transistors in PWM synchronous rectification configuration.

The Discrete DMOS Transistors

The DMOS discrete technologies are progressing at an amazing pace, making sure that the IC designers continue to avoid the temptation of integrating power and control on the same die. The latest crop of DMOS devices offers the flexibility and sophistication of transistors that are optimized for conduction losses compared to other transistors optimized for switching losses.

The low side driver in our charger application works a lot (has high duty cycle), and consequently it calls for a discrete transistor like the FDS6612A, a PowerTrench-based N-channel device optimized for low R_{DSON} and consequently low conduction losses (19 mΩ, SO8).

On the other hand, the high side driver needs a P-channel transistor that is optimized for switching losses. In this case, the NDS9435A—a "ten million cells" technology, in which discrete transistors are built as arrays of millions of cells; the more cells that are packed per square millimeter, the better the performance—is a good idea. It produces low switching losses while preserving R_{DSON} (50 mΩ, < 20 nC, SO8).

Finally, for a static P-channel transistor an NDS8435A (23 mΩ, SO8) may suffice.

The Smart Battery System

A smart charger allows for communication with a host microcontroller (Smart Battery System Level 3) via the SMBus (System Management Bus). See Figure 4–26.

Data Conversion

A wide range of charging currents and voltages is provided by the integration of three 8-Bit D-A converters. Precise charge termination voltage is provided by the integration of a very precise voltage reference.

Fast Charge

While the algorithm for charging resides elsewhere, typically in the smart battery, the smart charger can go a long way toward providing the right *hooks* in order to enable fast charging. Adaptive charging, in which all the available power from the AC adapter is constantly controlled and redirected between the notebook and the smart battery, can dramatically reduce the charging time.

In more traditional systems, the AC adapter current is limited to the maximum current tolerated by the smart battery. This means that the battery can only be charged by the difference between the total available current and the current needed by the notebook.

In an adaptive system, such a ceiling can be broken, and the AC adapter can be designed to provide sufficient power to allow for fast charge even when full power is provided to the notebook. The adaptive system guarantees that the battery is never exposed to more current than the maximum specified amount.

Battery Charger System

Figure 4–26 shows the Smart Battery System, with its three main components: the smart charger, smart battery, and microcontroller.

Figure 4–27 shows the power flow. The load (the regulator VREG powering the microprocessor) is powered either by the AC adapter or by the battery. The battery charges when the AC adapter is present.

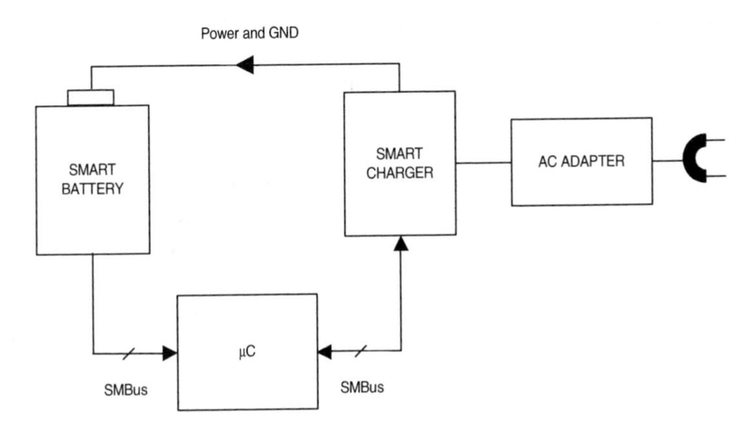

Figure 4–26 Smart battery system.

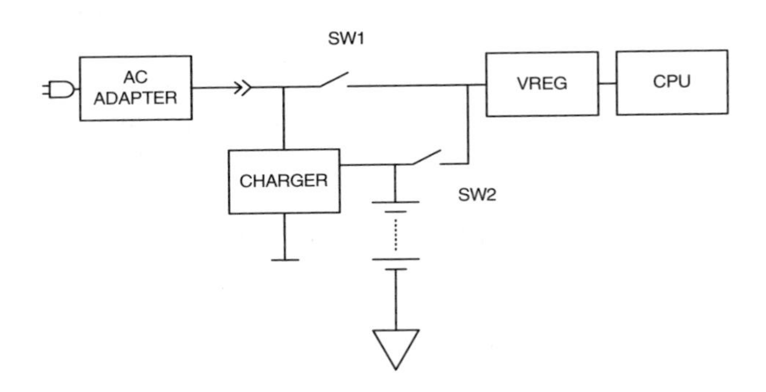

Figure 4–27 Power flow.

4.5 Digital Power

Control Algorithm of Modern Switching Regulators: Analog or Digital?

In the past few years, the pitch on digital control of switching regulators has been rising steadily. In a typical analog implementation, a *Pulse Width Modulated* (PWM) switching regulator is built around a modulator composed of a comparator with a periodic piecewise-linear (triangle or saw-toothed) modulation waveform on one end and the error signal at the other. As the quasi-stationary error signal falls between the minimum and the maximum of the modulation waveform, the comparator output produces a square wave at the heart of this modulation scheme. This system is "analog" simply because it is governed by the analog modulation waveform. This section explores the question: are there viable instances in which using true digital architectures in switching regulators is preferable to analog?

When people talk about digital control, they generally refer to one of the following three architectures:

1. The same analog engine as was just described, but equipped with digital peripheral functions like: a) serial communication (I2CBus, SMBus, etc.) b) Digital-to-Analog conversion (in CPU voltage regulators the digital inputs are called Voltage Identification codes (VIDs) and are essentially a digital means to vary on demand an otherwise constant reference) and/or c) small amounts of memory on-chip. I will refer to this first "digital" architecture as a *digitally controlled analog system.*
2. A microcontroller-based digital architecture is a useful architecture in terms of flexibility, especially in applications that require programmability as well as current and voltage profiling. As conventional digital algorithms are sequential in nature—requiring several clock cycles to execute an instruction—they are inherently slow and thus are not useful in applications requiring fast response. I will refer to this as a *microcontroller-based control architecture.*
3. A non-sequential machine, with hard-wired logic implementation that can produce a fast response comparable to an analog system will be referred to as a true *digital control architecture*. It follows that the challenge to the analog switching regulator dominance in fast response applications may only come from the true digital implementation described here.

Although true digital control is interesting technology, I have not yet heard a compelling case for it. Right now, this technology remains at the periphery and is not a mainstream architecture for the power industry.

But is there a relevant place for true digital control? One in which such control is not just more convenient, but fundamentally superior in performance?

To answer this question we must first look at the system we want to regulate.

If the system is truly linear, namely it is continuous and invariant or, *smooth* in its mode of operation; analog is the way to go. This is true in the case of a desktop CPU voltage regulator with an output voltage that must be continuously controlled by the same algorithm from no load to full load.

If, on the other hand, the system is non-smooth, namely discontinuous, and variable in its mode of operation, then digital may be the way to go.

For example, digital could be used in the case of a cell phone voltage regulator that, due to the necessity to save power at light loads, requires a mode change, typically from a PWM algorithm to *Pulse Frequency Modulation* (PFM). PFM is a mode in which the frequency adjusts with the load, thereby yielding lower frequencies and hence lower switching losses at lighter loads.

Such mode change in an analog system would require an abrupt commutation from one control loop (say PWM) to the other (PFM), typically at the time that the load is changing. This type of algorithm discontinuity would invariably lead to some degree of temporary loss of regulation of the output.

By contrast, a digital control is inherently equipped to handle discontinuities and thus would be capable of handling mode changes within a single control algorithm.

In conclusion, I believe that digital control may bring relief to its analog counterpart in non-smooth systems such as the one just described. In such systems, digital control may prevent risking loss of regulation and may save additional overhead in bill of materials that would be required in order to mitigate the effect of discontinuities in analog implementations.

Digital Power: Forward into the Past

Under the digital power umbrella falls a broad range of functions that go beyond regulation and include communications over a serial or parallel bus, power sequencing by means of state machines or microcontrollers, and digital algorithms for implementation of the servo, or feedback control loop.

The digitalization of power can be best understood when articulated at three different levels: chip design architecture, silicon processes, and board level design. The digitalization of power is progressing in each of these domains at different speeds, which causes much confusion.

Digital Power Chip Design

In recent years a number of startups have tried to crack the computing market space at the chip level with digital implementations of the traditional analog PWM modulator design, without much success. These implementations have found some space in high-end server blades applications that are low volume and tolerate higher cost.

In handsets, the increase in power levels is due to packing more elementary building blocks on-die. Such elementary building blocks have relatively low power consumption by themselves, but in large numbers add up to a considerable amount of power. The most complex power management units (PMUs) built today clearly show that LDOs and switchers remain primarily analog. On top of traditional analog, a good dose of digital is needed for communications and sequencing, and is implemented with architectures ranging from state machines to microcontrollers.

Digital Power IC Processes

As far as integrated circuits are concerned, the leading edge of the analog world long ago moved from pure analog to *mixed* analog and digital. Every VRM chip marries an on-die Digital to Analog Converter (DAC) to an analog switching regulator; the same way every PMU mixes analog and digital blocks. The move to mixed signal—the combination of analog and digital on the die—is a revolution that began around 1980. It started tentatively within the bipolar world with Integrated Injection Logic (I2L) logic gates and then fully blossomed with bipolar, CMOS, and DMOS (BCD) monolithic processes.

Today, the leading monolithic power companies have BCD mixed signal processes. These companies—as leaders often do—position themselves as solution-oriented, utilizing the most appropriate process, components, and techniques for the task at hand.

BCD processes can use bipolar for precision, CMOS for signal density, and DMOS for power density. Leading BCD is today in its 7th generation at 0.18 μm and soon it will be at 0.12 μm (BCD VIII): this is only two nodes away from the CPU roadmap at 0.65 nm by the end of 2005. Still, these companies understand the trade-off in terms of mask count and cost and, therefore, keep alive simpler or "traditional" analog processes that in specific applications—particularly single functional building blocks—may result in more cost-effective designs.

Board-Level Digital Power

Telecom and Datacom applications like *Point Of Load* (POL) are the areas where digital power may find the best niche. (Digital power in this case refers to power regulation equipped with a communication bus that allows for flexible set-output voltage, frequency compensation type, etc.) Leading power supply companies are battling for dominance of this new market.

Conclusion

At the board level, digital power seems to have found a hot niche in POL applications. Digital implementation of analog algorithms in silicon makes sense is some cases, like providing silicon in support of digital POL power, while in others it does not. The digitalization of power happened twenty years ago with BCD processes; today it is happening in the high-performance niches of computing power at the chip architecture level, and it will probably happen soon at the board level with POLs. As the saying goes, the next big thing—digital power in this instance—is already here. It is just not uniformly distributed. Perhaps more importantly for IC companies, a good process portfolio, which includes processes like BCD, takes us beyond the debate of digital versus analog and allows us to focus on solutions.

Fast Switchmode Regulators and Digital Control

The bulk of the CPU power regulation volumes are in the PC motherboard market, a fiercely competitive market dominated by the Taiwanese motherboard manufacturers operating on a relatively short-term horizon and driven by cost. Accordingly, these motherboards have the lowest possible bill of materials. It follows that the "sweet spot" for power is a voltage regulator built around some very resilient technologies based on the buck converter, which continues to reinvent itself (from buck to sync buck to multiphase to...) and thus far defies any new proposed architecture, and the electrolytic capacitor, which in its latest reincarnation, Aluminum-Polymer, keeps the emergence of ceramic caps at bay.

More precisely, huge amounts (mF) of "bulk" capacitors are employed in the design of buck converters to supply most of the energy during transient (the time it takes the feedback loop to respond) while a minimum number of ceramics are employed nearby the CPU socket for quasi-instantaneous response.

Modern specifications for CPU regulators require operation inside a tight voltage band (50 mV), while the source of degradation of the regulators is the Equivalent Series Resistance (ESR, in mΩ) of the output capacitors. Consequently at 50 A, the tolerable ESR has to be <50 mV/50 A = 1 mΩ.

Until now the $/mΩ figure of "merit" for electrolytic remains unsurpassed—namely the lowest—and this simple fact explains why any fast converter technology thrown at this niche does not stick, despite the promise to eliminate the "bulky" electrolytics.

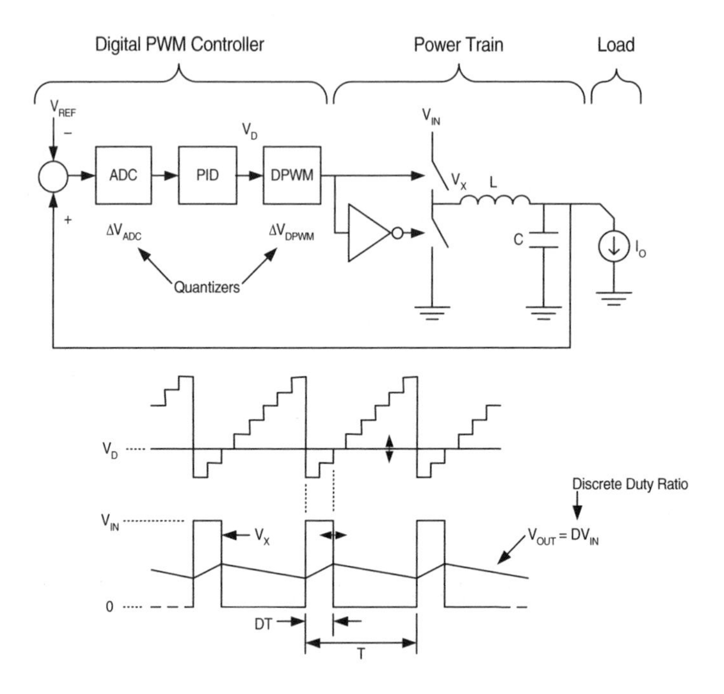

Figure 4–28 Digital power control loop.

Thanks to their requirement of desktop power packed inside thin form factors, other applications that currently have less commanding volumes, such as blade servers, offer a different value proposition and privilege size over cost. This niche has become a playground allowing a few companies to develop new and increasingly faster switchmode regulator architectures based on the more expensive but slimmer ceramic capacitors. The ultimate goal is to break the $/mΩ barrier by the design of switchmode controllers and power train filters that are fast enough to respond at or above the speed of the incoming current step di/dt (say 300 A/μs). Such a performance would go beyond the elimination of electrolytics and would reduce drastically the number of ceramics needed on the basis of plain ESR calculations. The underlying architecture would then finally defeat the established technology, with the entire regulation market being the prize. Fairchild is actively researching this field.

Digital switchmode control is a fledgling architecture testing itself against the abatement of the \$/mΩ barrier. In the process, digital control is regularly touted as an "inherently fast" technology. As conventional digital algorithms are sequential in nature, requiring several clock cycles to execute an instruction, there is nothing inherently fast about them. PWM digital control is all about going beyond the CPU's, or even the DSP's architectures, toward hard-wired logic that can respond at the speed of the process technology. Analog techniques, which are at the same process generation level, should be at least as fast.

Accordingly it is likely that at the core of future fast controllers, we will find a fast analog cell, may it be a "fast clamp," transient suppressor, or something similar. Around this fast cell we may find all kinds of bells and whistles, some digital and some analog.

What we need are *fast* architectures that deal effectively with the CPU voltage regulation—the rest is optional.

Chapter 5

Offline (AC-DC) Architectures

5.1 Offline Power Architectures

Introduction

System On a Chip (SOC) companies are claiming that the entire signal path (digital + analog + memory) and even a full *GSM* system—including power management—will be integrated in the next few years. However, the reality is that this up-integration march, fueled by nano-scale lithography (minimum features less than 100 nm), ends up defining the product's own technology boundaries: the higher the number of transistors on a chip, the lower their voltage and the more fragile their technology. At the 0.13 μm juncture, for example, the SOC processes work at voltages in the range of 1 V–2 V!

At the other end of the spectrum are the power chip companies creating technologies to deal with high voltages and high currents. Drawing power from the AC line down to an intermediate bus voltage requires robust devices capable of sustaining several hundred volts at several amperes. At the same time, the conversion from bus voltage to final load often requires low voltages at hundreds of amperes of current.

The way power conversion requirements are met in a PC application, from line *Power Factor Correction* (PFC) to intermediate bus voltage out of the silver box, down to the popular low voltages on the motherboard, nicely illustrates the new high-voltage and high-current silicon technologies and architectures. To describe this evolving power conversion technology, this chapter provides an application example of

Fairchild's single chip controller, the ML4803 PFC/PWM combo, and associated discrete transistors for the AC-DC conversion to intermediate voltage bus. Additionally, DC-DC conversion from bus to low voltage is demonstrated based on Fairchild's FAN5092 buck converter. Future trends in PFC/PWM and DC-DC converters are also discussed.

Offline Control

Harmonic Limits and Power Factor Correction

Optimum conditions for power delivery from the AC line are achieved when the electric load, a PC for example, draws current which is in phase with the input voltage (AC line) and when such a current is undistorted (sinusoidal). To this end, IEC 6100-2-3 is the European standard specifying the harmonic limits of various equipment classes. For example, all personal computers drawing more than 75 W must have harmonics at or below the profile demonstrated in Figure 5–1. With modern desktop *PSUs* drawing from 140 W to 250 W, all PCs shipped to Europe must comply. When it comes to compliance to IEC 6100-2-3, the rest of the world is following Europe's lead at varying paces.

Figure 5–1 illustrates one aspect of the European specification. Notice that the allowance grows stricter for the higher harmonics; however, these harmonics also have less energy content and are easier to filter. According to the specification, the allowed harmonic current does max out above 600 W, making it more challenging to achieve compliance at higher power.

Power Factor (PF) is a global parameter speaking to the general quality of the power drawn from the line. It is related to the input current *Total Harmonic Distortion* (THD) by the equation

$$PF = \frac{\cos\varphi}{(1 + THD^2)^{1/2}} \quad \text{Eq. 5–1}$$

where φ is the phase shift between line voltage and drawn current. With no phase shift ($\varphi = 0$) and no distortion ($THD = 0$) it follows that $PF = 1$. Since the numerator $|\cos\varphi|$ (bars indicate module or absolute value) is bounded between zero and one and the denominator is always higher or equal to one, it follows that $PF \leq 1$.

Since IEC 61000-3-2 specifies the harmonic components of *THD*, neither *THD* nor *PF* is a sufficient measure of performance. In reality, the harmonic distortion parameter to measure and comply with (as per

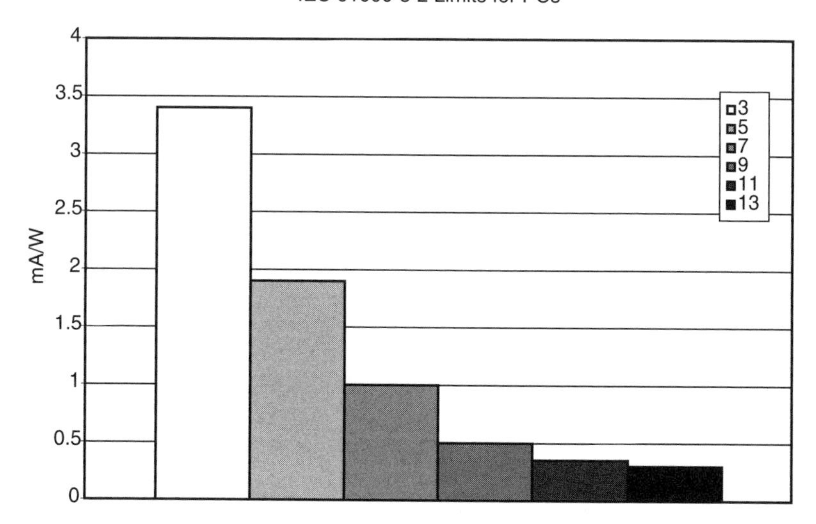

Figure 5–1 IEC 61000-3-2 harmonic current limits.

Figure 5–1), and the techniques to achieve that compliance generally are called *PFC*.

It is interesting to note that, in theory, the cosφ factor in the *PF* product can take on negative as well as positive values. Keep in mind that a negative cosφ value corresponds to the situation in which the load circuit is actually supplying real power to the line. In a rectifier circuit based on a diode bridge, this situation is impossible.

Harmonic Limits Compliance Constraints

The standard way to draw power from the AC line is via a diode bridge rectifier directly applied across the load (Figure 5–2).

If the capacitor is not present, the voltage and current are both rectified sinusoids with no distortion, no phase shift, and $PF = 1$ (see Figure 5–3). In this condition, the power delivered to the load consists of a waveform of double frequency, zero minimum (meaning in Figure 5–3 the lowest part of the waveform touches the horizontal axis corresponding to zero power) and instantaneous value of

$$P(t) = (V^2/R) \times \text{sen}^2 \omega t = (1/2) \times (V^2/R) \times (1 - \cos 2\omega t) \qquad \text{Eq. 5–2}$$

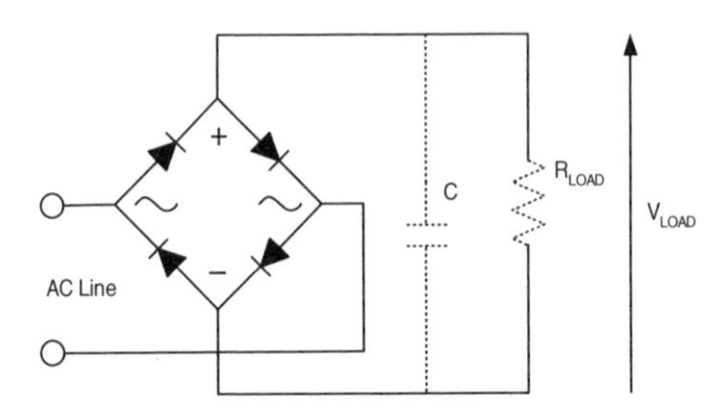

Figure 5–2 Diode bridge rectifier.

where V is the amplitude of the line voltage, R is the load, and ω is the line pulsation $2\pi f$, with $f = 50$ Hz or 60 Hz. From Eq. 5–2 the real or average power is

$$P_{AVE} = (1/2) \times V^2/R = V_{RMS}{}^2/R \qquad \text{Eq. 5–3}$$

with a time-varying zero average pulsating power of

$$P_{PULS} = -(1/2) \times (V^2/R) \times \cos 2\omega t \qquad \text{Eq. 5–4}$$

This simple example provides a model of an ideal rectification scheme as presented to the AC line. On the other hand, the scheme has no energy storage function, and the power delivered at the output of this rectifier has a double-line frequency component.

Continuing in this idealized framework, a typical load actually requires constant (DC) power. Thus, an inherent requirement is a bulk energy storage element, usually realized by an electrolytic capacitor, that handles the difference in power between $P(t)$, the input power, and P_{AVE}, the DC output power.

Adding a small capacitor C (the dashed line in Figure 5–2) to this scheme will naturally smooth the voltage across the load, reducing the ripple but also degrading the PFC, as the current waveform now drastically deviates from a sinusoid (see Figure 5–4).

The scheme of Figure 5–2 (with capacitor) represents the conventional, non-PFC architecture used in many commercial applications prior to IEC-61000-3-2.

PFC techniques are all about maintaining an input and output power match in the presence of low input harmonic current content and tightly regulated output voltage.

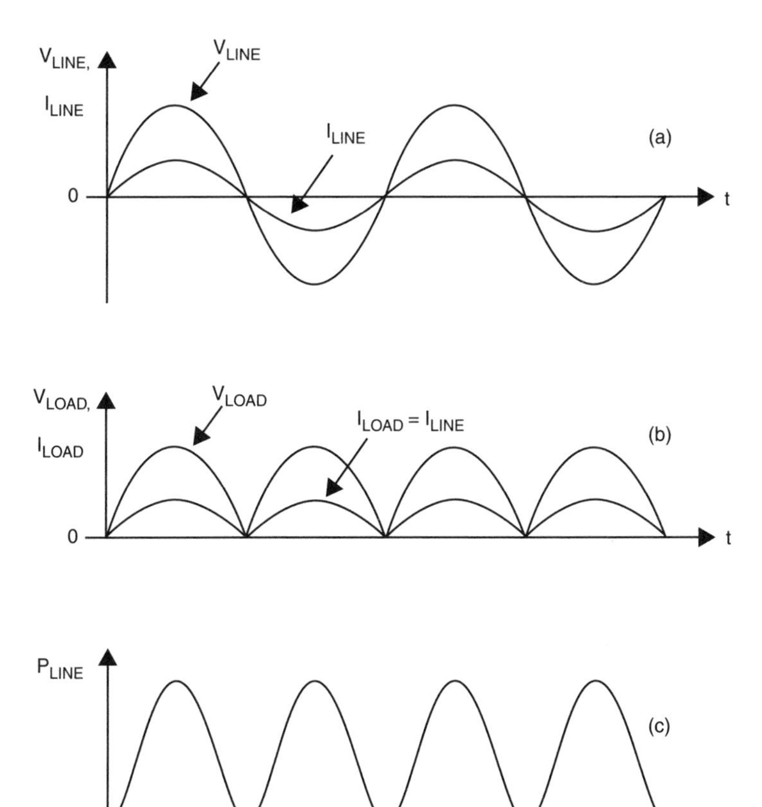

Figure 5–3 Power line ($P_{LINE} = V_{LINE} \times I_{LINE}$) has double frequency.

PFC Architecture

The general architecture for PFC is shown in Figure 5–5. As discussed in the previous section, a PFC stage will provide a good match between line voltage and current.

Assuming perfect balance ($PF = 1$), we find ourselves in the condition of Figure 5–3(a) on the AC line side. On the rectified side, the capacitor C will provide a reactive power

$$P_{CR} \cong -V_{CDC} \times C \times 2\omega \times V_{CRIPPLE} \times \cos 2\omega t \qquad \text{Eq. 5–5}$$

where V_{CDC} is the DC voltage across the capacitor, $V_{CRIPPLE}$ is its ripple peak, and $\omega = 2\pi f$ is the line voltage pulsation ($f = 50$ or 60 Hz). Notice that P_{CR} is analogous to P_{PULS} in the system from Figure 5–2 (no capacitor).

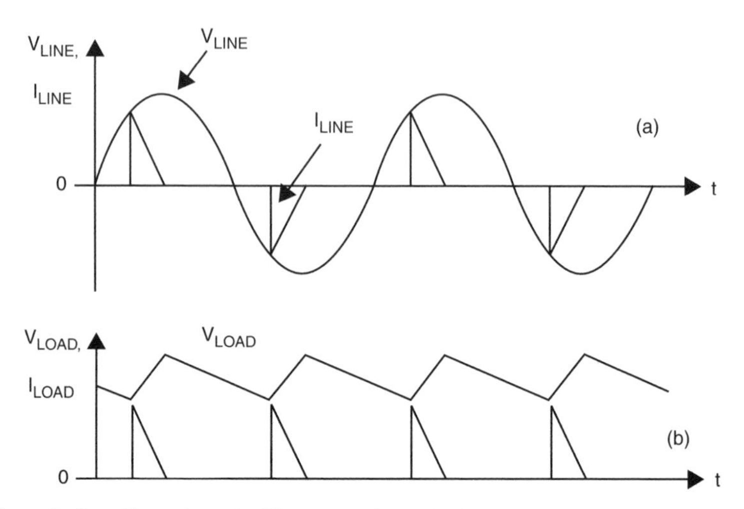

Figure 5–4 Capacitor C effect on voltage and current.

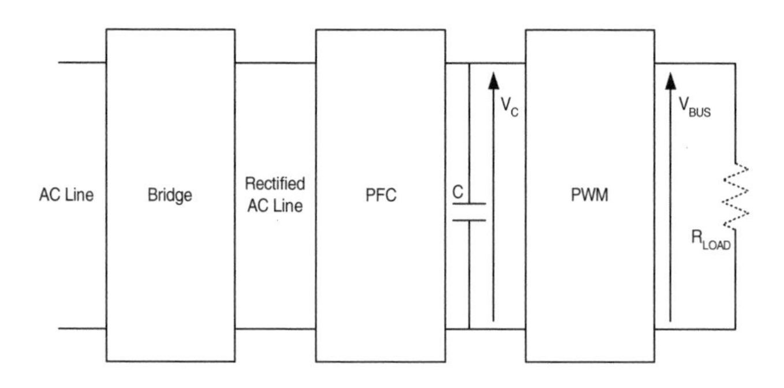

Figure 5–5 Example of PFC architecture.

From Eq. 5–5, we have

$$V_{CRIPPLE} \cong P_{CR(PEAK)}/V_{CDC} \times C \times 2\omega \qquad \text{Eq. 5–6}$$

This is a useful design formula showing the trade-offs between size of the capacitor C and its DC voltage and ripple values.

After the PFC stage has taken care of the line's harmonic content, the ripple across C is smoothed out by means of a DC-DC converter designed to have sufficient input ripple voltage rejection.

PFC and Pulse Width Modulation (PWM) Implementation

A high-level block diagram of the power conversion chain, from an AC line to an intermediate voltage bus V_{BUS} (for example, 12 V), is shown in Figure 5–6.

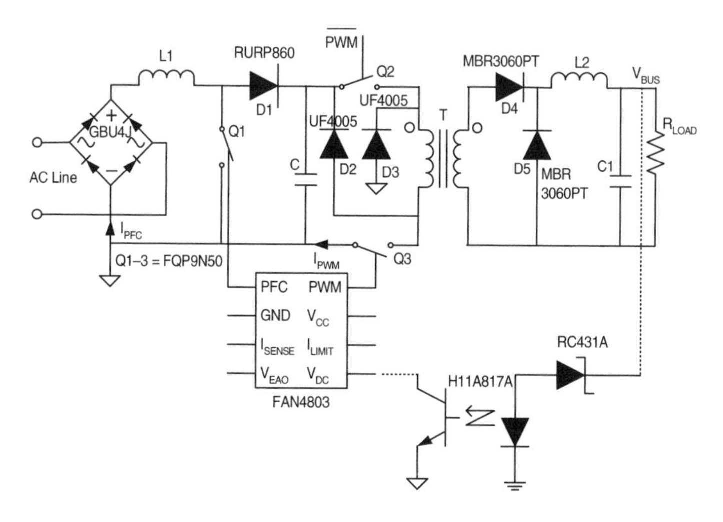

Figure 5–6 PFC and PWM chain based on FAN4803.

In Figure 5–6, the control is based on a product called the FAN4803, a very compact chip integrating two control loops on board. The inductor L1, switch Q1 (MOSFET), bulk capacitor *C*, and the diode D1 controlled by one half of the PFC/PWM controller FAN4803 (Figure 5–6), make up the PFC section. Next, the voltage across *C* is regulated down to the bus voltage by means of a *forward* converter. The forward converter includes switches Q2 and Q3, diodes D2–D5, passives L2 and C2, the second half of FAN4803 for primary side control, and RC431A for secondary side control. This conversion requires electrical isolation between the high input and the low output voltages. Isolation is accomplished via the utilization of a transformer T in the forward conversion path and an optocoupler H11A817A in the feedback path. Appendix B provides the data sheet of FAN 4803 for more technical details.

The Controller Architecture

The FAN4803 is powered (V_{CCPIN}) from the main transformer T via an auxiliary secondary winding transformer (not shown) yielding a relatively low voltage (15 V). Since every controller I/O pin sees voltages below 15 V, the chip is built in a low-voltage, dense BiCMOS process.

The top portion of Figure 5–7 shows the PFC control loop. The shaping function is accomplished by the continuous current mode architecture, which forces the current to follow the shape of the line voltage. In fact, on the small time period (15 μs) of the relatively fast clock frequency (67 kHz), when V_ε is roughly constant, the forced current is also constant. However, with an input voltage [V_{LINE}, Figure 5–3(a)] crossing zero twice per period (100 Hz or 120 Hz), the current in the inductor will collapse down to zero as well around the rectified line voltage dips [Figure 5–3(b)], yielding a current sufficiently close to the desired shape demonstrated by the I_{LOAD} waveform in Figure 5–3(b).

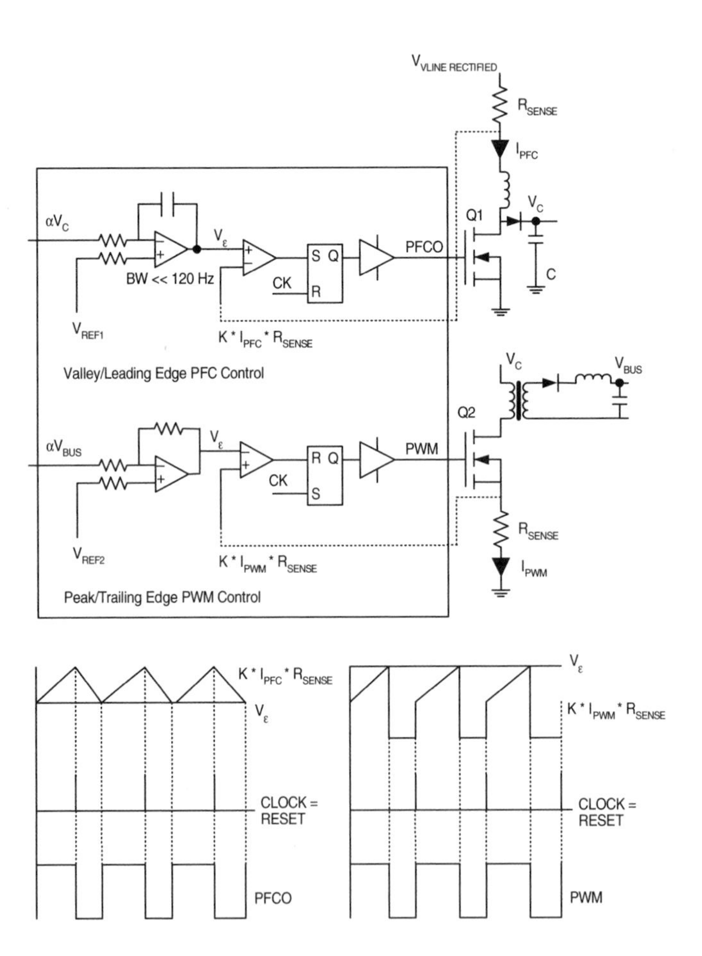

Figure 5–7 PFC and PWM control loops.

The very low bandwidth of the error amplifier assures control of the output voltage V_C according to Eq. 5–6. The PFC and PWM functions can be accomplished with minimum BOM when a synergistic mode of operation between the two sections is implemented. As illustrated in Figure 5–7, the PFC section is controlled with leading edge modulation. The MOSFET Q1 turns off on the clock edge, while turn-on, which corresponds to the leading, or rising edge of the PFC square wave, is under loop control. The PWM section is controlled with *trailing edge* modulation. The MOSFET Q2 turns on the clock edge while turn-off, which corresponds to the trailing, or falling edge of the PWM square wave, is under loop control. Consequently, with synchronized clocks the two transistors never draw currents concurrently; this further redistribution of the current results in minimum value of the high-voltage input capacitors.

Notice that while on the 50 Hz time scale, the waveforms look like the ones in Figure 5–3, on the 67 kHz (clock) scale the current will show ripples due to the chopping effects of the switching regulator. In Figure 5–8, I_L is the line current and R_{AMP} is the modulator ramp voltage shown on the 67 kHz scale.

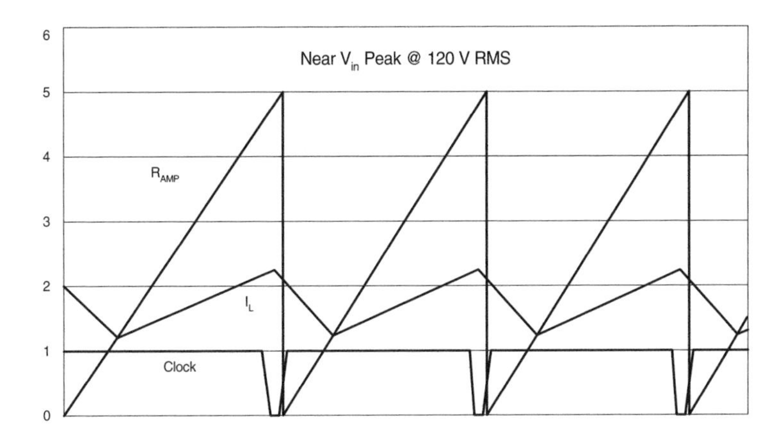

Figure 5–8 Ripple in the line current.

Offline Power Silicon

All the diodes and DMOS switches between the line and the primary of the transformer are high voltage devices. IEC 61000-3-2 specifies a voltage limit up to 240 V_{RMS} for single-phase (415 V_{RMS} for tri-phase) power line distribution. Accordingly, these components are able to withstand voltages in the 400–1000 V range.

The boost diode D1 in Figure 5–6 (RURP860) is a high-reverse voltage (600 V), low-forward voltage drop (1.5 V at 8 A), and ultra-fast recovery rectifier ($t_{RR} < 60$ ns). Its construction is shown in Figure 5–9. The other high-voltage components in Figure 5–6 are the ultra-fast UF4005 free-wheeling diodes, which are also able to stand 600 V, and the switches Q1–3 (FQP9N50). The three FQP9N50 transistors in Figure 5–6 are 500 V N-channel enhancement MOSFETs built with planar stripe DMOS process, a process yielding high switching speed and very low "on" resistance (0.73 Ω at 10 V of V_{GS}). Figure 5–10 shows a cross section of the DMOS transistor. Finally, Figure 5–11 shows the picture of a silver box.

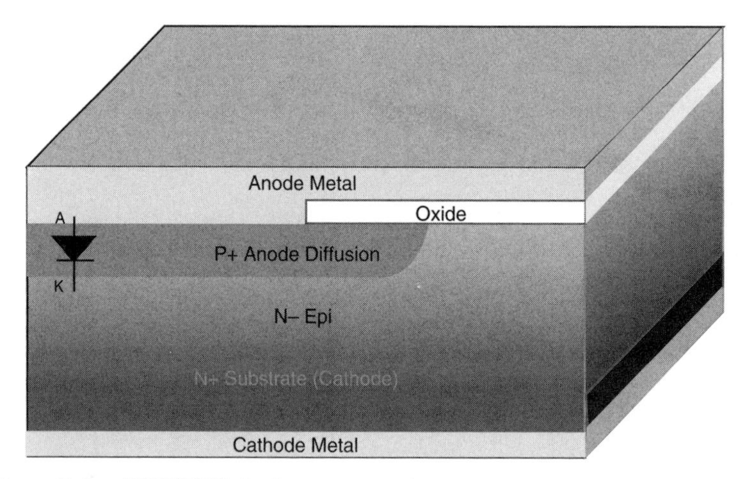

Figure 5–9 RURP860 device cross section.

DC-DC Conversion Down to Low Voltage

The bus voltage V_{BUS} (12 V in Figure 5–12) is distributed and reduced to the popular 3.3 V, 2.5 V, 1.8 V, or V_{CPU} by means of switching regulators, typically synchronous buck converters.

The FAN5092 step-down (buck) is a two-phase interleaved buck converter switching up to 1 MHz per phase, thanks to its leading edge valley control architecture. This IC is able to directly drive the discrete DMOS transistors' high side Q1–3 (FDB6035AL) and low side Q2–4 (FDB6676S), with integrated drivers exhibiting the lowest impedance in the industry (1 Ω).

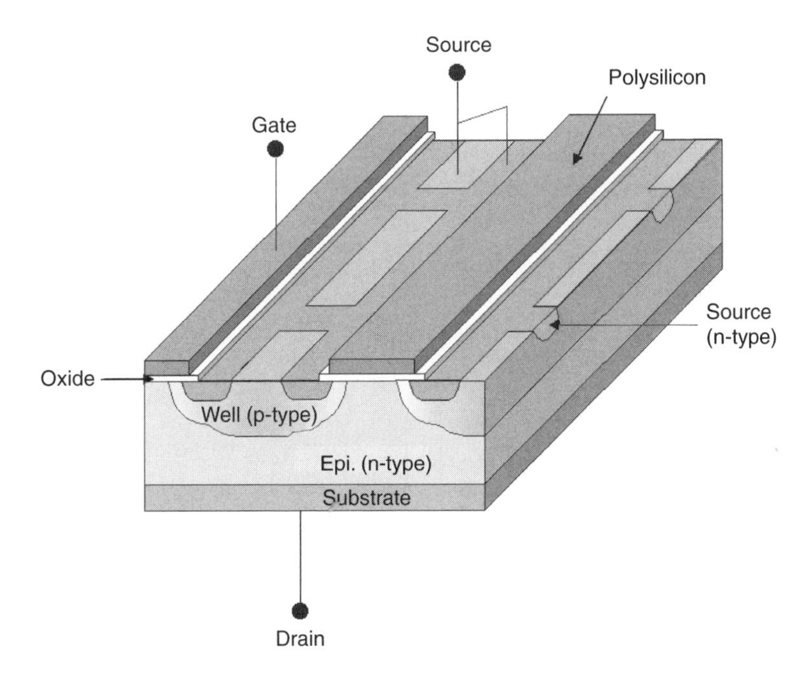

Figure 5–10 Cross section of high voltage DMOS transistor.

Figure 5–11 Typical silver box.

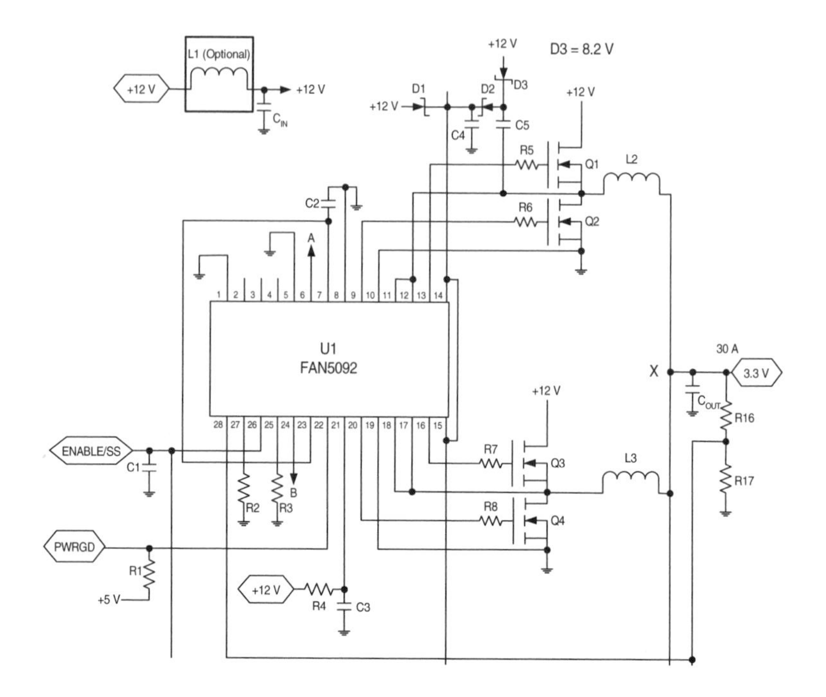

Figure 5–12 Buck converter: from V_{BUS} (12 V) down to 3.3 V.

Future Trends

Active power correction allows us to meet easily IEC 6100-3-2 power factor specifications but unquestionably requires a relatively heavy bill of materials. The state-of-the-art FAN4803 helps reduce the silicon complexity by integrating two controllers on the same die. These two controllers each require a full set of passive components to do their part of the job. Ideally, what is needed in the future is a true, single-stage PFC/PWM controller that will cut in half—or less—such complexity. Integration of the PFC/PWM function is in its infancy. In the future, slick new architectures will be developed that will significantly cut the BOM of current implementations.

As far as power distribution trends are concerned—DC-DC conversion from V_{BUS} to low voltage—the dominant architecture today is based on the resilient, interleaved synchronous buck converter. The challenge will be to reduce the bank of output capacitors by means of fast architectures that can respond quickly to load changes. Advanced work in these areas is intense, but the prize for such breakthroughs will be as big as the entire power conversion market.

5.2 Power AC Adapter: Thermal and Electrical Design

Thermal and electrical design techniques satisfy new requirements for AC adapters.

Introduction: The Challenge

The power management industry makes a tremendous effort to reduce the power dissipated by modern appliances, such as cell phones. A top priority is to find ways to extend the battery life of such devices. This narrow focus on extending untethered operation has generally limited the power management effort to the consumer side of the appliance, leaving the other side—the one concerned with wall power (as in the case of a cell phone's AC adapter)—relatively neglected.

Energy trends and regulations, however, such as the EPA's Energy Star® initiative that focuses on single voltage external AC-DC power supplies, are pushing for devices, including AC adapters, to meet or exceed specific active and no-load mode requirements in order to claim compliance to these initiatives and associated labels. Active mode refers to the device—for example a charger—providing power to an active load. A battery under charge would be an example of active load. A charged battery, even if connected to a charger, would not draw power and hence would represent a case of no load.

In addition to being efficient in both light and full load operation, an AC adapter also should be as small as possible for ergonomic reasons. Such minimum size (and maximum power density) is, in turn, defined by the amount of heat that an AC adapter *cube* can dissipate while maintaining reasonable temperatures.

AC Adapter Power Dissipation

The AC adapter brick transfers power from the line to the load with a certain efficiency such that

$$\eta = P_{OUT}/P_{IN} = P_{OUT}/(P_{OUT} + P_D) \qquad \text{Eq. 5–7}$$

where

η = efficiency

P_{OUT} = power delivered to the load

P_{IN} = input power drawn from the AC line

P_D = power dissipated inside the AC adapter

Inverting equation Eq. 5–1 yields the relation between dissipated power and output power

$$P_D = P_{OUT} \times (1 - \eta)/\eta \qquad \text{Eq. 5–8}$$

From Eq. 5–2, we see that a switching regulator with an efficiency of 80 percent inside the adapter will lose an amount of power equal to 25 percent of the delivered power, while a linear regulator with 50 percent efficiency will lose an amount of power equal to the one delivered to the load, or half of the power drawn from the line. In this example, the linear regulator dissipates four times (1/0.25) more power than the switching regulator in operation. Accordingly, a 5 V/620 mA AC adapter delivering a peak power of 3 W will leave, inside the adapter box, 750 mW in switch-mode and 3 W in linear mode.

AC Adapter Case Temperature

AC adapters generally are required to have a max case temperature below 75°C. The heating of the case is proportional to the power dissipation and to the ambient temperature (assume max 45°C). The amount of heat that can be dissipated inside an enclosed box is governed by the thermodynamics laws for heat convection and radiation. A simple model of a plastic box was analyzed with *ANSYS*, a thermal simulator based on the finite element method. The box had the following dimensions:

$$V = h \times w \times l = 0.5 \times 1 \times 2 = 1 \text{ in}^3 \qquad \text{Eq. 5–9}$$

where

h = height

l = length

w = width of the box, including a heat source

The box was heated with a power source and temperature profiles were obtained for the box surface. This first-order simulation showed that it would take 1 W of power dissipation inside the box to produce a peak temperature on the box surface—in the spot closest to the heat source—of roughly 74°C (45°C ambient).

Accordingly, the switching regulator discussed previously could be placed comfortably inside such a box without overheating it, while the linear regulator would certainly exceed the maximum allowed temperature limits.

Active and No-load Operation

The ENERGY STAR specification for single voltage external AC-DC and AC-AC power supplies took effect on January 1, 2005.

To meet Energy Star efficiency criteria for active mode, the 3 W AC adapter in our example will need to have efficiency above 60 percent. In no-load mode the same device should consume less than 0.5 W. State-of-the-art designs can go as low as 0.1 W unloaded. However, such levels of performance cannot be met by traditional and generic solutions.

Development of a Solution

Fairchild's performance offerings for AC adapters are based on a solid, high-voltage mixed BCD process. They offer a highly integrated, monolithic flyback architecture that already has reduced the number of components needed to build the AC adapter, making it a cost-effective solution even when compared to dis-integrated solutions. The following will discuss what it takes to design an integrated circuit (FAN210 in Figure 5–13) suitable to implement an AC adapter (the entire circuit in Figure 5–13 is the full AC adapter) to meet light load and no-load operation.

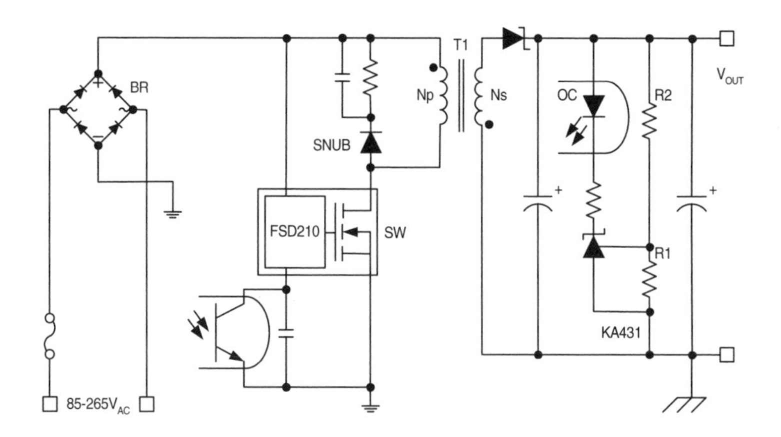

Figure 5–13 AC adapter simplified block diagram.

Power dissipation at no-load has many contributors to losses, including:

- IC power consumption
- *snub* network
- transformer
- bridge rectifier

All the losses associated with these elements have to be cut down substantially to stay within the allowed budget of 0.3 W.

A flyback architecture is just fine for full load operation, but it would not satisfy the no-load requirements. However, burst mode operation can be implemented in silicon (FSD210) to achieve the no-load objective. By virtue of gating the clock frequency and stopping it under light load conditions, the AC adapter is able to operate at the nominal frequency only for brief bursts and then "sleep" for the rest of the cycle—effectively reducing the frequency of operation down to a few kHz during no-load or light load operation (Figure 5–14). As most of the losses listed above vary proportionally to the frequency, burst mode reduces each substantially, allowing the device to easily meet the desired no-load power budget.

Figure 5–14 shows the output variations around the reference voltage, V_{FB} illustrates the mechanism of entering/exiting burst mode, and I_{DS} and V_{DS} illustrate the bursts of current and voltage, respectively, associated with the DMOS integrated power transistor.

Implementation of the burst mode operation in both silicon and board design can happen very quickly. To speed the process, Fairchild makes available the "FPS™ Design Assistant," a simple and effective software design tool that is also available online on Fairchild's website (see application note AN4137 online at *www.fairchildsemi.com* for details).

The result of the demand for better power dissipation is the Fairchild Power Switch (FPS™) FSD210, an off-line power switcher (see Figure 5–13). This device combines a SenseFET lateral DMOS transistor (LDMOS) for current driving and sensing (700 V minimum breakdown rating) with a voltage mode PWM IC—a combination that minimizes external components, simplifies the design, and lowers power dissipation and cost in targeted power saving or, *green mode* AC adapter applications. Figure 5–15 shows a compact traveler adapter with FSD210 circled. Appendix C provides the data sheet of FSD210 for more technical details.

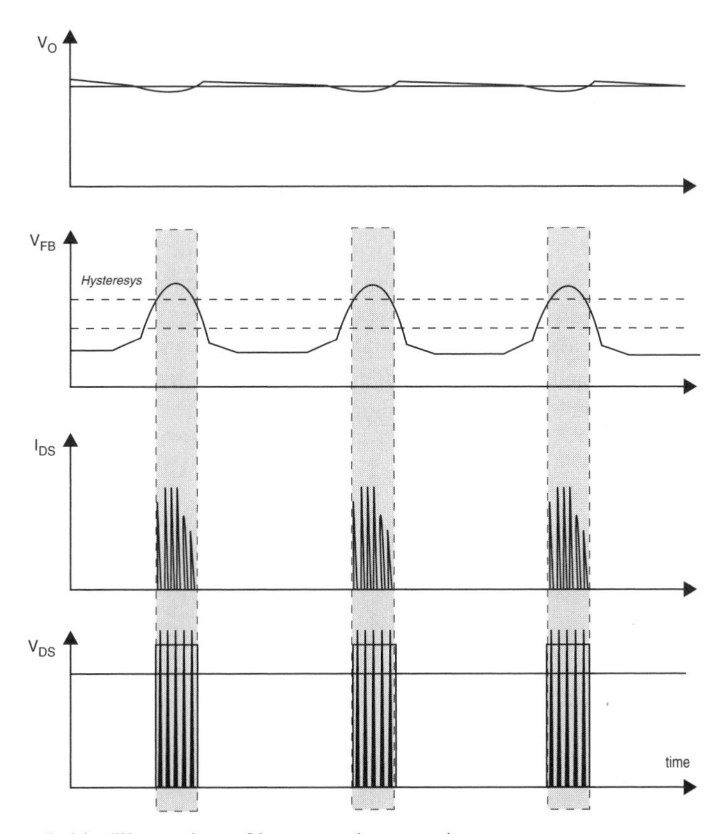

Figure 5–14 Illustration of burst mode operation.

Figure 5–15 A compact traveler adapter with FDS210 circled.

Conclusion

Miniaturization trends of modern electronic appliances, and their market diffusion by the billions, fuel a keen interest in more efficient designs. This is evident by the many protocols and initiatives already in place. Power requirements are pushing technology advancements beyond the traditional, cost-oriented model of minimizing the appliance's BOM. To meet these demands, AC adapter performance can be adequately met with proper thermal and electrical design techniques. Even with rising power requirements of today's smart phones—convergent devices that deliver all the data voice and video features imaginable—it seems safe to predict that the AC adapter will not be the bottleneck for power delivery when designs are based on efficient switching architectures.

Chapter 6

Power Management of Ultraportable Devices

6.1 Power Management of Wireless Computing and Communications Devices

Cellular telephone technology is one of the best success stories of recent years for its ability to keep the user working untethered for the entire day, with a single overnight recharge. The ultimate vision for this technology is the *smart phone*, which would have the advanced functionality of a handheld computing device, a digital still camera, a global positioning system, a music player, a portable television set, a mobile phone, and more in a convergent device. Reaching such a level of functionality without compromising the usage model will present enormous challenges as well as opportunities for the electronics industry and in particular for power management.

The Wireless Landscape

The wireless landscape is, and will remain for many years, very fragmented along both geographical and communications standards lines.

Three generations of digital cellular technologies—second (2G), third (3G), and in-between (2.5G)—already coexist (see Table 6–1).

Japan is ahead of the pack with 3G (W-CDMA and CDMA2000 flavors), while as I write the United States is building the infrastructure to provide 2.5G technology. Europe and Asia are somewhere in between.

Table 6–1 Common Cellular Standards

Generation	Symbol	Type	Description	Speed
Second	2G	GSM	Global System Mobile	14.4 Kbps
Two and a half	2.5G	GPRS	General Packet Radio Service	25–40 Kbps
Third	3G	EDGE	Enhanced Data Rate GSM Evolution	>144 Kbps[a]
Third	3G	CDMA, W-CDMA	Wideband Code Division Multiple Access	>2 Mbps

a. 2 Mbps standstill, 384 Kbps walking, 144 Kbps.

The Japanese typically do not own home computers and rely increasingly on their phones to exchange text messages as well as access email and the Internet. If this behavior takes hold elsewhere, the future of smart phones is assured.

The real possibility that smart phones will become the next *disrupting technology*—meaning that the success of smart phones will threaten almost every other established consumer technology, including PCs and notebooks—seems to be confirmed by the recent entrance into the wireless arena of powerful novices like Intel and Microsoft.

Power Management Technologies for Wireless

The majority of cellular and handheld devices are powered today by single cell Lithium-Ion batteries. The wireless semiconductor smart ICs in the signal path, following an established industry-wide trend, are mostly designed in sub-micron, low voltage, and high density processes. Consequently the power management ICs are—with a few exceptions—low voltage devices themselves, bridging the gap between the power source voltage range (2.7–4.2 V) and the operational voltage of the signal ICs (1–3.5 V). Such low operational voltages in conjunction with the necessity of low quiescent currents for long standby times have established low voltage CMOS (0.5 μm minimum feature at his juncture) as the process of choice for wireless voltage regulators. Since in these applications the space is premium, these voltage regulators come in very small packages (see Figure 6–1), from leaded to lead-less to *chip scale* varieties.

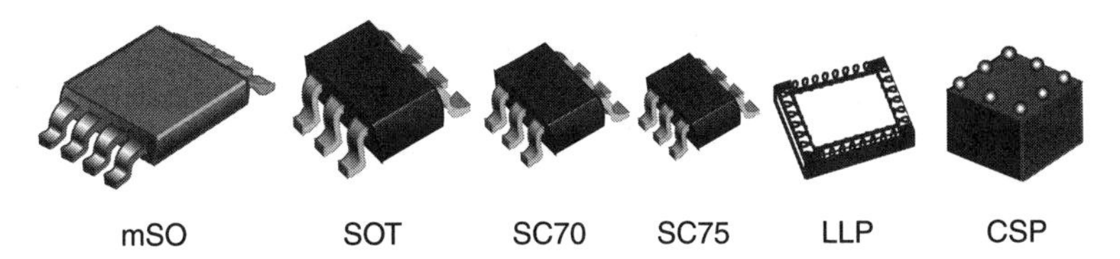

Figure 6–1 Small form factor packages for voltage regulators.

Cellular Telephones

For the next few years, the cellular telephone will remain the dominant wireless device, accounting for 80 percent of total units. Handheld devices and smart phones will account for the remaining 20 percent in roughly equal amounts.

Figure 6–2 shows the typical block diagram of a 2.5G digital cellular telephone, in the class of the T68 mobile phone by Sony Ericsson.

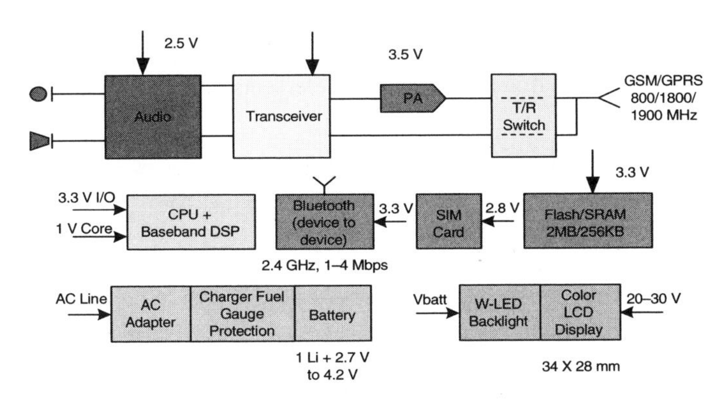

Figure 6–2 Block diagram of 2.5G mobile phone.

Each block requires a specialized power supply. The RF section is particularly sensitive to noise and is best served with low noise linear regulators, while other sections will be served by either linear or switching regulators based purely on architectural and cost constrains.

Figure 6–3 illustrates a possible strategy for the configuration illustrated in Figure 6–2.

The battery can directly power the audio LDO since its output voltage (2.5 V) is below the minimum operational battery voltage (2.7 V). The rest of the LDO outputs fall somewhere inside the battery range of operation (2.7–4.2 V) and consequently, need a higher supply voltage, in this case provided by the boost converter. The DSP core at 1 V will need a dedicated buck

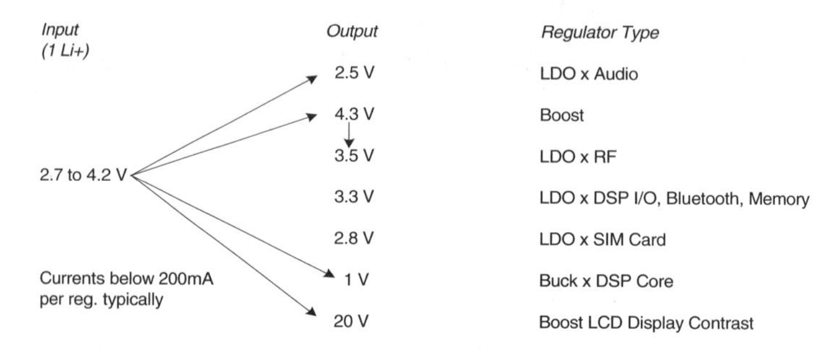

Figure 6–3 Power management strategy for a cellular telephone.

converter, while the LCD display contrast at 20 V will need a dedicated boost converter.

Figure 6–4 shows the block diagram of the power management system. In this case, a total of seven voltage regulators are necessary to power this device.

Finally, Table 6–2 shows a wide selection of chips, classified by function, from which to draw for each of the elements in Figure 6–4.

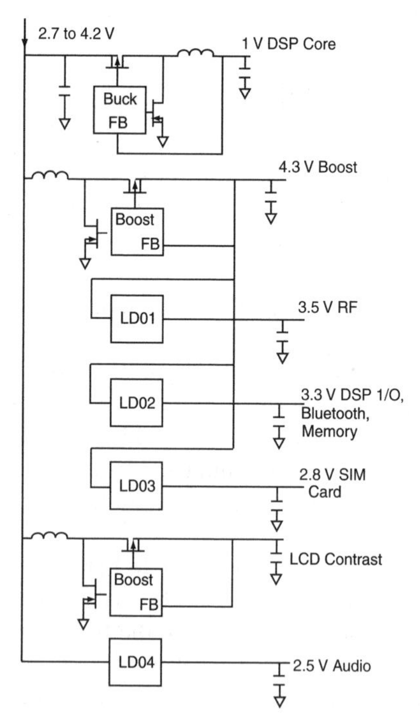

Figure 6–4 Mobile phone power management system.

Table 6–2 Semiconductor Building Block for Wireless Applications

LDO (RF)	**LDO**	**DC-DC Converter**	**Charge Pumps**
ILC7010 ILC7011 ILC7071 ILC7080/ILC7081 ILC7082 ILC7080 Dual	FAN2502/3/10/11/12/13 FAN2534/5 FAN2544 Dual FAN2558 Low Voltage FAN2591 with reset	ML4854 Boost ML4855 Boost FAN5301 Buck FAN5303 Buck ILC6363 Buck-Boost FAN2321 Buck-Boost	FAN5660 FAN5601 regulated FAN5602 regulated FAN5603 regulated
LDC & Backlight Boost ILC6383 FAN5377 (Main + LCD bias)	**Audio Amplifier** FAN7000D Headphone FAN7005 Headphone FAN7021 8Ω Speaker FAN7023 8Ω Speaker	**PA Controller** KM4112 MS4170/4270 KM7101	**Wall Adapter** FSDH0165 Ka5L/M/H0165RN Ka5L/M/H0165RVN Ka5L/M/H0265RN Ka5L/M/H0265RVN
White LED Driver FAN5611/2/3/4 Matched FAN5610 Any LED FAN5620 Serial	**Dual MOSFET** FDW2501/3NZ (TSSOP8) FDS9926A (SSOP8)	**Charge Controller** FAN7563 **Tiny Logic™** NC7S/SZ/SBxx	**Supervisory** FM809/10 FM1233 FM6332/3/4 ILC803/9/10

Wireless Handheld

A lot of activity is going into wireless handheld devices, thanks to their potential to intercept and take over a share of the cellular market. Figure 6–5 shows the typical block diagram of a 2G wireless handheld, in the class of the recently announced Palm i705.

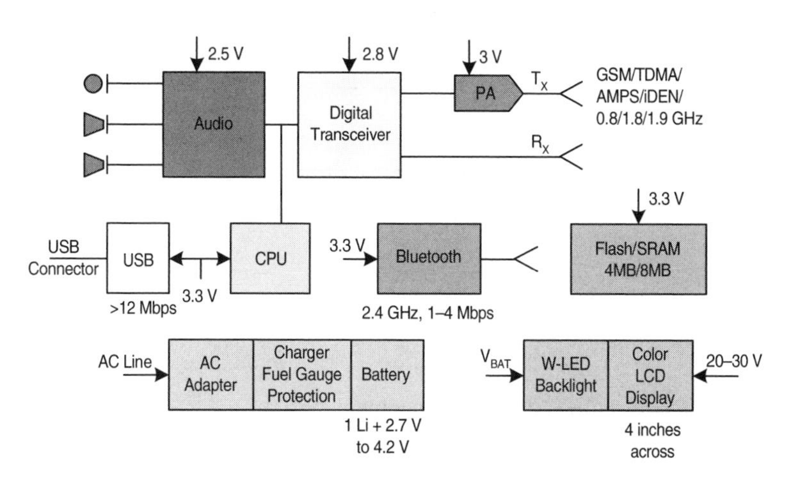

Figure 6–5 Block diagram of a 2G handheld computer.

Here again each block requires a specialized power supply, but due to the absence of a DSP and of the SIM card, the power management is a bit leaner than for the case illustrated in Figure 6–2.

With similar considerations to those used for Figure 6–4, Figure 6–6 shows the strategy chosen for the handheld power management and Figure 6–7 shows the implementation, obtained with a total of five regulators.

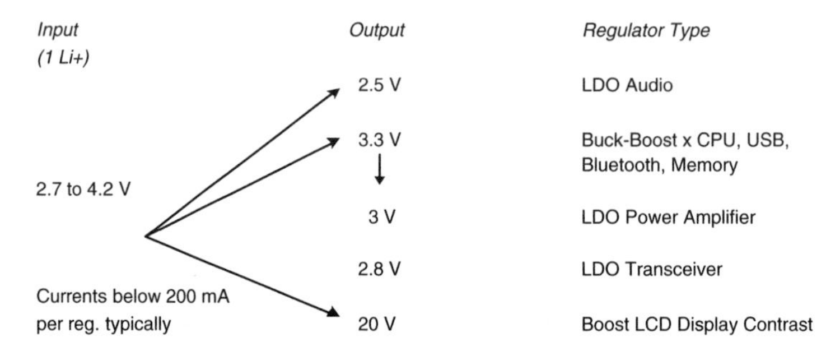

Figure 6–6 Power management strategy for wireless handheld.

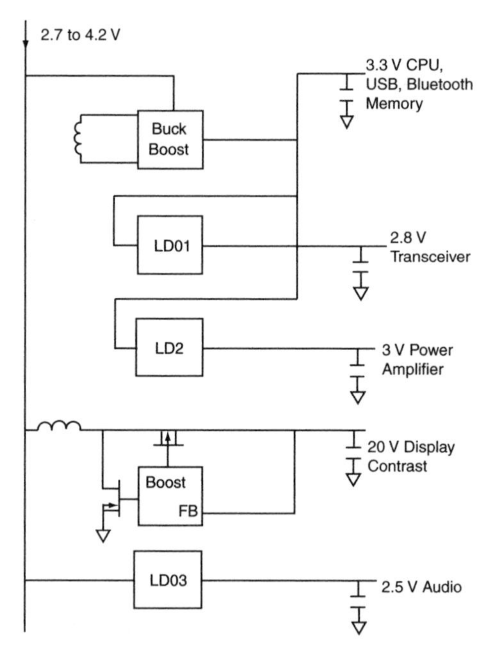

Figure 6–7 Handheld power management system.

Here again the specific components can be drawn from Table 6–2.

Charge

Important elements of both cellular phones and handheld devices are the external AC adapter and the internal charger. Many AC adapters on the market are very simple implementations based on a transformer, a bridge rectifier, and a resistive current limit. More sophisticated controls can be obtained with integrated controllers such as those indicated in Table 6–2.

The Lithium-Ion charger is a constant-current/constant-voltage regulator that is either implemented with specialized controllers (see Table 6–2 for an example) or by PWM of a pass transistor controlled directly by the CPU.

Protection and Fuel Gauging

This section deals briefly with the in-battery electronics, namely that section of power management residing inside the Li^+ cell.

Lithium-Ion cells' energy density makes them dangerous elements that need a very precise protocol for charge and handling. Overcharge needs to be prevented as well as undercharge, which leads to reduced energy storage.

To this end, protection electronics measures the battery voltage and opens a pass transistor as soon as the charge voltage threshold is crossed.

Fuel gauging is necessary to be able to display the state of charge of the battery and predict the residual time of operation in battery mode. This is an interesting and challenging feature because residual time of operation matters to the user only toward the end of the battery charge, exactly when the accuracy of the prediction begins to falter. In fact, no matter how precise the measurement system, eventually the residual time of operation will translate into an amount of residual charge that is of the order of magnitude of the system precision, leading to increasing prediction errors as the battery approaches the empty state. This results in the requirement of amazing levels of precision in the current measurement. The measure of current over time is also referred to as Coulomb counting. Analog front-end amplifiers are called to resolve micro-Volts of voltage drops across small sense resistors, followed by 10-bit or higher order A-D (Analog to Digital) converters. The actual processing of the row data today—at the 2 to 2.5G juncture—is generally done in the central processing unit. With 3G systems and above and with smart phones, we expect that the taxing of the DSP—or its successor—will be such that the fuel gauge data processing will be decentralized, leading to smart fuel gauge devices incorporating compact 8-bit microcontrollers.

Figure 6–8 shows an example of an integrated combo fuel gauge and protection control IC that utilizes Fairchild's dual MOSFET FDW2508D as the pass transistor for the protection section.

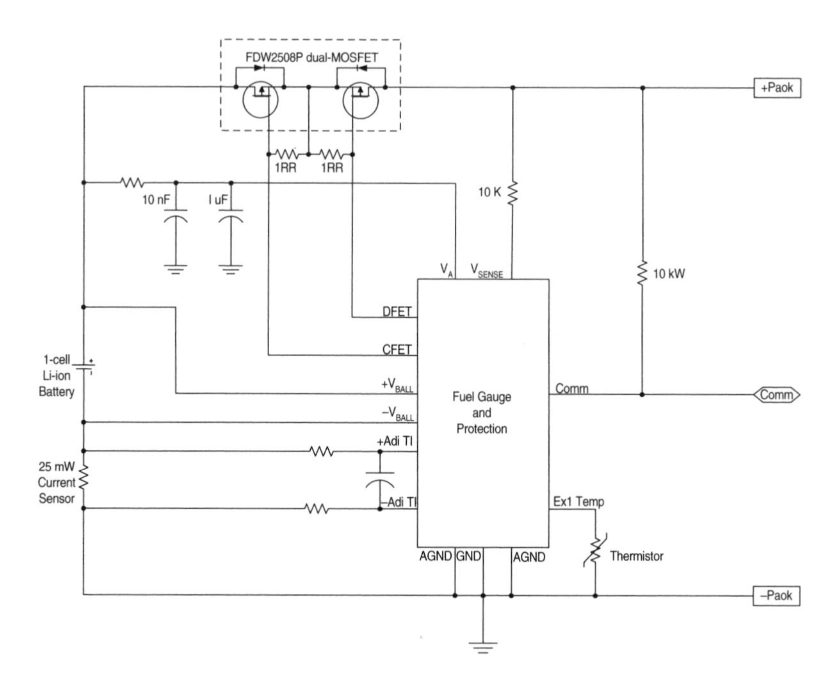

Figure 6–8 Fuel gauge and protection with FDW2508P as the pass transistor.

Convergence of Cellular Telephone and Handheld

By studying the block diagrams in Figure 6–2 and Figure 6–5, it becomes obvious how similar the two systems are. Both rely on the same radio technologies and frequency ranges, Bluetooth for device-to-device networking, single Lithium-Ion for power source, etc.

In fact, it is easier to point out the differences between the two devices. The DSP, present in the cellular phone only, is a key differentiator allowing for voice processing. Otherwise, it really comes down to size. The handheld typically will have a bigger screen and more memory (Flash memory for operating system, phone book, and files storage and SRAM for temporary data storage) as well as stereo audio for music player emulation (MP3).

A few examples of smart phones, namely handheld devices with DSP on board, already are appearing in the market. One early example is the Blackberry 5810, a handheld PC that can be transformed into a cellular phone by means of a hands-free module connected to the device via a 2.5 mm jack.

As pointed out at the beginning of this chapter, the challenge for the electronics industry is in part technological (will the smart phone be able to deliver the same standby and talk times to which the cellular customers are accustomed?) and in part cultural (will the Japanese model of connectivity illustrated at the beginning of this chapter prevail?).

Future Architectures

At the 3G juncture, the system complexity for cellular and smart phones is such that one DSP is not enough and an additional DSP or ASIC often is necessary to support video and audio compression. In turn, this increases power consumption and reduces battery operation time.

Adaptive Computing Machines (ACMs) are a new class of ICs appearing on the horizon that promise to solve the power dissipation problem by means of a flexible architecture that optimizes software and hardware resources.

Power management in wireless devices is a pervasive issue that encompasses every element in the signal as well as in the power path. With system complexity increasing dramatically from one technology generation to the next, the long duration of untethered operation in wireless devices can be preserved only through the introduction of new breakthrough technologies. New architectures along the lines of the aforementioned ACMs as well as the conversion of large scale ICs from bulk CMOS to *Silicon On Insulator* (SOI) should go a long way toward reducing the power dissipation of the electronic load. At the other end of the equation, new and more powerful sources of power, such as fuel cells, should be able to provide higher power densities inside the same cell form factor. The entire technology arsenal should be able to continue to provide more features at no compromise.

Power conversion technologies already have achieved impressive efficiencies, reaching peaks of 95 percent. Accordingly, they are a critical element of the power management equation but not its bottleneck. The holdups are at the process and power source levels, and eliminating these bottlenecks will require new process technologies that lead to chips with reduced power dissipation and new power sources with higher power density. The analog building blocks for effective power management of a wireless device in its present and future incarnations are already in place. It is not uncommon today to find these building blocks assembled inside custom combo chips. In this case, such combo chips integrate the entire power management function on board a single IC. Voltage regulators—today designed with 0.5 μm minimum features—will continue to follow the CMOS minimum feature reduction curve, staying only a few technology generations away from the loads they are powering (state-of-the-art 0.13 μm minimum features). Accordingly, they will continue to be able to sustain adequately the power, performance, and cost curve that will be required to power future generations of wireless devices.

6.2 Power Management in Wireless Telephones: Subsystem Design Requirements

Trends in power management are driven by a demand for products loaded with features. Convergent wireless devices, such as smart phones that combine the features of cell phones, PDAs, digital still cameras (DSCs), music players (MPs), and global positioning systems (GPSs), stretch many technology boundaries, including those of power. This section discusses the latest power management products being used in today's most sophisticated cellular phone designs.

Smart Phone Subsystems

A state-of-the-art smart phone system (with handset and AC adapter/charger) can be divided into up to five main board constituents: display board, baseband main board, keypad board, Li^+ in-battery board, and AC adapter board. Additional modules may be present for DSC, Bluetooth, or other functions. Accordingly, power management breaks down along these five subsystems. Figure 6–9 illustrates such system partitioning. We will review these subsystems with respect to their power management chips' content, both integrated and discrete, including Light Emitting Diodes (LEDs). These power management chips are:

1. LED driver ICs and four white LEDs in the display board
2. LED drivers and eight white or blue LEDs in the keypad board
3. Power management ICs in the main board
4. Lithium-Ion protection and fuel gauge ICs and MOSFETs in the battery pack
5. Offline regulator ICs in the AC adapter board

Figure 6–10 shows the corresponding block diagram for a cell phone in the class of Nokia's 7650, which integrates a digital camera.

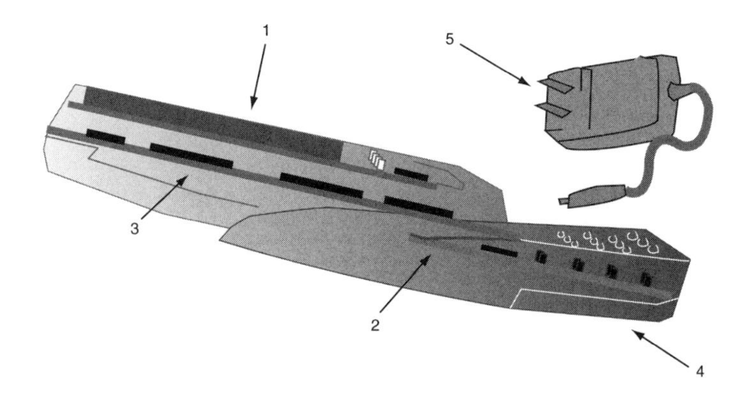

Figure 6–9 System partitioning of a state-of-the-art smart telephone system.

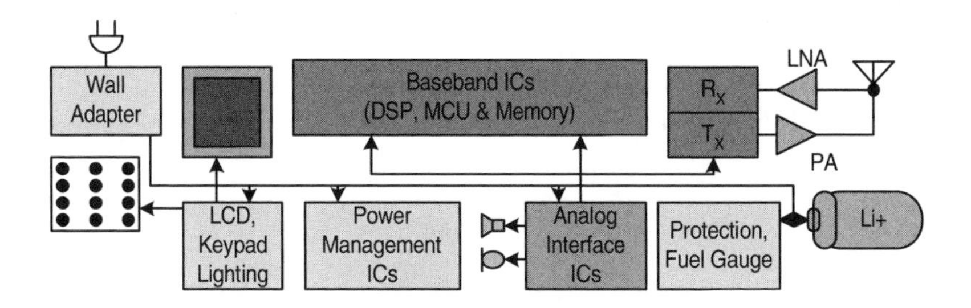

Figure 6–10 Block diagram of a smart phone system.

Display Board

In monochromatic displays, the backlight can be made up of different colors, which generally are obtained with four LED lamps of the same color. In smart phones the color, *Thin-Film Transistors* (TFT) *Liquid-Crystal Display* (LCD), necessarily requires only white backlighting. White LED diodes have low forward voltage (around 2.7 V) and require a simple DC current to produce light. Accordingly, a low DC power source (V_{DD} in Figure 6–11), as low as 3.1 V, is necessary to bias these devices. Thanks to such low bias voltage, the four LEDs can operate directly off a single cell Lithium-Ion, the power source of choice in cellular telephones.

A monolithic quadruple LED driver, such as Fairchild Semiconductor's FAN5613 (shown in Figure 6–11), can be housed in a tiny MLP 8-lead package and provide up to 40 mA bias for each diode.

As continuous-time current control may result in poor color consistency, the LEDs can be excited with a pulse width modulated source via the ON/OFF pin to achieve higher color fidelity.

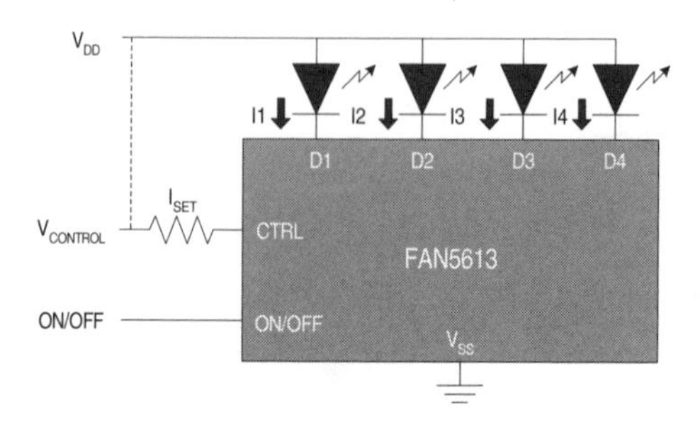

Figure 6–11 Bias scheme for White LED driver with FAN5613.

Keypad Board

Similar to the display configuration, the keypad is also illuminated by LED lamps. In this case, eight white or blue LEDs generally are utilized. For blue backlight, eight QTLP601C-EB InGaN/Sapphire surface mount chip LEDs (shown in Figure 6–12) can be used, with two Fairchild FAN5613 LED Driver ICs driving them.

Figure 6–12 QTLP601C-EB low VF blue LED lamp.

Main Board

The main board contains the vast majority of the electronics, from the baseband DSP and application MCU to the transceiver and analog interface. Each of these blocks is powered by a dedicated voltage regulator. The growing complexity of smart phones requires strict management of the power source. This is obtained by means of a "power manager" inside the baseband processor, communicating to the outside world via logic signals. On the power source side, the voltage regulators are able to receive such logic signals and react accordingly.

In some instances, all the regulator is required to do is to enter into a light load operation or sleep mode or into a shutdown mode via direct logic signals. In other instances, such as in powering the baseband processor, power management is more sophisticated and requires a voltage source that varies with the task at hand, delivering just enough power as necessary and no more. In this case, a voltage regulator, coupled with a D-A converter and a serial bus with the ability to communicate with the host microcontroller, is required. While this technique may sound exotic, such power management schemes are commonplace in notebook computing, battery operated devices that long ago crossed the threshold of complexity that today's smart phones have just now reached. (Examples of popular power management techniques in notebooks are SpeedStep™ from Intel and PowerNow™ from AMD.)

Figure 6–13 illustrates an example of distributed power management on the main board. The combination of a simple buck converter and an SMBus serial-to-parallel interface, such as the FM3570 by Fairchild, allows the CPU to drive the V_{core} supply with a 5-bit D-A converter resolution. A combination of switching and linear regulators assures a good compromise between simplicity and performance, and all the devices can be shut down via a dedicated logic pin.

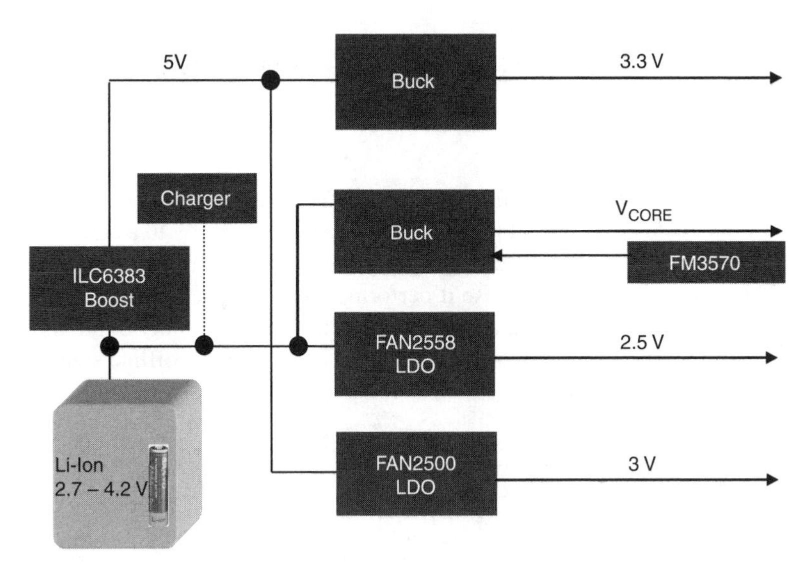

Figure 6–13 Example of distributed power management system for the main board.

Battery Pack

The power management inside the battery pack consists mainly in the Lithium-Ion protection and fuel gauge ICs and MOSFETs. The protection electronics measures the battery voltage and opens a pass transistor as soon as the charge voltage threshold is crossed. Fuel gauging is necessary to display the battery's state of charge and to predict the residual time of operation in battery mode. Figure 6–14 shows an example of in-battery electronics that uses Fairchild's dual MOSFET FDW2508D as the pass transistor for the protection section.

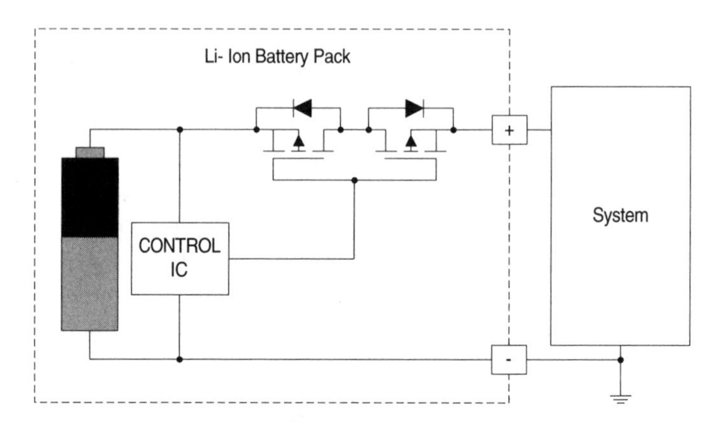

Figure 6–14 In-battery fuel gauge and protection with FDW2508D.

AC Adapter

The AC adapter board rectifies the AC line and converts it down, either to a low DC voltage manageable by the main board, or directly to a constant-current/constant-voltage charging algorithm required by the single Lithium-Ion cell, in which case it performs both the functions of adapter and charger. A charger on the main board will be required only in the first case. In Figure 6–15 the AC adapter/charger is based on an offline switch-

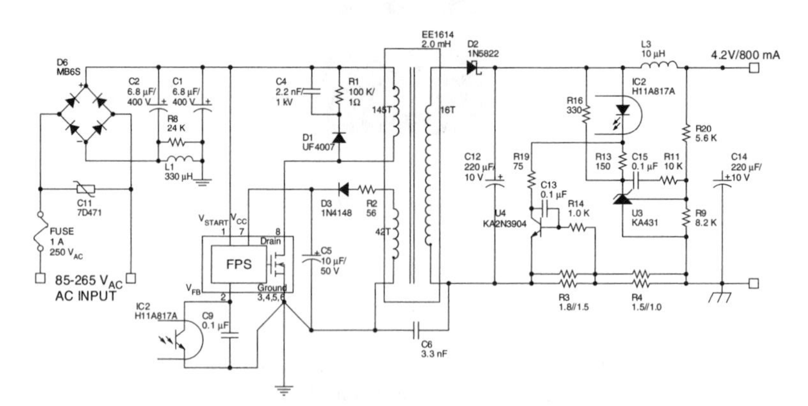

Figure 6–15 AC adapter for single cell Lithium-Ion.

ing architecture for the best efficiency. In this example, the high voltage product called the FSDH0165 is powered directly by the AC line and integrates the power into the DMOS transistor for minimum complexity. Later in this chapter, we will discuss in more detail the AC adapter/charger subject.

The power management of state-of-the-art wireless telephones typically breaks down along five main subsystems, each following its own integration dynamics. The main board requires many low voltage sources and lends itself to higher levels of integration. At the opposite end is the AC adapter (or adapter/charger) with its requirement of high voltage (600–800 V) and galvanic isolation with respect to the low voltage side. The keypad board and display can be serviced by the same class of technologies, namely LEDs and LED drivers, while the in-battery electronics is another unique domain where true mixed signal technologies are needed for fuel gauging and protection.

It seems safe to say that the natural boundaries of these subsystems and the diversity of the technologies needed in each of them will assure a plurality of technologies, solutions, and players in the power management of wireless devices.

6.3 Powering Feature-Rich Handsets

Market trends show that devices incorporating color screens, camera phones, and Personal Information Management (PIM) applications are growing steadily. For example, market data by major market research companies like Dataquest and iSuppli point to the possibility that in 2006 the number of smart phones will be larger than the number of the notebook computers shipped that year and will far outnumber single function devices like digital still cameras and PDAs.

With this in mind, we have little doubt that newly emerging applications in cell phones and handhelds, such as video streaming and high quality digital media playback, will soon become legitimate in high-end handsets and will later be embraced by the mainstream.

In this section, we will look at the challenges that such complex devices pose, with a special focus on power management. We will also discuss new solutions and future trends.

Growing Complexity and Shrinking Cycle Time

Today's *OEMs* play in complex markets, spanning across different platforms—second generation or 2G platforms such as GSM, Time Division Multiple Access (TDMA), and CDMA, and 3G platforms such as W-CDMA and CDMA2000—and each proposed in different models. See Table 6–1 for a review of these acronyms.

For the best time to market, the reference design for a single platform typically will rely on a relatively rigid "core" chipset, while a more flexible periphery will accommodate a model's differentiation within the given platform. In other words, it takes time, sometimes up to a year, to develop new chips that incorporate new features. Consequently a product can go on the shelf faster if it can rely on a rigid core inherently undifferentiated with the differentiation—new features, better performance, etc.—that is accommodated with add-ons. The resulting product may be less integrated and more bulky but goes to market faster. Hence, the final architecture of a product, like the one in Figure 6–16, is influenced by technical factors as well as by time. In the system illustrated in Figure 6–16 the power management—the section of interest for us—is accomplished with a core PMU tightly integrated inside the chipset and an auxiliary PMU servicing the add-on features.

Figure 6–16 illustrates the core chipset, with the baseband section, including the application *MCU* handling the data, the *DSP* for voice, FLASH memory, the RF section (with its receiving RX and transmitting TX blocks), and the power management unit section.

A number of add-on modules, such as Bluetooth for untethered data transfer on a short distance, cameras, and LCD modules surround such a core chipset. These blocks require additional power provided by an auxiliary PMU, represented by the PMU add-on block in Figure 6–16.

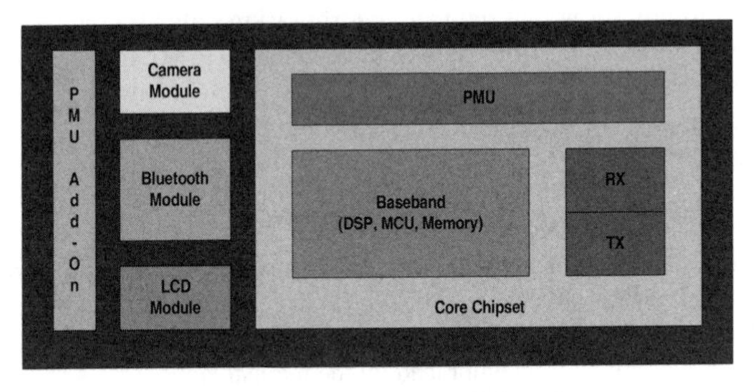

Figure 6–16 Block diagram of the handset mainboard.

Power Management Unit

The increasing number and performance of smart loads supported by the power management unit demands an increasingly sophisticated PMU, capable of going well beyond providing the basic functions of voltage regulation, charging, and fuel gauging.

In sophisticated systems, the PMU may need to be programmable in order to become platform-specific via software implementation of the protocol. To this end the PMU must be capable of communicating with the host CPU via a serial interface (I2CBus or similar). This is to adjust the power delivery mode to the load demand (heavy, light, or intermediate) and to take responsibility for many critical functions, such as power sequencing, at a time when the communication bus is disabled.

Such PMU can be implemented with varying levels of integration, perhaps initially starting with a solution based on multiple chips for fast time to market, and subsequently up-integrating to a single package (Multi-Chip Package or MCP) or even a single IC, depending on the volumes and other considerations.

Low Dropouts (LDOs)

In Figure 6–17, a microcontroller-based power management architecture provides all the hardware and software functions, as discussed above, in a multi-chip implementation. When defining this unit, many trade-offs need to be considered. The Li^+ low voltage (3 V typical) power source is conducive to a high level of integration on standard CMOS. However, this choice hits a snag if a charger, interfacing with an external AC adapter, needs to be integrated, in which case the process technology needs to withstand voltages well above the standard 5 V of CMOS.

Ultimately, if the cost structure allows for its high mask count, a powerful mixed signal BCD process can enable a true single chip solution capable of handling high voltage, high current, and high gate count. As illustrated in Figure 6–17, each subsystem in the handset requires its own specific version of power delivery—low noise LDOs in the RF section and low power LDOs elsewhere. Each subsystem also requires an efficient buck converter for the power consuming processors, a boost converter in combination with LED drivers for the LED arrays, and a linear charger interfacing the Li^+ battery with the external AC adapter during charge.

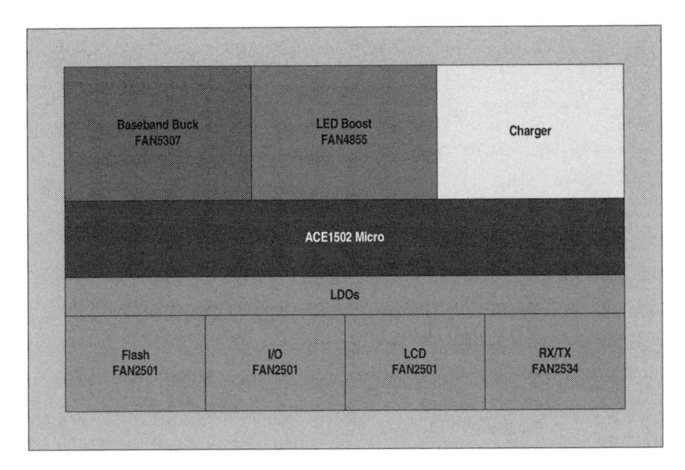

Figure 6–17 Power management unit.

6.4 More on Power Management Units in Cell Phones

The power management of cell phones is one of the most dynamic areas of growth for integrated circuits. Recently, a number of PMUs for cell phone applications have appeared on the market. Handset PMUs range from low integration (single function building blocks) to medium integration (three to ten integrated regulators) to high integration (entire power management, audio power amplifiers, etc.). Interestingly enough, a systematic study of numerous recently released handsets shows no clear trend with respect to levels of integration. Different business models and privileging aspects, from performance to size to cost to time-to-market, explain this fragmented picture. This section focuses on complex PMUs and the capabilities that drive successful product offerings in this area.

Ultraportable devices are feeding the up-integration trend due to shrinking handset dimensions and increasing capabilities such as color displays and the convergence of cell phones with PDAs, DSCs, MP3s, GPSs, and the like. This phenomenon is somewhat similar to what happened in the early 1990s when the shrinking of Hard Disk Drives (HDDs) went from 3.5" to 2.5" to 1.8" and below, which pushed up-integration despite increased costs. In that case, size reduction was paramount (i.e., it didn't matter if the single die would cost more than the dis-integrated solution, due to the inherent lower yield of a bigger die). At that time, many power management companies, lagging on the large-scale integration power process curve needed for that function (Bipolar-CMOS-DMOS), had no choice but to exit the market.

One important difference between the HDDs of the early 1990s and the cellular telephones of today is that, due to its low power and low voltage, the power management IC up-integration in cell phones does not require specialized processes. Instead, this integration can often be accomplished with run-of-the-mill 0.35–0.25 μm CMOS homegrown technology, or with processes easily available at the foundry houses.

Because today's process technology does not appear to be a big barrier, we are witnessing the emergence and participation of fab-less power semiconductor startup companies, a business model not seen before. These are companies with no fabrication facilities that rely on external foundries for chip production. Hence, contrary to what happened with HDDs, we are witnessing an abundance of players at the starting line of the cell phone up-integration race.

Barriers to Up-Integration

The power section in a cell phone, including the power audio amplifiers and charger, is relatively simple; it consists mostly of an array of low-power linear regulators and amplifiers. The complexity comes from managing these functions, which require reliable data conversion and the additional integration of digital blocks such as SMBus for serial communication and state machines, or microcontrollers, for correct power sequencing. Such levels of complexity on board a single die bring their own set of problems, like interference from cross-talk noise.

This new class of power management devices requires technical skills, as well as IP and CAD tools, which go beyond the traditional power team's skill set and cross into logic, microcontroller, and data conversion fields. Such an extension of the capability set in the power management space can be a barrier to entry for traditional analog power companies, while cost competitiveness will likely be a barrier with which the fab-less startups will have to contend.

PMU Building Blocks

Highly integrated power management units are often complex devices housed in high pin count packages. Available devices range from 48 to 179 pins. Such units either can be monolithic, with perhaps a few external transistors for heavy-duty power handling, or multi-chip solutions in a package (MCP). The complexity effectively makes these units custom devices. Because of the custom nature of these units, the following section will discuss the architecture (Figure 6–17) and fundamental building blocks of a PMU in generic terms rather than focusing on a specific device. For the same reasons, building blocks will be illustrated by means of available stand-alone ICs.

Figure 6–17 illustrates a generic microcontroller-based power management architecture, providing all the hardware and software functions, as discussed above. Many trade-offs need to be considered when defining this unit. Some of the regulators, like the charger, are required to provide a continuously rising level of power, which may be difficult to accommodate on board a single CMOS architecture. For example, an external P-channel DMOS discrete transistor, such as Fairchild's FDZ299P, housed in an ultra-small BGA package can help solve the problem. As illustrated in the figure, each subsystem in the handset requires its own specific flavor of power delivery. Low noise LDOs like Fairchild's FAN5234 are used in the RF section and low power LDOs like FAN2501 are used elsewhere. This architecture also requires an efficient buck converter for the power consuming processors as well as a boost converter in combination with LED drivers for the LED arrays.

CPU Regulator

Figure 6–18 shows the die of the FAN5307, high-efficiency DC-DC buck converter; the big V-shaped structures on the left are the integrated P- and N-channel MOS transistors, while the rest of the fine geometries are control circuitry. The FAN5307, a high efficiency low noise synchronous PWM current mode and Pulse Skip (Power Save) mode DC-DC converter, is designed specifically for battery-powered applications. It provides up to 300 mA of output current over a wide input range from 2.5 V to 5.5 V. The output voltage can be either internally fixed or externally adjustable over a wide range of 0.7 V–5.5 V by an external voltage divider. Custom output voltages are also available.

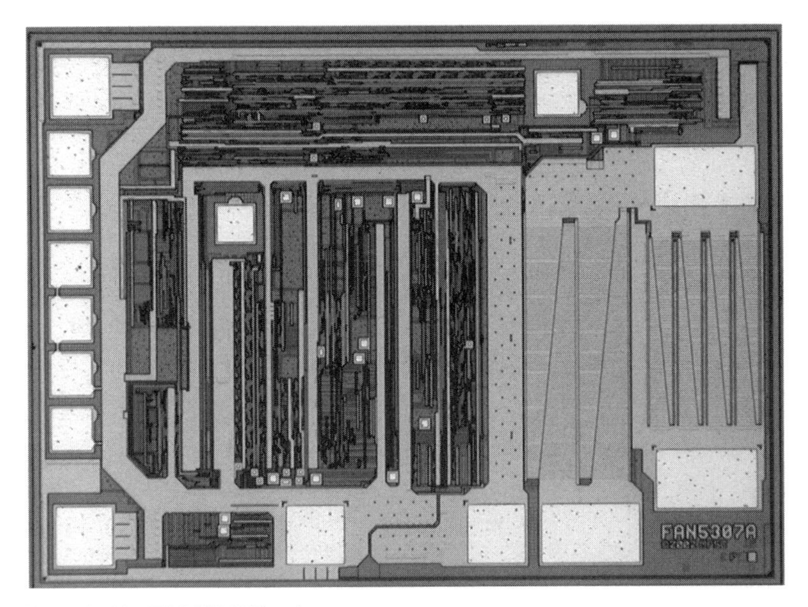

Figure 6–18 FAN5307 buck converter.

Pulse skipping modulation is used at moderate and light loads. Dynamic voltage positioning is applied, and the output voltage is shifted 0.8 percent above nominal value for increased headroom during load transients. At higher loads, the system automatically switches to current mode PWM control, operating at 1 MHz. A current mode control loop with fast transient response ensures excellent line and load regulation. In Power Save mode, the quiescent current is reduced to 15 μA in order to achieve high efficiency and to ensure long battery life. In shut down mode, the supply current drops below 1 μA. The device is stand-alone and is available in 5-lead SOT-23 and 6-lead 3 × 3 mm MLP packages.

Figure 6–19 shows the voltage regulator application complete with external passive components. The integration of the power MOS transistors leads to a minimum number of external components, while the high frequency of operations allows for a very small value of the passives. Appendix D provides the data sheets of FAN5307 for more technical details.

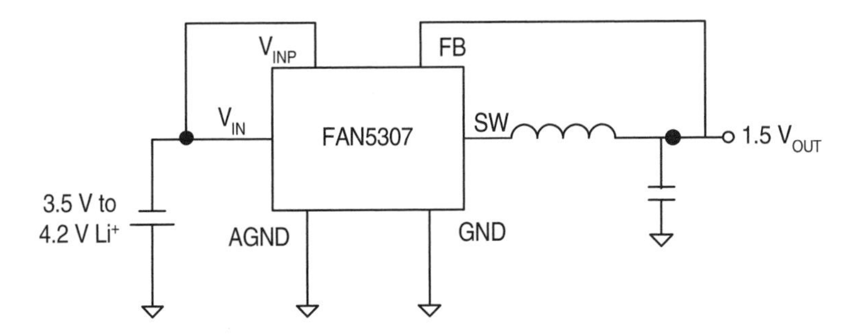

Figure 6–19 FAN5307 application.

Low Dropout Block

Due to the relatively light loads (hundreds of mA rather than hundreds of Amperes as in heavy-duty computing applications), low voltages (one Li^+ power source or 3.6 V typical), and often low input-to-output dropout voltages, simple linear regulators are very popular in ultraportable applications. Figure 6–20 shows the die of the FAN2534 low dropout (180 mV at 150 mA) regulator: a state-of-the-art CMOS design that targets ultraportable applications and is characterized by low power consumption, high power supply rejection, and low noise. Here again, the V-shaped structure is the P-MOS high side pass transistor and the rest of the fine geometries are the control logic.

In this section, we have discussed the evolution of complex PMUs in cell phones, illustrating the benefit of using the microcontroller in sophisticated applications such as a handset illumination system. We have reviewed the breadth of mixed-signal technologies and architectures coming into play, focusing on fundamental building blocks of the PMU: the microcontroller, the buck converter, and the LDO. These, and other building blocks like LED drivers, chargers, and audio power amplifiers, can all be integrated monolithically or in multi-chip package form to implement a modern handset power management unit.

From this discussion, it should be clear that the likely winners of the race for the PMU sockets will be the companies with the broadest combination of skills and capabilities to meet the technical hurdles and the stringent cost targets imposed by this market. The successful companies will

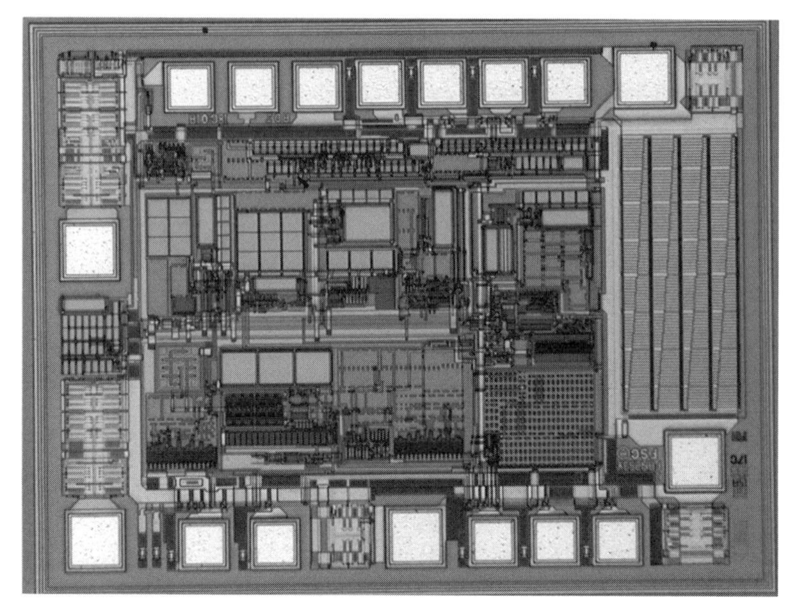

Figure 6–20 FAN2534 LDO die photo.

need to have knowledge of ultraportable systems, power analog and digital integration experience, and the ability to mass-produce these chips.

The Microcontroller

As discussed in the last section, the microcontroller, a block diagram of which is shown in Figure 6–21, is the basis of a feature-rich, or smart phone, power management unit. Fairchild's ACE1502 (Arithmetic Controller Unit) family of microcontrollers, for instance, has a fully static CMOS architecture. This low power, small-sized device is a dedicated programmable monolithic IC for ultraportable applications requiring high performance. At its core is an 8-bit microcontroller, 64 bytes of RAM, 64 bytes of EEPROM, and 2/k bytes of code EEPROM. The on-chip peripherals include a multi-function 16-bit timer, watchdog and programmable under-voltage detection, reset and clock. Its high level of integration allows this IC to fit in a small SO8 package, but this block can also be up-integrated into a more complex system either on a single die or by co-packaging.

Another important factor to consider when adding intelligence to PMU via microcontrollers is the battery drain during both active and standby modes. An ideal design will provide extremely low standby currents. In fact, the ACE1502 is well suited for this category of applications. In halt mode, the ACE1502 consumes 100 nano-amps, which has negligible impact on reduction of battery life. Appendix E provides the data sheet of ACE1502 for more technical details.

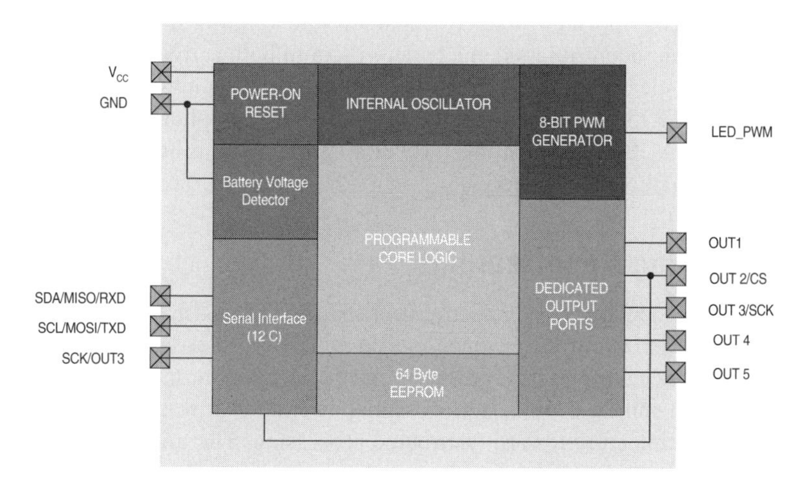

Figure 6–21 Microcontroller architecture.

The Microcontroller Die

The microcontroller is often the basis of a feature-rich, or smart phone power management unit. Fairchild's ACE1502 microcontroller die is shown in Figure 6–22. This IC fits in a small SO8 package, but this block can also be up-integrated in a more complex system, either on a single die or by co-packaging. .

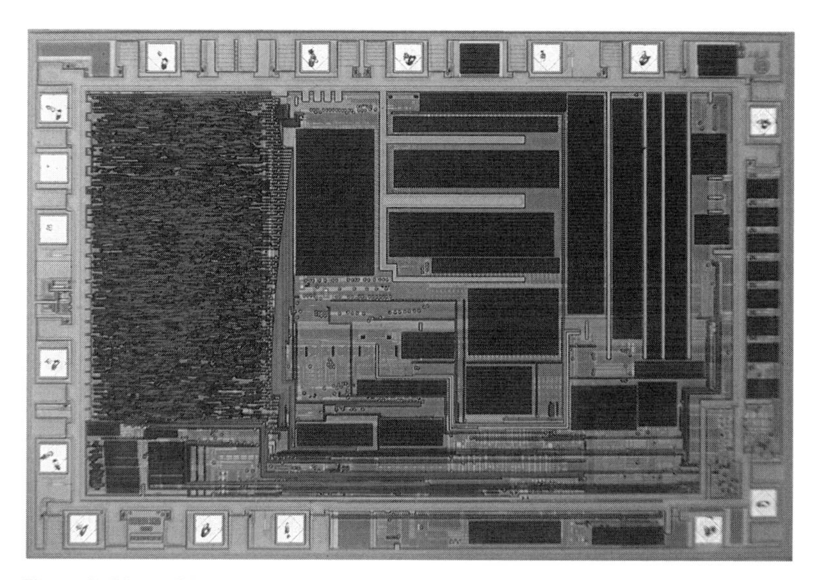

Figure 6–22 ACE1502 microcontroller die.

Another important factor to consider when adding intelligence to PMU via microcontrollers is the battery drain in both active and standby modes. An ideal design will provide extremely low standby currents. In fact, the ACE1502 is well suited for this category of applications. In halt mode, the ACE1502 consumes 100 nano-amps, which has negligible impact on reduction of battery life.

Processing Requirements

As the trend continues toward convergent cell phone handsets, development of software and firmware becomes an increasingly complex task. In fact, as the systems tend toward larger displays and the inclusion of more functions, such as 3-D games, a phone's processing power and software complexity drive its architecture toward distributed processing. The microcontroller adds further value in off-loading the power management tasks from the main CPU, thus freeing it to perform more computing intensive tasks.

The application of "local intelligence," via a microcontroller, can assume various levels of sophistication, such as the recent trend of *feature phones*. For example, it is common to find phones with digital cameras built into them. However, the lack of a photoflash limits the use of the phone's camera to brightly lit scenes. To address this problem, it is now possible to include a flash unit built from LEDs. The addition of a flash requires several functions such as red-eye reduction and intensity modulation, depending on ambient lighting and subject distance as well as synchronization with the CCD module for image capture. These additional functions can be easily off-loaded to a peripheral microcontroller. Such architecture leads to optimized power management and simplifies the computing load on the main CPU.

Microcontroller-Driven Illumination System

A complex LED based illumination system is illustrated in Figure 6–23. Typically, an array of four white LEDs is needed for the color display backlighting, while another array of four white or blue LEDs implements the keyboard backlighting. White LEDs, typically assembled in a quad package, are needed for the camera flash. And finally, an RGB display module provides varying combinations of red, green, and blue flashes for lighting effects. As mentioned earlier, the sequencing and duration of all the illumination profiles are under micro control.

Figure 6–24 demonstrates the lighting system described previously, with all the elements of the system excited at once. The back light and display light locations are obvious. The flash is the top light and the RGB is the one in the middle.

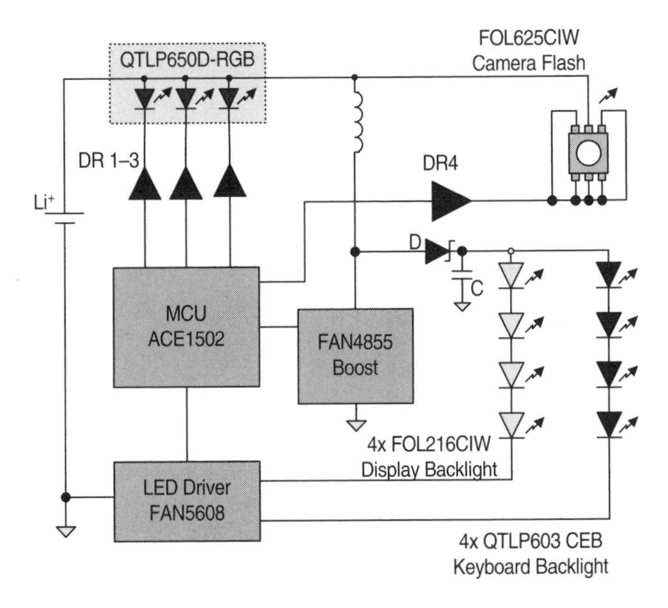

Figure 6–23 Handset illumination system.

Figure 6–24 Lighting system demonstration.

Figure 6–25 shows the typical waveform generated by the microcontroller to drive the lighting system. The oscilloscope waveforms are:

- A1 FLASH LED cathode signal
- A2 primary back light intensity control via 8-bit PWM signal
- 2 secondary back light intensity control via 8-bit PWM signal
- 3 RGB LED Module: Red channel controlled using 4-bit PWM signal
- 4 RGB LED Module: Green channel controlled using 4-bit PWM signal
- 5 RGB LED Module: Blue channel controlled using 4-bit PWM signal:

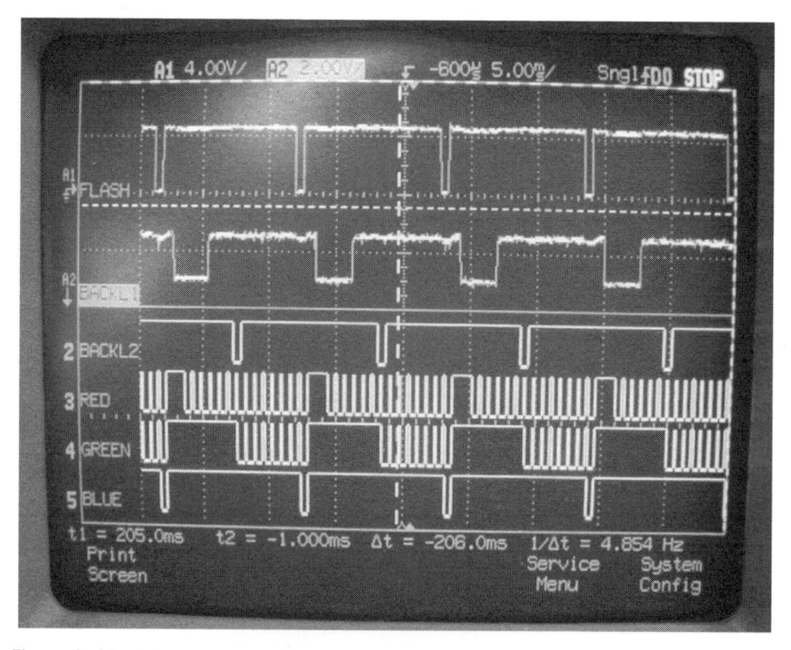

Figure 6–25 Lighting system waveforms.

6.5 Color Displays and Cameras Increase Demand on Power Sources and Management

One of the most amazing recent trends in ultraportable technology is convergence. With smart phones representing the convergence of PDAs, cell phones, digital still cameras, music players, and global positioning systems. With Audio Video Recorders (AVRs) converging camcorders, DSCs, audio players, voice recorders, and movie viewers into one piece of equipment.

While some of these convergences will take time to materialize in the mainstream, others are improving rapidly. One of these rapidly improving areas is the convergence of two very successful ultraportable devices: DSCs and color cell phones, into a single portable device.

This section reviews the DSC first and then dives into the integration of this function into cell phones. Finally, the implications in terms of power consumption and power sources are discussed.

Digital Still Camera

Digital still cameras have enjoyed a brisk growth in the past few years and today there is more of a market for DSCs than notebook computers. One third of these DSCs are high resolution (higher than three megapixels); today top of the line cameras exhibit close to five megapixels with seven on the horizon.

Figure 6–26 illustrates the main blocks of a DSC and the power flow, from the source (in the example one Li^+ cell) to the various blocks.

The key element in a DSC is its image sensor, traditionally a charge coupled device (CCD) or more recently a CMOS integrated circuit that substitutes the film of traditional cameras and is powered typically by a 2.8–3.3 V, 0.5 W source.

A Xenon lamp powered for the duration of the light pulse by a boost regulator converting the battery voltage up to 300 V, produces the camera flash. The lamp is initially excited with a high voltage (4–5 kV) pulse ionizing the gas mixture within the lamp. The pulse is fired by a strobe unit composed of a high voltage pulse transformer and firing IGBT like the SGRN204060.

The color display backlight can be powered by four white LEDs via an active driver like the FAN5613 which allows duty cycle modulation of the LED bias current to adjust the luminosity to the ambient light, thereby minimizing the power consumption in the backlight.

The focus and shutter motors are driven by the dual motor driver KA7405D and the Li^+ battery can be charged by the FSDH565 offline charger adapter.

Finally powering the DSP will be accomplished by a low voltage, low current (1.2 V, 300 mA) buck converter.

As an example, the peak power dissipated by a palm sized DSC (1.3 megapixels) during picture taking can be around 2 W and 1.5 W (or 500 mA at 2.4 V) during viewing. Two rechargeable alkaline cells in series with 700 mAh capacity can then sustain close to one hour of picture taking and viewing.

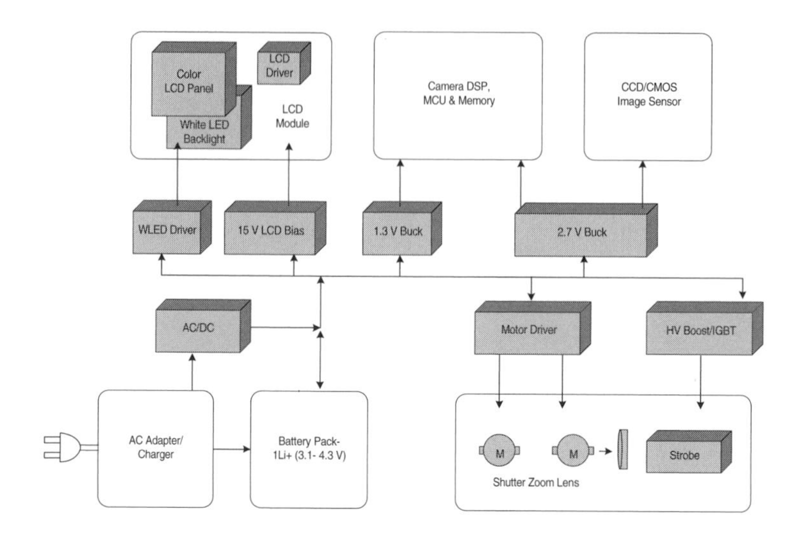

Figure 6–26 Generic DSC and power distribution.

Camera Phones

If DSCs are doing well, camera phones are sizzling. It is expected that soon the number of camera phones will surpass the number of DSCs and by 2007, one forth of all cell phones produced will have integrated cameras.

The Japanese have been leading the demand of high-end camera phones equipped with mega-pixel, solid-state memory cards and high-resolution color displays.

At the time of this writing, a number of camera phones are being announced in Japan with a resolution of 1.3 megapixels, matching, at this juncture, the performance of low end DSCs. Not surprisingly, forecasts for DSCs are starting to exhibit more moderate growth rates.

Cameras for current cell phones are confined inside tiny modules and generally meet stringent specifications, including one cubic centimeter, 100 mW power, and 2.7 V power source and cost ten dollars.

Right now, a big technology battle is going on regarding image sensors. Cell phone manufacturers are willing to allocate 100 mW or less of power dissipation to image sensors. CCDs are currently close to that limit, while CMOS typically require half.

While at the lower resolutions, CMOS image sensors seem to have won out over CCD thanks to their lower power dissipation, at the higher resolutions (greater than one megapixel) CDD is in the lead.

Camera phones that are currently available have resolutions in the 0.3 megapixels range and consume pretty much the same peak power levels (below 1.5 W) in call and picture mode.

Current camera phones, like DSCs, come with 8 to 16 MB memory stick flash memory for storage. The new solid-state memory cards, dubbed Mini SDs (Security Data), will go up to 256 MB by the end of 2005.

Based on the DSC example discussed earlier, a 1.3 megapixel camera phone could exhibit peaks of power consumption in picture mode (2 W) higher than in call mode (1.5 W).

Such state of the art camera phones typically equipped with a 3.6 V, 1000 mAh Li^+ cell should warrant up to two hours of call and picture mode.

Figure 6–27 shows the picture of a GSM camera phone main board and Figure 6–28 shows the disassembled battery powering a CDMA2000 camera phone, both courtesy of Portelligent.

The trade-off for all these features is a reduction of the cell phone talk time ability, from six hours for regular cell phones to one or two hours for the new camera phones.

The attacks on talk time will continue as the pressure for a higher number of pixels, higher resolution displays and more features incorporated into the cell phone increases.

With one to two hours of operation, the camera phone finds good company in its bigger relative, the notebook PC: both devices badly in need of new technologies capable of extending their untethered operation time. As both rely on the same display (LCD) and battery (Li^+) technologies, it is no surprise that they also suffer from the same problem, namely short operation time in mobile mode. For the notebook to achieve its goal of eight hours of operation and the cell phone to go back to its initial talk time of six hours, we need new technologies to come to bear. Fuel cells, electrochemical devices converting the energy of a fuel like methanol directly into electricity, have the potential to store ten times the energy of current battery technology, and it is likely that they will be ready for prime time in a couple of years.

On the display front, emissive technologies like Organic LEDs (OLEDS) clearly need to take over from current transmissive LCD technology, thereby eliminating the power-hungry backlight outfits. The first OLED display-based camera phone was announced in March of 2003. Since it appears that it is more difficult to produce reliable large sized OLED displays, this technology will probably penetrate the ultraportable market first, before moving to the notebook and beyond.

Finally, it is worth mentioning that White LEDs are moving beyond backlighting applications and enabling the use of flash in phone cameras,

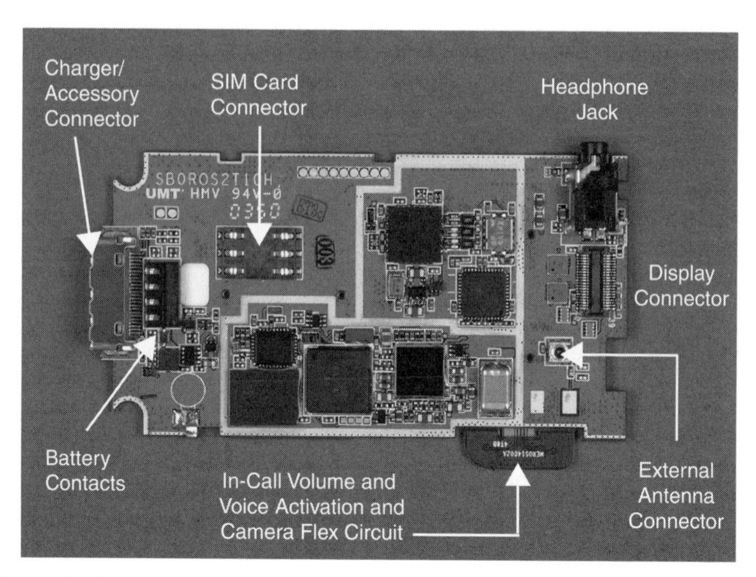

Figure 6–27 Camera phone mainboard example. *(Courtesy of Portelligent)*

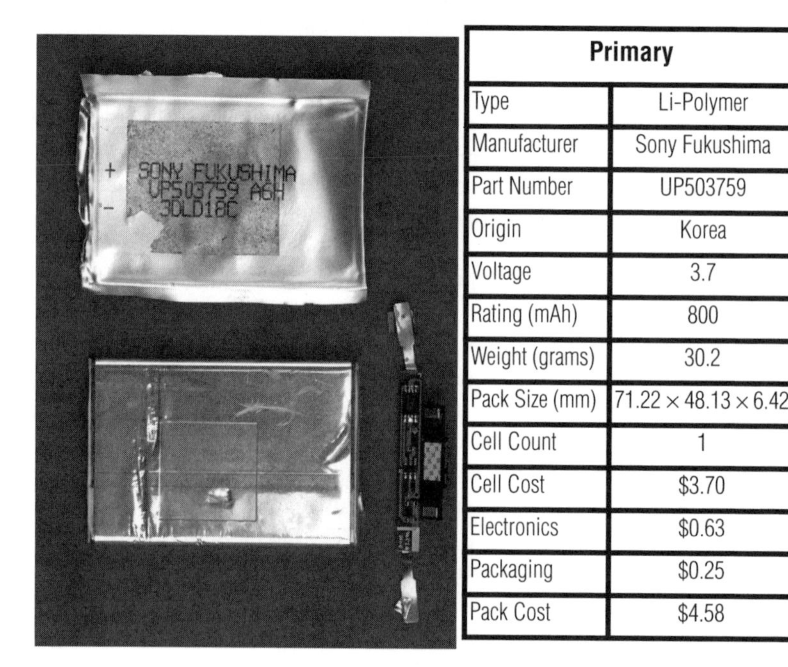

Primary	
Type	Li-Polymer
Manufacturer	Sony Fukushima
Part Number	UP503759
Origin	Korea
Voltage	3.7
Rating (mAh)	800
Weight (grams)	30.2
Pack Size (mm)	71.22 × 48.13 × 6.42
Cell Count	1
Cell Cost	\$3.70
Electronics	\$0.63
Packaging	\$0.25
Pack Cost	\$4.58

Figure 6–28 Battery and electronics daughterboard disassembled. *(Courtesy of Portelligent)*

thanks to their greater efficiency and simplicity of operation compared to xenon lamps.

No doubt the convergence phenomenon will continue. If high-resolution displays, cameras, and storage cards have been the drivers so far, no less compelling applications are on the horizon, like video on a handset, GPS, and more.

Fortunately, new technologies are coming along that are capable of both taming the escalation of power consumption (White LEDs, OLEDs) as well as breaking the current bottlenecks (fuel cells).

Is there an upper limit to the power consumption? Higher power consumption translates directly into higher temperatures in the gadgets we all love. Again, look at the notebook for an answer—in the near future we will likely be called to bear in our hands similar temperatures to those that we currently endure from our laptops. We expect then that our handsets will become as hot as possible without crossing the threshold of discomfort, as cooling down is an expensive and bulky proposition.

Power Minimization

The battle for power waste-minimization extends to the signal path as well. The logic gates, operational amplifiers, and data conversion devices used extensively in ultraportable applications are all specifically designed for ultra low power dissipation and are housed in space efficient packages.

For example, the Ultra Low Power (ULP and ULP-A) TinyLogic® devices, such as Fairchild's NC7SP74, a D flip-flop, and the NC7SP00 dual NAND gate, operate at voltages between 3.3 V and 0.9 V and have propagation delays as short as 2.0 ns, consuming less than half as much power as existing high performance logic.

Untethered Operation

Recent high-end handsets exhibit amazing features such as dual color LCD displays, camera, video on demand, and audio on demand. An 800 mAh Li^+ battery (corresponding to a 2.4 Wh at 3 V average output) can sustain heavy-duty activities like playing games, taking pictures, or recording and viewing videos—assuming each activity consumes power at a rate of 1.4 W for less than two hours. Such figures of merit are getting better, thanks to the power management methods discussed previously, but they remain a far cry from the desired performance of 6–8 hours of untethered operation as in more basic handsets.

The two technologies on the horizon promising to improve this situation are organic LEDs, which do eliminate the power consuming backlights, and fuel cells; electrochemical devices capable of extracting electricity directly from fuels like methanol. Fuel cells already promise to flank Li^+, for example as untethered chargers, and then to progressively substitute Li^+ technology.

Alternative power sources, such as fuel cells, will require even more sophisticated power management. This increased management will necessitate further proliferation of local intelligence to manage tasks (i.e. additional microcontrollers,; including sophisticated mixed signal capabilities to perform supervisory functions.

Digital still cameras with OLEDs are already commercially available and this technology is expected to take a wider hold in the next three to five years. Fuel cells are a proven technology but difficult to miniaturize and they may come to larger devices like notebooks before trickling down to handsets. Prototype handsets, some powered by, and others simply charged by fuel cells, have been demonstrated and are expected to become commercially viable in the same timeframe as OLEDs.

Power management techniques are adapting and evolving to keep up with the increased complexities of today's systems. These techniques include traditional cell library regulation elements as well as untraditional digital functions, such as bus interfaces, data converters, and microcontrollers.

Feature-rich handsets and smart phones are clearly the devices pushing the edge of every technology, including power, and more features will be coming in the future. For example, it is conceivable that a series of "plug and play" standards will be debated and then adopted to allow for mix-and-match of add-on peripherals (camera, GPS modules, etc.) from various sources, as well as promote the re-use of peripherals that a user already owns. The addition of microcontrollers in power management applications will become an increasingly important theme in the ICs that provide system power for these platforms.

This "smartening" of power management electronics, combined with the increasing maturity of new technologies for energy storage and displays, promises to keep these feature-rich devices on a steep growth curve for the foreseeable future.

Chapter 7

Computing and Communications Systems

7.1 Power Management of Desktop and Notebook Computers

Power management of PCs is becoming an increasingly complex endeavor. Figure 7–1 shows the progression of Intel PC platforms, from the launch of the Pentium in 1996 to current times. The Pentium brand CPU opens up the modern era of computing however, the birth of the CPU goes back as far as 1971 to the Intel 4004 CPU. In Figure 7–1 below each Pentium generation and associated voltage regulator offered by Fairchild Semiconductors, we find the year of the platform launch, the voltage regulation protocol (VRMxxx), the minimum feature (minimum line width drawn) of the transistors at that juncture in micro-meters, and the current consumption of the CPU.

Before Pentium, CPUs required relatively low power and could be powered by linear regulators. With Pentium the power becomes high enough to require switching regulators, devices distinctively more efficient than linear regulators. With Pentium IV the power becomes too high to be handled by a single phase (1Φ)—just to grasp the concept, think of a single piston engine trying to power a car—regulator, and the era of interleaved multiphase regulation (the paralleling and time spacing of multiple regulators) begins. At the VRM10 juncture, the breathtaking pace of Moore's law has slowed down somewhat, as exemplified by the unusual longevity of this platform. At the VRM11 juncture, the rate of increase in CPU power consumption has been slowed down with sophisticated techniques such as back biasing of the die substrate, to

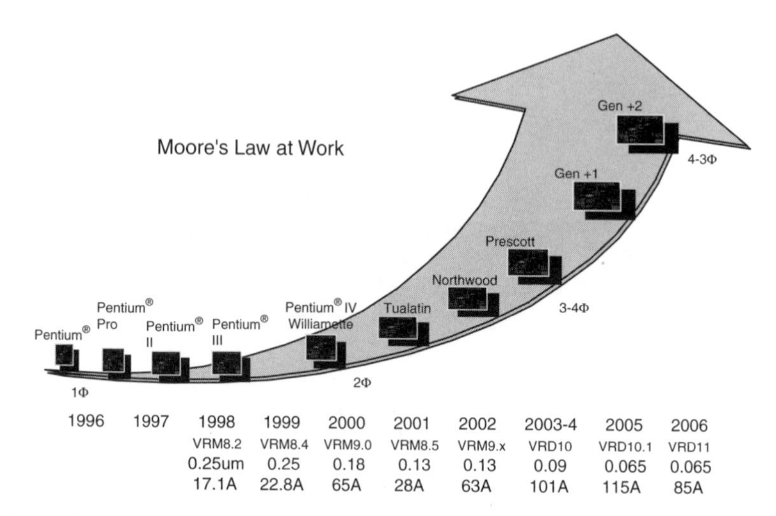

Figure 7–1 Progression of CPU platforms according to Moore's law.

reduce leakage, new dielectric materials to reduce switching losses and strained silicon, a technique that stretches the silicon lattice resulting in wider passages for the electrons and hence lower ohmic resistance. In the following session we will discuss first a Pentium III platform, covering most of the basic power management technology needed for the PC. Then we will cover a Pentium IV platform, focusing on new features specific to this platform, such as interleaved multiphase and extending the discussion to notebook systems as well.

Power Management System Solution for a Pentium III Desktop System

In this section we will review in detail the power management for a Pentium III system, while in the next section we will focus on Pentium IV. With PIII, the PC power management reaches a very high level of complexity in terms of power management architecture. The subsequent PIV platform doesn't change much except for the fact that more powerful CPUs require more hefty CPU voltage regulators.

Nine Voltage Regulators on Board

With the PIII platform the motherboard needs nine distinct regulated voltages, none of which are directly generated by the ATX silver box and all of which consequently need to be generated locally on the motherboard. The voltage types are as follows:

The Main Derived Voltages

These voltages all come from the 5 V silver box or 3.3 V silver box Mains

- DAC controlled CPU voltage regulator
- 2.5 V clock voltage regulator
- 1.5 V V_{TT} termination voltage regulator
- 3.3 V or 1.5 V Advanced Graphics Port (AGP) voltage regulator
- 1.8 V North Bridge (now renamed Micro Controller Hub or MCH) voltage regulator

The Dual Voltages

The dual voltages are managed according to the ACPI (Asynchronous Computer Peripheral Interface) protocol and powered from the silver box Mains during normal operation, or from the silver box 5 V standby during "suspend to RAM" state.

- 3.3 V Dual voltage regulator (PCI bus power)
- 5 V Dual voltage regulator (USB power)

The Memory Voltages

Memory voltages turn off only during "soft off" state.

- 3.3 V SRAM voltage regulator
- 2.5 V RAMBUS voltage regulator

The motherboard is becoming too crowded to be able to make room for nine separate power supplies. The best architecture is one in which the number of chips, and consequently the total area occupancy, are minimized. Figure 7–2 shows an architecture in which four of the five Main derived voltages are controlled by a single chip, while the four ACPI voltages are controlled with a second chip (effectively a dedicated quad linear regulator with ACPI control). The ninth regulator (North Bridge regulator) is provided separately for maximum flexibility.

The CPU Regulator

The CPU regulator is by far the most challenging element of this power management system. The main tricks of the trade employed to deliver high performance with a minimum bill of materials are discussed in detail in this section, including the handling of ever increasing load currents and input voltages in conjunction with decreasing output voltages, voltage positioning, and FET sensing techniques.

The Memory Configuration

Intel's recommended configuration for memory transition from SDRAM to RDRAM is 2 RIMM modules and 2 SIMM modules—this points clearly toward an architecture for the ACPI controller in which both RAMBUS and SRAM voltages are available at the same time, as opposed to an architecture in which a single adjustable regulator provides one or the other.

The RC5058 + RC5060 power management chipset shown in Figure 7–2 is proposed as an example of a complete motherboard solution.

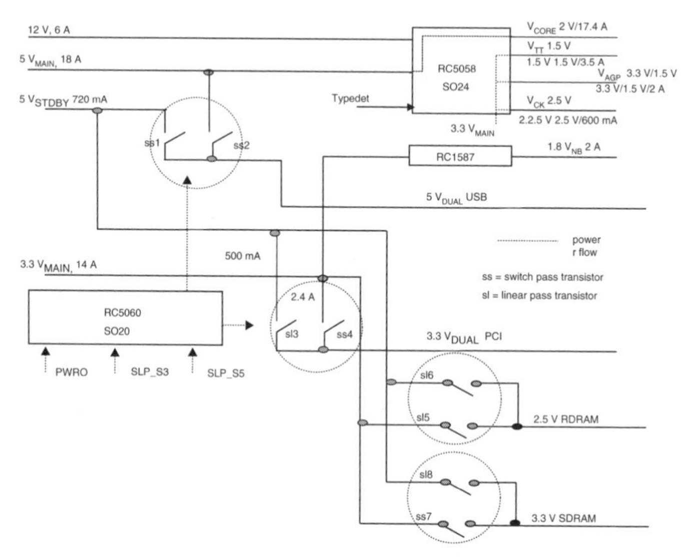

Figure 7–2 Pentium III system power management.

Power Management System Solution for Pentium IV Systems (Desktop and Notebook)

This section reviews the main challenges and solutions for both desktop and notebook PCs.

Introduction

Personal computers, both desktop and notebook, play a central role in the modern communication fabric and in the future will continue that trend toward a complex intertwining of wired and wireless threads (Figure 7–3)

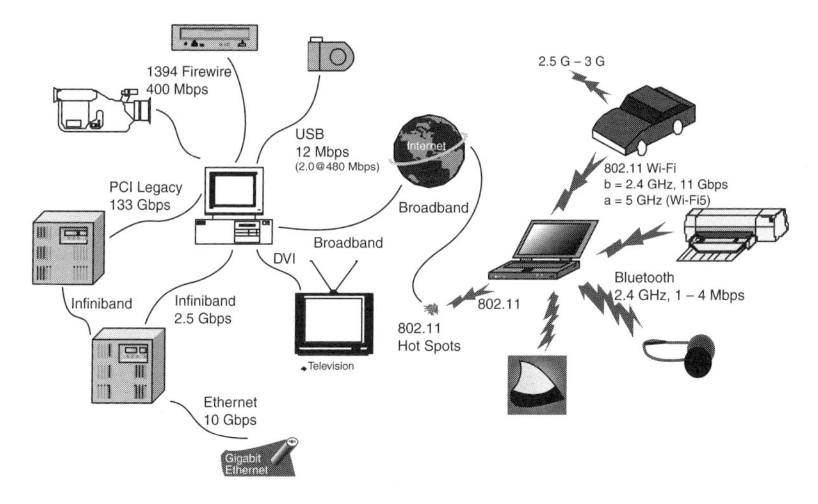

Figure 7–3 PCs and notebooks at the center of the communication fabric.

all contributing to the ultimate goal of computing and connectivity anywhere anytime.

To make the challenge even more daunting, this goal must be accomplished in conjunction with another imperative, "performance without power dissipation."

This chapter focuses on the latter challenge. Today's state-of-the-art technology is illustrated through the discussion of power management for a desktop system and a notebook system. We will also discuss future trends toward the achievement of both goals mentioned above.

The Power Challenge

Moore's law has two important consequences regarding power:

It creates a technology hierarchy with the CPU at the top, produced with the smallest minimum feature (today 0.13 μm) and requiring the lowest supply voltages (1–1.5 V) available thus far. Consequently, the previous CPU generation infrastructure for 0.18 μm gets recycled down the "food chain" for memory, which is powered at voltages around 2.5 V. The cycle goes on and on with lower minimum features and lower voltages continuously generated. The end result is a downward proliferation of power supply voltages from 5 V down to 3.3 V, down to 2.5 V, down to 1.5 V, etc. This phenomenon fuels the proliferation of "distributed power." Every new generation motherboard has more functions and requires more voltage regulators than the previous one, while at the same time the motherboard form factor is shrinking to meet the new demands for slick form factors.

By stuffing more and more transistors in a die (a Pentium IV CPU in 2004 has 42 million transistors versus 2300 transistors in 1971 for the Intel 4004 CPU) modern devices create tremendous problems of heat and power dissipation which must be resolved by us, the power specialists. This phenomenon has generated a tremendous migration from inefficient power architectures like linear regulators (believe it or not, the first CPUs were powered by Low Drop Output regulators, or LDOs, notorious for their high losses) to more efficient ones like switching regulator architectures. The workhorse for switching regulators continues to be the buck, or step down converter, continuously renewing itself into more powerful implementations, from conventional buck to synchronous to multiphase, in order to keep up with the CPU growing power.

Both trends compound the same effect, a phenomenal concentration of heat on the chips and on the entire motherboard that cannot continue untamed. More on this subject will be discussed later.

Desktop Systems

Figure 7–4 and Figure 7–5 illustrate a modern desktop system in its main components: the microprocessor, the Memory Channel Hub (MCH, also referred to as North Bridge), and the I/O Hub (IOH or South Bridge), connecting to the external peripherals. While the silver box can only provide the "row" power (5 V, 3.3 V, and 12 V), most of the elements in the block diagram need specialized power sources, to be provided individually and locally. Going from wall to board, an impressive slew of processes and technologies come to bear, from high voltage discrete DMOS transistors to Bipolar and Bi-CMOS IC controllers, from power amplifiers to linear and switching regulators. They all fall under a centrally orchestrated control providing energy in the most cost effective and power efficient way possible.

Powering the CPU

By far the most challenging load on the motherboard is the CPU. The main challenges in powering a CPU are:

Duty Cycle

High input voltages (12 V) and low output voltages (1.2 V typically) for the regulator, leading to duty-cycles of 10 percent ($V_{OUT}/V_{IN} = 0.1$). This means that useful transfer of power from the 12 V source to the regulated output happens only during 10 percent of the time period. For the remaining 90 percent of the time the load is powered only by the output bulk capacitors (tens of thousands of microFarads).

Figure 7–4 Desktop PC motherboard.

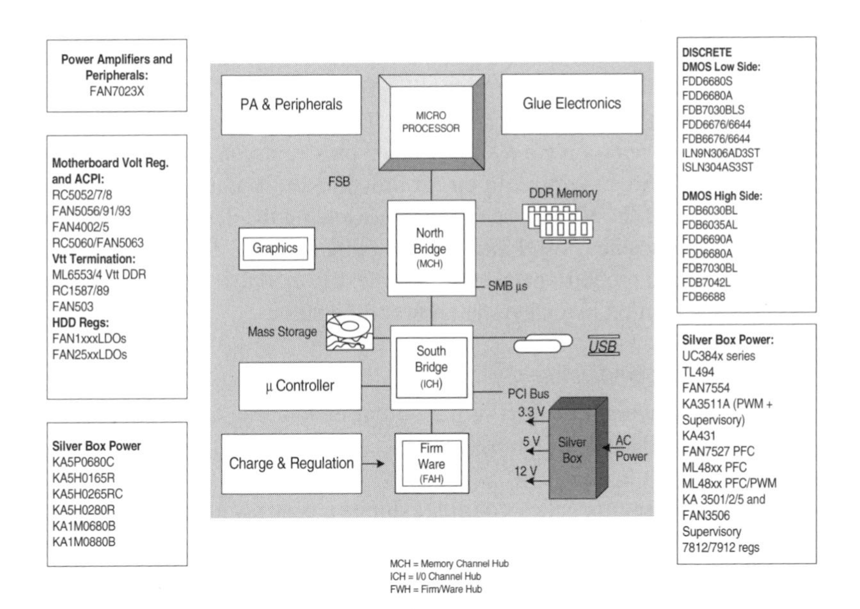

Figure 7–5 Desktop system.

So far, the introduction of interleaved multiphase buck converters has helped reduce the number of both input and output bulk capacitors. Still, the amount of capacitors used today is huge and the current trend is racing toward faster architectures and technologies capable of reducing the size of the passive elements without compromising the efficiency of the regulator, which must remain near 80 percent. This efficiency requirement also puts the discrete technology on the front line: it will take a new generation of discrete DMOS transistors with sensibly lower switching losses to achieve this goal. The desirable endpoint is elimination of electrolytic capacitors—the current workhorse of bulk capacitors—in favor of an all-ceramic solution.

Tight Regulation

Output voltage is tightly regulated. The voltage across the CPU is allowed to vary over production spreads and under wide and steep load transients (30 A/μs) for only a few tens of millivolts.

One industry standard practice, in order to alleviate the problem of eating up the voltage margins due to voltage spikes during load transients, is the utilization of a controlled amount of output "droop" under load. The waveform in Figure 7–6(a) shows the normal behavior of a regulated output in absence of droop, exhibiting a total deviation of $2 \times ESR \times I$, where I is the current and ESR is the series resistance of the bulk capacitor C. By adding a droop resistor R_{DROOP} in the indicated position and of value equal to ESR ($R_{DROOP} = ESR$) the total deviation is reduced to $ESR \times I$ as the (b) waveform illustrates. It follows that the waveform (b) will have a total deviation equal to the waveform (a) at twice the ESR value, corresponding to half the amount of output bulk capacitors. The technique has been illustrated here with passive droop via R_{DROOP}. Since passive droop is dissipative, the best practice is to do "active droop" or drooping of the output by controlled manipulation of the VRM load regulation, yielding the desired reduction in BOM at no efficiency cost. Finally, as the output voltage droops under load, less voltage and proportionally less power is delivered to the load, leading to a sensible reduction in total system power dissipation.

Dynamic Voltage Adjustment

Dynamic voltage adjustment of the output is done via D-A converter on the order of hundreds of nanoseconds to accommodate transitions to and from low power modes.

Our architecture, valley control, exhibits a very fast transient response and hence fits this type of application very well.

Figure 7–7 illustrates valley current-mode control based on leading-edge modulation. The error amplifier forces V_{OUT} to equal V_{REF} at its input. However, contrary to the standard peak control technique, now its output voltage, V_{ε}, is compared to the low-side MOSFETs current (I_L) times R_{DSON}. When $I_L \times R_{DSON}$ falls below the error voltage, the PWM comparator goes high. This sets the flip-flop, initiating the charge phase by turning on the high-side driver and terminating the discharge phase by turning off the low-side driver. The charge phase continues until the next

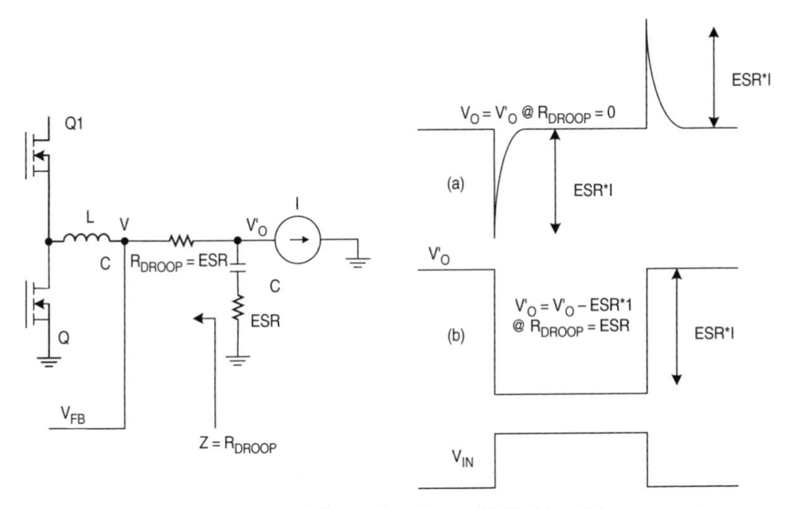

Figure 7–6 Output voltage "droop" reduces BOM by fifty percent.

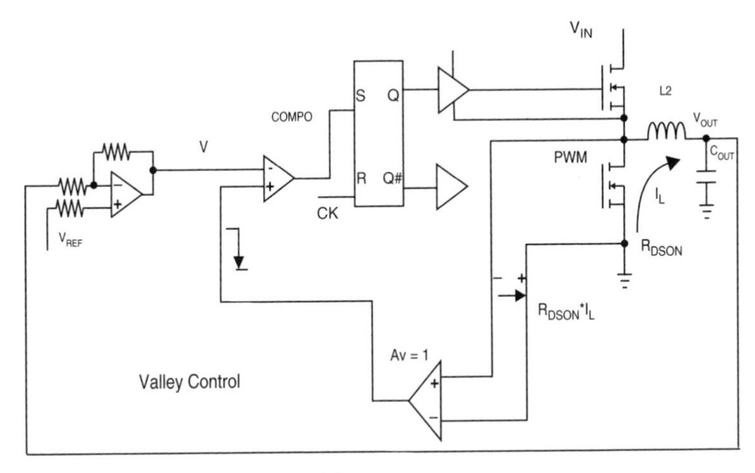

Figure 7–7 Valley control architecture.

clock pulse resets the flip-flop, initiating a new discharge phase. The advantage of this architecture is that it easily senses the current on the low side driver, where the current is present for 90 percent of the time in a 10 percent duty cycle application like this. For example if the clock frequency is 300 kHz, the high side pulse is only 330 ns, whereas the low side pulse is 2.97 μs. Consider also that sensing a 330 ns pulse on the low side driver would correspond to operate the VRM at a frequency of 2.7 MHz, a measure of how fast valley control can operate compared to peak control.

The turn-on of the high side driver is instantaneous and asynchronous as opposed to peak control in which the turn-on can only happen at every clock edge. It follows that standard peak control has inherently a delay of one clock period (say 3.3 μs at 300 kHz), whereas valley control has fast response (200 ns) independently of the clock.

This architecture is proposed both in Fairchild's line of fully integrated converters, having controllers and drivers on board (Figure 7–8), as well as in the new line of controllers and separated drivers (Figure 7–9).

Current Sensing

In modern voltage regulator modules precision current sensing is critical for two reasons: without precision current sensing there is no accurate "active droop" and there is no good current sharing in interleaved multiphase controllers.

The easiest way to accomplish precise current sensing would be to utilize a precise current sense resistor but because of cost and power dissipation issues, this isn't a practical solution.

Mainstream solutions today accomplish current sensing in a "lossless" fashion by measuring current across the drain-source ON resistance of the discrete DMOS transistor (Figure 7–7). This method eventually will run out of steam because of the temperature dependency of this resistance (more than sixty percent over 100°C roughly.) Other methods like the one measuring current on the basis of the inductor parasitic resistance are no better over temperature.

A few brute force techniques are starting to appear in response to this problem, including the use of external thermistors, diode temperature sensors, etc. There is a simple way to accomplish precise current sensing, namely the ratioed Sense-FET technique. This technique exploits the cellular nature of a modern DMOS discrete transistor in order to isolate a small portion of it into a separate source capable of reflecting current in a predictable amount with respect to the main transistor. This technique has not taken over in VRMs yet because until now it was not needed and because an earlier attempt at an industry standardization of this device failed. Probably the time has come to revisit this technology.

Powering the Entire Motherboard ACPI

Advanced Configuration and Power Interface (ACPI) is an open industry specification co-developed by Compaq, Intel, Microsoft, Phoenix, and Toshiba. ACPI establishes industry-standard interfaces for OS-directed configuration and power management on laptops, desktops, and servers. The specification enables new power management technology to evolve

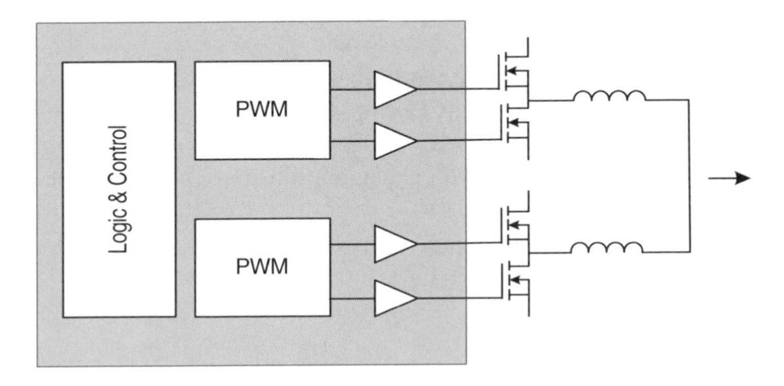

Figure 7–8 FAN5093/FAN5193 two-phase monolithic controller and driver.

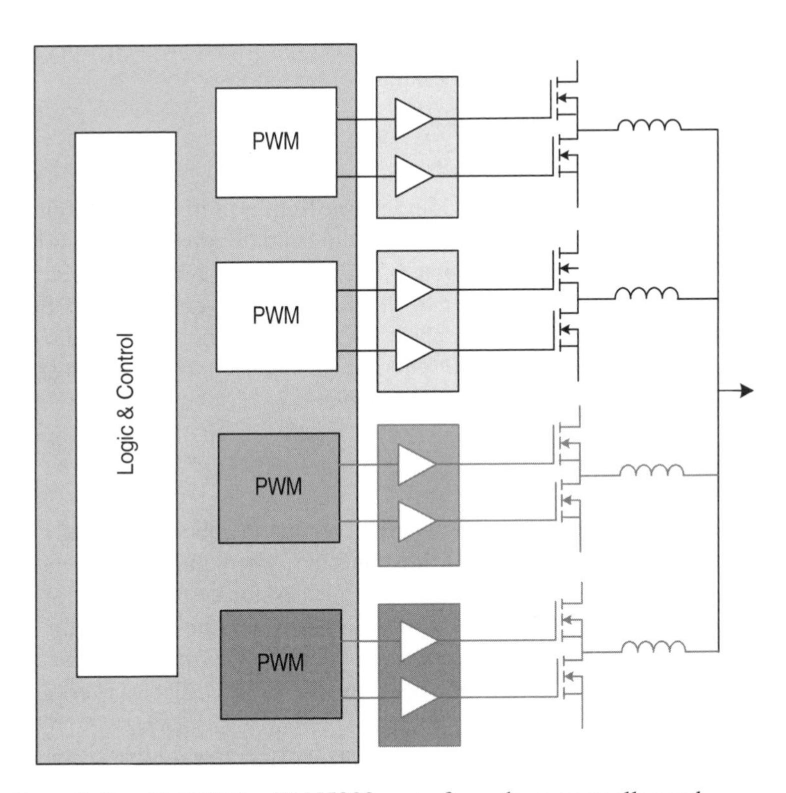

Figure 7–9 FAN5019 + FAN5009 up to four-phase controller and separate drivers.

independently in operating systems and hardware while ensuring that they continue to work together.

Figure 7–10 illustrates the power management of the entire desktop motherboard. In this case our ACPI chip, FAN5063, in coordination with the rest of the chips and under ACPI control, assures smooth operation of hardware and software. This block diagram illustrates power for the following functions:

- CPU regulator V_{CORE}
- BUS Termination V_{TT}
- DDR Memory
- graphics chip AGP
- PCI
- clock

The illustration in Figure 7–11 reports typical currents and voltages for a modern Pentium IV class desktop PC.

Powering the Silver Box

In line with the mission of providing power from wall to board, Fairchild also provides all the electronics necessary to build the silver box, the brick inside a PC box providing the main 5 V, 3.3 V, and 12 V. The coverage goes from the bridge rectifier to the power factor correction (PFC) switcher to the PWM main switcher. The electronics in the forward path are covered, as are the opto-electronics in the feedback loop including the opto-coupler H11A817, as illustrated in Figure 7–12.

Notebook Systems

Figure 7–13 is the picture of a modern notebook mainboard (courtesy of Portelligent), while Figure 7–14 illustrates the system diagram of a modern notebook system; by comparison with the desktop system (Figure 7–5) we can see many architectural similarities between the two systems as well as some important differences like the presence of a battery as the main source of power for the notebook and the AC-DC adapter necessary for battery recharge.

Battery life is one of the most serious barriers toward the vision of "computing and connectivity anywhere anytime." The industry needs to move from the current two hours of effective battery life to six or eight hours! This will require the optimization of every technology from battery to CPU, to display, to passives.

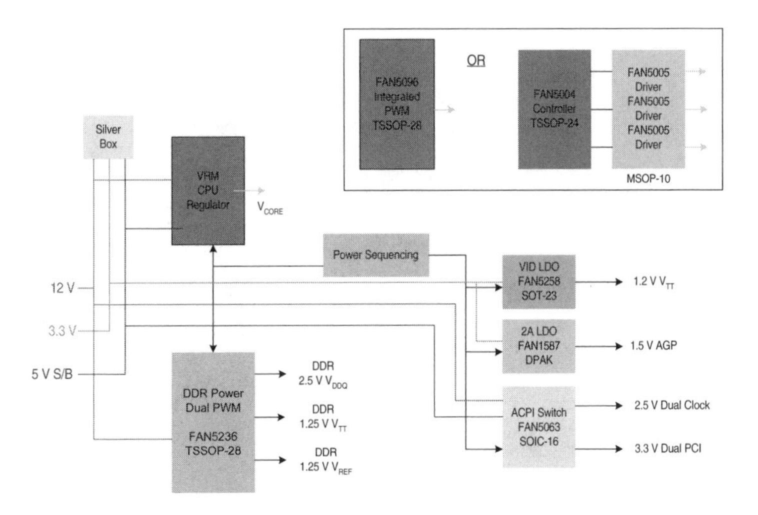

Figure 7–10 Motherboard power management.

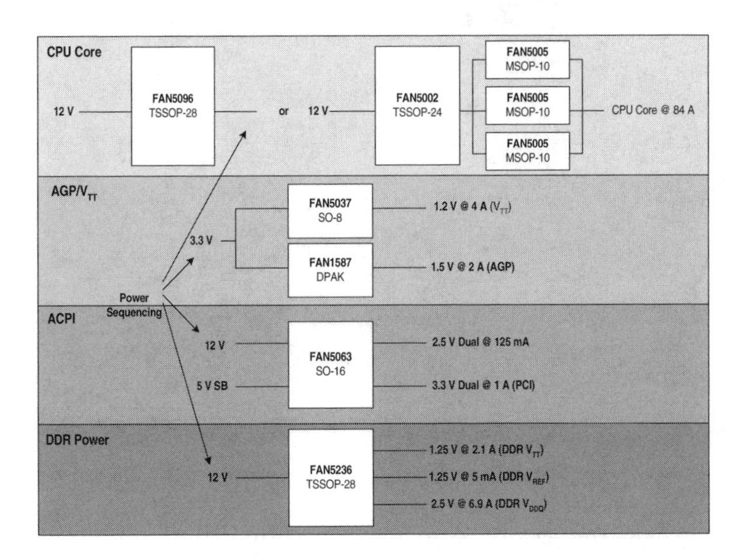

Figure 7–11 Currents and voltages for modern Pentium IV class desktop PCs.

Form factor is another important differentiator between desktop and notebook. The growing trend toward light and thin notebooks calls for thin, surface mountable components.

For these reasons, despite the initial similarities, the amount of desktop technology that can be reused in notebooks is limited.

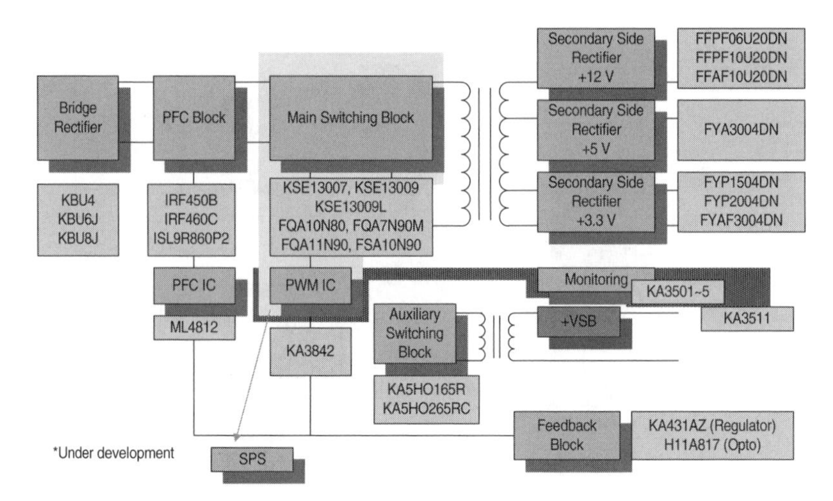

Figure 7–12 Silver box power.

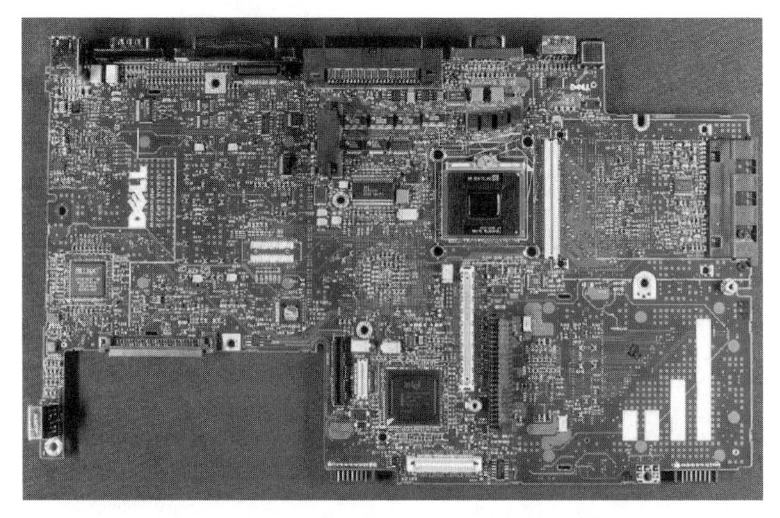

Figure 7–13 Notebook PC motherboard. *(Courtesy of Portelligent)*

Intel Mobile Voltage Positioning (IMVP™)

IMVP consists of a set of aggressive power management techniques aimed at maximizing the performance of a mobile CPU with the minimum expenditure of energy. Such techniques are similar to those discussed in the desktop CPU power management section but go well beyond. The additional power management techniques for notebooks are:

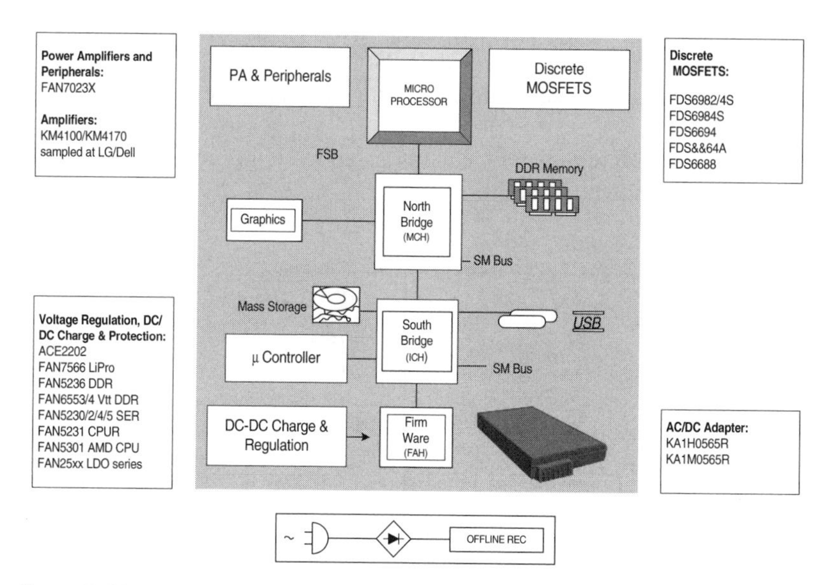

Figure 7–14 Notebook system.

Light Load Operation

At light load the switching losses become dominant over ohmic losses. For this reason, frequency of operation is scaled down at light load. This is done either automatically, commuting to light load mode below a set current threshold, or under micro control, via a digital input toggling between the two modes of operation.

Clock Speed on Demand

One of the most effective ways to contain power in notebooks is to manage the CPU clock speed and supply voltage as power dissipates with the square of the voltage and in proportion to the frequency (CV^2f). Different CPU manufacturers offer variations of this technique. SpeedStep™ is Intel's recipe for mobile CPU power management while PowerNow™ is AMD's version. The bottom line is that for demanding applications, such as playing a movie from a hard disk drive, the CPU gets maximum clock speed and highest supply voltage, thereby yielding maximum power. Conversely, for light tasks like writing a memo, the power is reduced considerably.

Accordingly, when customers buy a 1.2 GHz PIV machine, they really get a CPU that may sometimes run at peak clock speed of 1.2 GHz.

Powering the Entire Mobile Motherboard

Figure 7–15 illustrates the power scheme for the entire mobile motherboard.

In this case the system runs under ACPI, as illustrated by the blocks under S3 (Suspend to RAM) and S5 (Soft Off) control, as well as under one of the mobile specific power saving schemes like Intel IMVP discussed earlier.

Figure 7–16 reports typical currents and voltages for a modern Pentium IV class notebook PC.

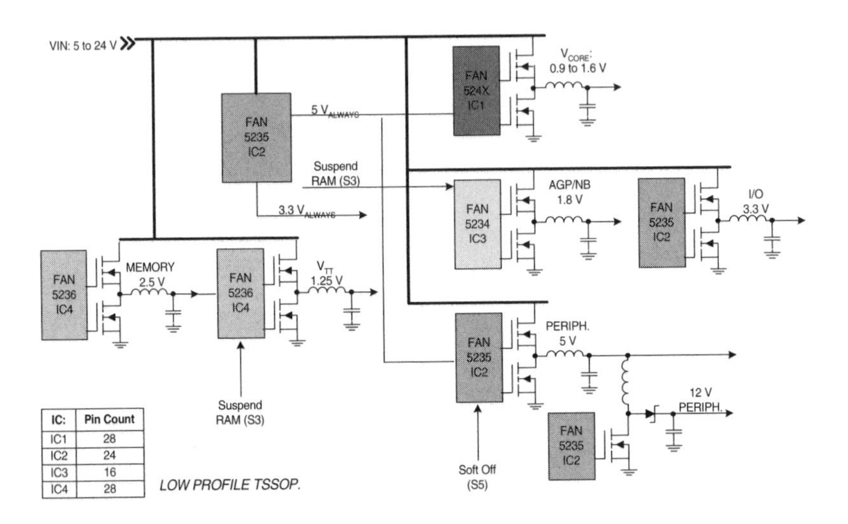

IC:	Pin Count
IC1	28
IC2	24
IC3	16
IC4	28

Figure 7–15 Powering the mobile motherboard.

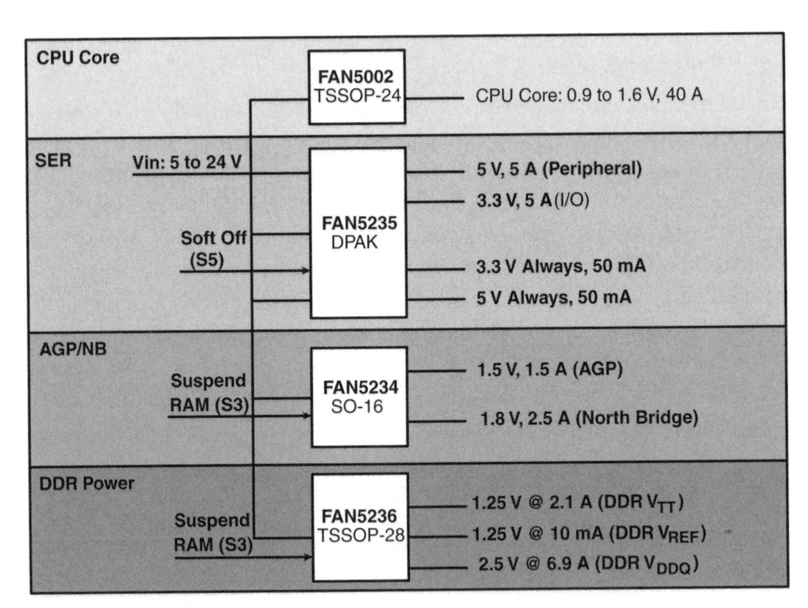

Figure 7–16 Current and voltages on the notebook motherboard.

Notebook AC-DC Adapter

Figure 7–17 shows the simplified application of a notebook AC-DC adapter based on Fairchild components, from the bridge rectifier to the offline converter (FSDH0165) to the opto loop (H11A817A opto coupler and KA431 voltage reference), which, together with the transformer T provides electric isolation between the high voltage on the AC line side and the low voltage on the load side. The control is a constant current/constant voltage (CC/CV) implementation geared toward the charging of lithium batteries. The output voltage can be easily adjusted for 3s + 2p (three series, two parallel cells pack) or 2s + 3p packs by simple scaling of a resistor divider (not shown in the figure).

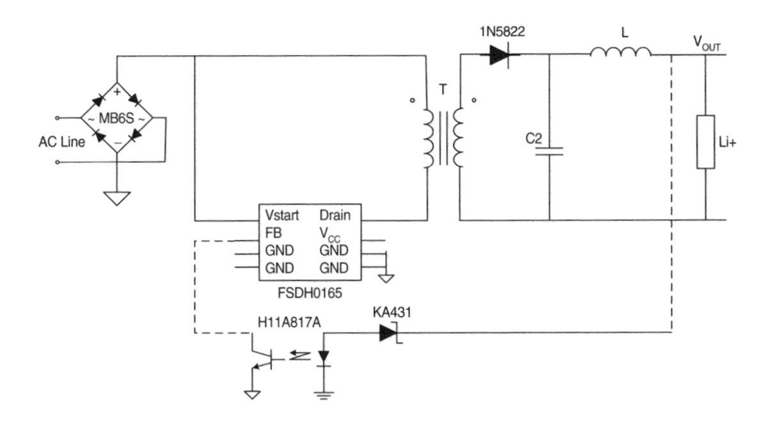

Figure 7–17 AC-DC adapter for notebook.

Future Power Trends

The pressure will not relent for the motherboard power management designer in the future. With performance and complexity increasing and shrinking form factors, the challenge will move from handling rising power to handling rising power density! The future motherboard will pack more and more power in less and less space, calling for new power technologies delivering unprecedented power densities obtainable only by a new generation of discrete MOSFETs able to work at many MHz of clock speed without appreciable losses in conjunction with all ceramic input and output bulk capacitors and smart PWM controllers capable of working reliably at very low duty cycles and high clock frequencies.

7.2 Computing and Data Communications Converge at the Point of Load

The convergence of computing and communications—"*Comm*puting"—is happening both on the signal and the power path, creating new opportunities as well as challenges for designs. Silicon integration of computing, communications, and wireless functions on the same die, or on the same process technology, is blurring the lines between computing and communications. Powering this silicon requires a good understanding of a new environment that does not conform well to traditional schemes and classifications.

The Proliferation of Power Supplies

The two end points in the power chain are the "load" and the "wall power." The load end of the chain is in constant evolution, resulting in continuous changes that generate new opportunities and new challenges. The technology driver on the load side is ultimately Moore's law. Doubling the number of transistors per a given area every 18 months creates a technology hierarchy by which the CPU—at the top of the food chain—is designed with the smallest minimum feature (90 nm in 2004–5) and requires the lowest supply voltages (1–1.5 V). Consequently, the previous generation fab infrastructure at 130 nm gets recycled down the food chain for the next high protein product—say memory—that gets powered at voltages around 2.5 V or lower. This cycle goes on and on. As the performance of such loads (e.g., CPUs, memories, chipsets) tends to go up while voltage decreases, the end result is an increase in power (Watts) demand.

When such a load, say a CPU, is part of a computing system, it gets powered by a voltage regulator. The voltage regulator typically is referred to as voltage regulator module if the power supply is a module plugging into a socket on the motherboard, or Voltage Regulator Down (VRD) if the same circuitry is built-in permanently "down" on the motherboard. When the same load is part of a communications system, it will be powered essentially by the same regulation electronics, now called the Point of Load regulator or simply POL.

On the wall power side we have two different power distribution systems: the 48 V power for telecom systems and the AC line (110 V or 220 V AC) for computing. Figure 7–18 illustrates and compares the two systems.

Telecom Power Distribution

Traditionally, telecom systems (Figure 7–18a) have distributed DC power (–48 V typically) obtained from a battery backup that is charged continually by a rectifier/charger from the AC line. This is the case for the power

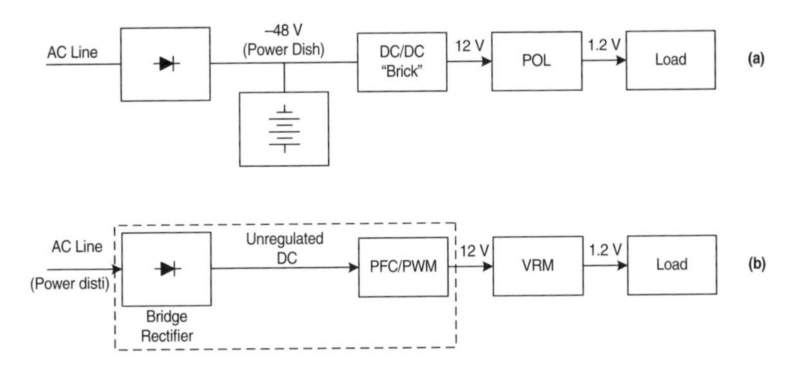

Figure 7–18 (a) Telecom versus (b) Computing power distribution system.

distribution in land telephones for example. Subsequently, this 48 V (in reality a voltage spreading from 36 V to 72 V) is converted into various low positive DC voltages (Figure 7–18a shows 12 V only for simplicity). This down conversion generally is accomplished by isolated DC-DC converters referred to as *bricks*, although non-isolated buck converters are sometimes used in telecom.

Isolation in bricks is driven by a number of technical factors, including cleaner ground loops, ease of handling the wide input to output voltage ratios (easily 10:1) by means of the transformer turn ratio, and inherently good over-voltage protection of the load due to the low voltage at the output of the transformer. Such a 12 V (or 5 V) bus may then be reduced down to the final voltage rails (3.3 V, 2.5 V, 1.2 V, etc.) by means of a DC-DC converter for each rail or even one for each single load, depending on the overall power management scheme. This type of low voltage DC-DC converter is referred to as point of load in telecom systems.

Computing Power Distribution

In a typical computing system (Figure 7–18b), such as a desktop PC, the power is drawn from the AC line. After rectification (AC to unregulated DC voltage conversion), the high input voltage is "bucked" down to the standard 12 V, 5 V, and 3.3 V busses by the PFC and PWM block. The silver box inside the PC box performs the down conversion. These voltages are then delivered by a cable to the motherboard, where they are reduced to the final voltage rails by VRMs, VRDs, and other types of voltage regulators.

Multiphase Buck Converter for POLs and VRMs

POLs and VRMs essentially are modules and come in a number of more-or-less standard form factors. Standardization and modularization differentiate these elements and make them specific to the application at hand, but at their heart they are powered by similar technologies and architectures. Their similarity derives from the fact that they are powering similar or identical loads from similar or identical input voltages.

The most popular architecture for step-down regulators, from 12 V or less to any voltage down to 1 V or less, is the non-isolated, *multiphase interleaved buck converter.*

The buck converter is a very popular and resilient architecture, thanks to its simplicity and effectiveness. Interleaved multiphase is the feature that has given a new lease on the life of this architecture. *Multiphase* refers to paralleling of two or more buck converters, and *interleave* signifies the time spacing of the clock cycles between the converters. Figure 7–19 refers to two buck converters (or two slices or phases of a multiphase buck converter) working in opposition of phase.

In Figure 7–19(a), the two clocks in phase opposition are generated at the output of a *logic device* (see glossary), starting from a master clock (MCK). The currents in each phase have very high DC components and a small ripple on top of such DC interleave is all about reducing that ripple amplitude even more, as the ripple represents noise or deviation from an ideal direct current waveform. Hence in this discussion we ignore the DC components and focus only on the variable content, or ripples called $I_{\Phi 1RIPPLE}$ and $I_{\Phi 2RIPPLE}$ in Figure 7–19(b). In the same Figure 7–19(b), $I_{2\Phi RIPPLE}$ is the resulting ripple current after the currents in the two slices are summed. The interleave produces these fundamental benefits:

Effective operation at twice (or n times for n slices) the single slice frequency without the switching losses associated to high frequency of operation.

Smaller output ripple as demonstrated graphically in Figure 7–19(b) by the smaller amplitude of the resulting ripple compared to the amplitude of the ripple components. On the other hand, instead of working toward a smaller ripple, this architecture can be utilized to maintain a specified ripple with smaller and hence cheaper output components (inductors and capacitors).

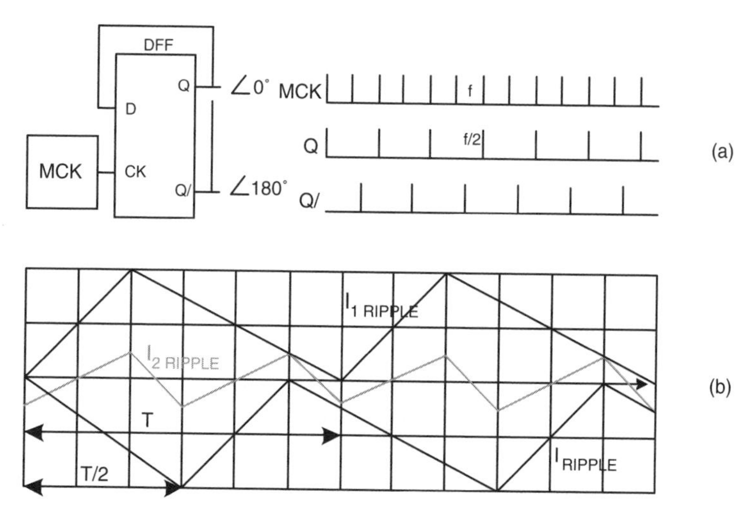

Figure 7–19 Two-phase interleaved (a) buck converter and (b) output ripple current waveforms.

The higher duty cycle and higher on-time is demonstrated in Figure 7–19(b) by the longer duration of the positively sloped segments in the resulting ripple compared to the ripple components. Higher on-time means lower peak currents within a clock cycle. Since the on-time is the time during which current is drawn from the input capacitors, lower peak currents lead to savings in input capacitors as well.

An issue that needs attention when it comes to interleaved schemes is phase current balancing, or the need to assure that all phases carry an identical amount of current. This can be accomplished in many ways, from simple ballast schemes to active current sensing and balancing.

Conclusion

The convergence of computing and communications brings together two cultures and, in fact, two separate power universes, each with their own language, systems, and classifications. Power distribution at the source starts very differently for these two fields, but at the point of load there is clear convergence. When we dig below the surface of VRMs and POLs, we find the same technologies and architectures at play and between the latter, the interleaved buck converter rules.

7.3 Efficient Power Management ICs Tailored for DDR-SDRAM Memories

Introduction

A new type of *Single Data Random Access Memory* (SDRAM), *Double Data Rate* (DDR) DDR-SDRAM for short, has gained popularity in desktop and portable computing thanks to its superior performance (initially 266 Mbps data rate versus 133 Mbps data rate for plain SDRAM) and low power dissipation at a competitive cost when compared to competing memory technologies. Subsequently, the DDR data rate has increased to 400 Mbps. A second generation DDR, or DDR2 (JESD79-2A), has been introduced recently, extending the data rate from 400 up to 667 Mbps. DDR memories require a new and more complex power management architecture in comparison with the previous SDRAM technology.

This chapter reviews the power requirements for DDR-SDRAM memories, covering static, transient, and stand-by modes of operation. Alternative schemes of power management are discussed and an example of a complete power management system, based on efficient switching voltage regulation, is provided. Finally, future trends in power management for DDR-SDRAM memories are examined.

DDR Power Management Architecture

Figure 7–20 illustrates the basic power management architecture for first generation DDR memories.

In DDR memories the output buffer is a push-pull stage, while the input receiver is a differential stage. This requires a reference bias midpoint V_{REF} and, consequently, an input voltage termination capable of sourcing, as well as sinking, current. This last feature (sourcing and sinking current) differentiates the DDR V_{TT} termination from other terminations present in the PC motherboard, noticeably the termination for the *Front System Bus* (FSB), connecting the CPU to the *Memory Channel Hub* (MCH), which requires only sink capability due to termination to the positive rail. Hence, such DDR V_{TT} termination cannot reuse or adapt previous V_{TT} termination architectures and requires a new power design.

In first generation DDR memories the logic gates are powered by 2.5 V. Between any output buffer from the chipset and the corresponding input receiver on the memory module, typically we find a routing trace or stub, that needs to be properly terminated with resistors RT and RS as indicated in Figure 7–20. When all the impedances, including that of the output buffer are accounted for, each terminated line can sink or source ±16.2 mA. For systems with longer trace lengths between transmitter and receiver, it may be necessary to terminate the line at both ends, doubling the current.

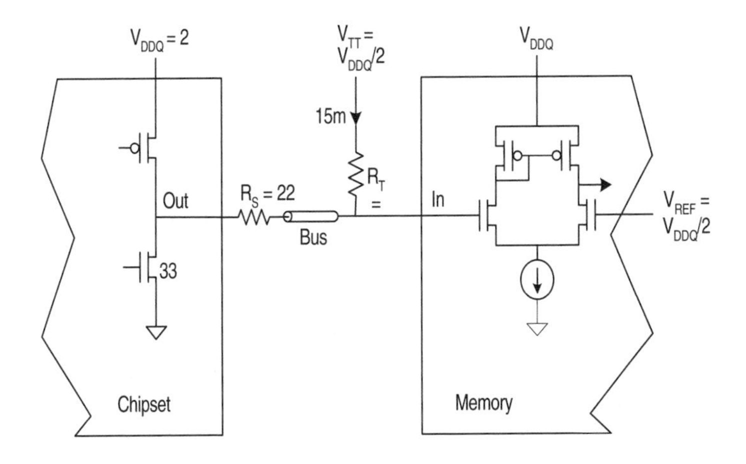

Figure 7–20 Illustration of DDR power supply architecture.

The 2.5 V V_{DDQ} required for the DDR logic has a tolerance of ±200 mV. To maintain noise margins, DDR termination voltage, V_{TT}, is required to track V_{DDQ}. It must be equal to $V_{DDQ}/2$, or approximately 1.25 V, with an accuracy of plus or minus three percent. Finally, the reference voltage, V_{REF}, must be equal to V_{TT} to +40 mV. These tracking requirements, plus the requirement that V_{TT} can both sink and source current, are the features that present the unique challenges of powering DDR memory.

Worst Case Current Consumption

V_{TT} Termination

Assuming the following structure for a 128 Mbyte memory system:

128 bit wide bus

8 data strobe

8 mask bits

8 V_{CC} bits

40 address lines (two copies of 20 addresses)

192 lines

With each line consuming 16.2 mA we have a maximum current consumption of

$$192 \times 16.2 \text{ mA} = 3.11 \text{ A peak}$$

V_{DDQ} Power Supply

V_{DDQ} sources current during the phase in which V_{TT} sinks current. It follows that the current for V_{DDQ} is unipolar and its maximum equals V_{TT}'s maximum value of 3.11 A.

Average Power Consumption

A 128 Mbyte memory system is typically made up of 8 × 128 Mbit devices and consumes an average power of 990 mW excluding the V_{TT} termination power.

It follows that the average current I_{DDQ} drawn from V_{DDQ} will be

$$I_{DDQ} = P_{DDQ}/V_{DDQ} = 990 \text{ mW}/2.5 \text{ V} = 0.396 \text{ A}$$

Similarly, the average power P_{TT} consumed by the termination resistors is 660 mW.

It follows that the current I_{TT} drawn from V_{TT} will be

$$I_{TT} = P_{TT}/V_{TT} = 660 \text{ mW}/1.25 \text{ V} = 0.528 \text{ A}$$

Finally the V_{REF} current, I_{REF} is selected of a value high enough for the V_{REF} supply to exhibit low enough impedance to yield good noise immunity (< 5 mA).

In summary, the main static parameters for the design of a 128 Mbyte DDR memory power management system are

$$V_{DDQ} = 2.5 \text{ V}, I_{DDQ} = 0.396 \text{ A average, 3.11 A peak (source)}$$

$$V_{TT} = V_{DDQ}/2 = 1.25 \text{ V}, I_{TT} = 0.528 \text{ A average, 3.11 A peak (source and sink)}$$

$$V_{REF} = V_{DDQ}/2 = 1.25 \text{ V}, I_{REF} = 5 \text{ mA}$$

Naturally if V_{DDQ} is utilized to power other loads besides the termination load then its sizing must be increased accordingly.

Transient Operation

The governing documents for DDR memory, JEDEC, JESD79, and JESD 8-9, specify that the V_{TT} voltage must be equal to half the V_{DDQ} voltage with a tolerance of plus or minus three percent. This tolerance should include load transients on the bus caused by the lines transitioning. However, two items necessary for evaluation of the capacitor requirements for the V_{TT} supply are missing from this: the JEDEC spec does not say with what bandwidth V_{TT} must track V_{DDQ}, nor does it specify the maximum load transient that V_{TT} can have.

In practice, it appears that the intent of the spec was to maximize noise margins. Thus, while it is not mandatory for V_{TT} to follow half of V_{DDQ} at all times, the greater the bandwidth with which it does so, the more robust the system. For this reason, a wide-bandwidth switching converter is desirable for generating V_{TT}.

For the V_{TT} load transient, conceivably the current could step from +3.11 A down to –3.11 A, from sourcing to sinking current. This 6.22 A step with a 40 mV window would require an output capacitor with an ESR of only 7 mΩ. Two practical considerations moderate this requirement, however. The first is that actual DDR memory doesn't really draw 3.11 A. Measurement shows typical current in the range of 0.5–1 A. Secondly, the transition between sinking and sourcing current occurs very quickly, so quickly that the converter doesn't see it. To go from positive maximum current to negative maximum current would require that the bus go from all 1s to all 0s and then remain in that state for a time at least equal to the inverse of the converter bandwidth. Since this is something on the order of 10 μs, and since the bus runs at 100 MHz, it would need to stay at all 0s for a thousand cycles! In practice, then, the output capacitor for V_{TT} need be only about 40 mΩ ESR.

Standby Operation

DDR memory supports standby operation. In this mode, the memory retains its contents, but it is not being actively addressed. Such a state may be seen, for example, in a notebook computer in standby mode. In standby the memory chips are not communicating so the V_{TT} bus power can be turned off to save power; V_{DDQ}, of course, must remain on in order for the memory to retain its contents.

Linear versus Switching

As noted earlier the average power dissipation of a DDR system is

$$P_{DDQ} = 990 \text{ mW}$$

$$P_{TT} = 660 \text{ mW}$$

for a total of

$$P_{TOTDDR} = 990 \text{ mW} + 660 \text{ mW} = 1650 \text{ mW}$$

while a comparable SRAM system consumes 2040 mW.

If a linear regulator were used to terminate V_{TT}, then the P_{TT} power is processed with 50 percent efficiency according to the ratio $V_{OUT}/V_{IN} = V_{TT}/V_{DDQ} = 0.5$. This means that an additional 660 mW of power is dissipated in the V_{TT} regulator, raising the total average power dissipation to 1650 + 660 = 2310 mW. Such a figure now exceeds the corresponding power dissipation figure for SDRAM, wiping out one of the advantages of the DDR memories, namely lower power dissipation.

As far as P_{DDQ} goes, most of the power advantage comes from having a V_{DDQ} of 2.5 V, as opposed to 3.3 V for conventional SDRAM. However, in a typical PC environment, the 3.3 V is provided by the silver box, while the 2.5 V is not available and needs to be created on the motherboard. Here again, unless an efficient regulation scheme is utilized to generate V_{DDQ}, the power dissipation advantage is lost. So, it follows that switching regulation should be the preferred means of processing both P_{DDQ} and P_{TT} power for DDR memories.

Second Generation DDR—DDR2

With DDR2, V_{DDQ} is reduced from 2.5 V down to 1.8 V and V_{TT} from 1.25 V down to 0.9 V with a sink/source drive capability of ±13.4 mA. Accordingly DDR2 memories end up consuming much less than first generation DDR. For example, a DDR2-533 ends up consuming roughly half of the power consumed by a DDR-400. All the static and dynamic observations made in the previous sections for DDR also apply to DDR2. The termination scheme for DDR2 is slightly different from the one for DDR shown in Figure 7–20 and the termination resistors are on the chip, not the motherboard; however, an external V_{TT} termination voltage is still necessary. At the much lower levels of DDR2 power consumption, linear regulators for V_{TT} can be utilized, especially if simplicity and cost are a prevailing consideration over power consumption minimization.

FAN5236 for DDR and DDR2 Memories

There are a variety of DDR power ICs; for example, Fairchild Semiconductor has the ML6553/4/5 with integrated MOSFETs, the FAN5066 for high power systems, and the recently released FAN5068, a combo DDR and ACPI. But the Fairchild FAN5236 (Figure 7–21) is specifically designed for all-in-one powering of DDR memory systems. Integrated in this single IC are a switcher controller for V_{DDQ}, a switcher controller for V_{TT}, and a linear buffer for V_{REF}. The switcher for V_{DDQ} runs off any voltage in the range from 5 V to 24 V. The switcher for V_{TT}, however, is different; it is designed to run from the V_{DDQ} power and switches synchronously with that switcher. Both switchers' outputs can range from 0.9 V to 5.5 V. Since the bus lines are driven with 2.5 V (DDR) or 1.8 V (DDR2) for V_{DDQ}, and are terminated to 1.25 V (DDR) or 0.9 V (DDR2) for V_{TT}, the power to some extent is circulating between V_{TT} and V_{DDQ}. Drawing V_{TT} from V_{DDQ} minimizes total circulating power, and thus circulating power losses. The V_{TT} switcher can also be shut down for standby mode. Figure 7–21 shows the typical application and Table 7–1 the associated BOM for a 4 A continuous, 6 A peak V_{DDQ} application. This circuit can easily be modified to set V_{DDQ} at 1.8 V (via divider R5/R6) and V_{TT} to 0.9 V for DDR2 applications. Appendix F provides the data sheets of FAN5236 for more technical details.

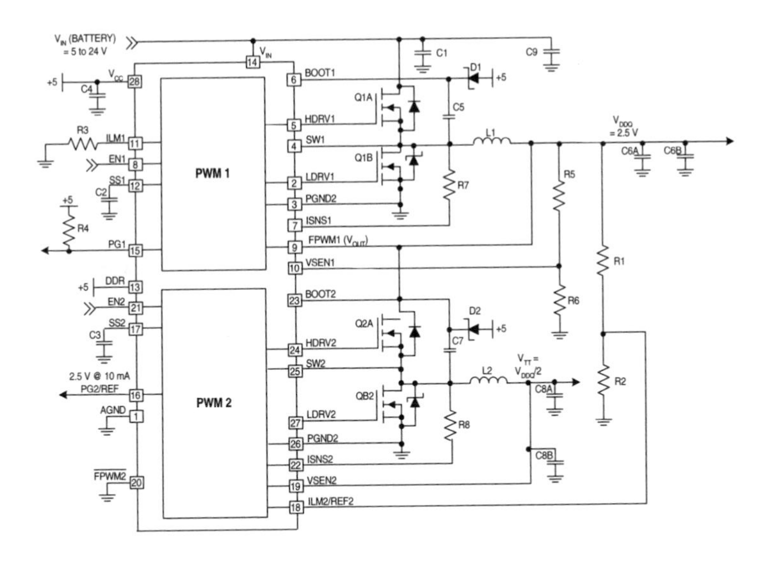

Figure 7–21 FAN5236 powering V_{DDQ} and V_{TT}.

Table 7–1 DDR Regulator BOM for a 4 A continuous, 6 A Peak V_{DDQ} Application

Description	#	Ref.	Vendor	Part Number
Capacitor 68 μF, Tantalum, 25 V, ESR 150 mΩ	1	C1	AVX	TPSV686*025#0150
Capacitor 10 nF, Ceramic	2	C2, C3	Any	
Capacitor 68 μF, Tantalum, 6 V, ESR 1.8 mΩ	1	C4	AVX	TAJB686*006
Capacitor 150 nF, Ceramic	2	C5, C7	Any	
Capacitor 180 μF, Specialty Polymer, 4 V, ESR 15 mΩ	2	C6A, C6B	Panasonic	EEFUE0G181R
Capacitor 10,008 μF, Specialty Polymer, 4 V, ESR 10 mΩ	1	C8	Kemet	T510E108(1)004AS4115
Capacitor 0.1 μF, Ceramic	1	C9	Any	
1.82 kΩ 1% Resistor	3	R1, R2, R6	Any	
56.2 kΩ 1% Resistor	1	R3	Any	
10 kΩ 5% Resistor	1	R4	Any	
3.24 kΩ 1% Resistor	1	R5	Any	
1.5 kΩ 1% Resistor	2	R7, R8	Any	
Schottky Diode 30 V	2	D1, D2	Fairchild	DAT54
Inductor 6.4 μH, 6 A, 8.64 mΩ	1	L1	Panasonic	ETQ-P6F6R4HFA
Inductor 0.8 μH, 6 A, 2.24 mΩ	1	L2	Panasonic	ETQ-P6F0R8LFA
Dual MOSFET with Schottky	2	Q1, Q2	Fairchild	FDS6986S
DDR Controller	1	U1	Fairchild	FAN5236

Future Trends

As has been the trend for many years, customers will demand more and more memory to run their ever larger software applications. Systems such as the Intel boards for servers are already being designed with large amounts of DDR memory; some systems contain as much as 16 GB. DDR's decreased power requirements may still not be adequate to power such systems, hence the move toward DDR2 memory technologies. While we are just at the beginning of the DDR2 cycle, the industry is already buzzing about the next generation memory technology for PCs, DDR3 memories, which are not expected to reach the market until 2007 or later.

7.4 Power Management of Digital Set-Top Boxes

The Digital Set-Top Box (DSTB) market is one of the fastest growing applications for semiconductors. The market in millions of units is bigger and is expanding faster than the notebook market, offering tremendous opportunities for digital and analog semiconductor manufacturers. In this section, we will focus on the power management ICs that power the digital set-top box.

Set-Top Box Architecture

DSTBs control and decode compressed television signals for digital satellite systems, digital cable systems, and digital terrestrial systems. In the future, DSTBs will be an important means of access to the Internet for web browsing.

Figure 7–22 shows the main elements of a set-top box, from the video and audio processing sections to the CPU, memory, and power management sections.

Contrary to the PC architecture, which is well established and dominated by a few players, the set-top application is still going through an exciting phase of evolution and creativity. Today, there are many architectures and many implementations on the market. They range from a classic PC-like architecture based on Athlon or Pentium CPUs with associated chipsets, to embedded architectures with varying degrees of integration, all the way up to very large scale integrated circuits that include all but tuner, modem, and memory functions (see Figure 7–22).

In each case, power to each element of the architecture must be delivered readily and efficiently.

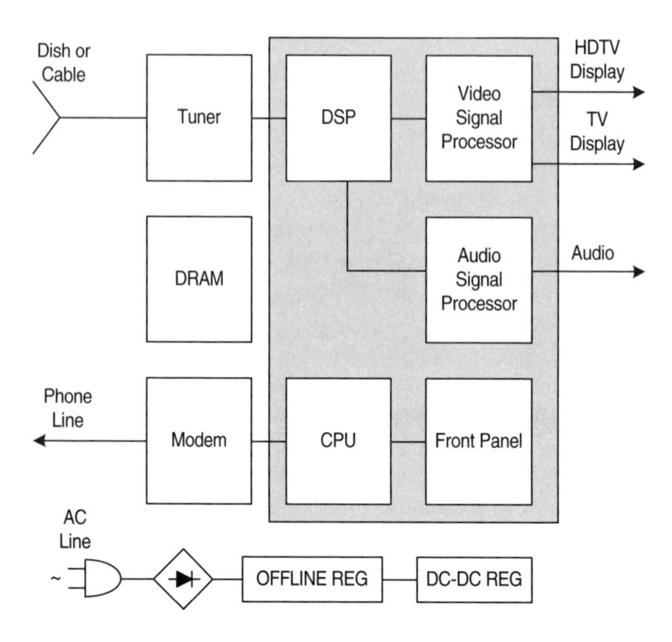

Figure 7–22 Digital set-top box block diagram.

Power Management

The strategies for powering set-top boxes are as diverse as their architectures. However, the underlying digital technologies are common to sister applications like PCs and handheld computers. Such commonalties allow the power system designer to draw from a rich portfolio of Application-Specific Standard Product (ASSP) ICs in order to power these devices, at least at the current stage of the game. As volumes increase and architectures solidify around a few leading core logic chipsets, it will become increasingly necessary to develop specific power management solutions for this market.

Here, however, we will reduce the discussion to two major cases: high performance and high power set-top boxes, which consume 50–240 W and require Power Factor Correction (PFC), and low power set-top boxes, below 50 W.

High Power Set-Top Boxes

In this section we will discuss a typical power management system for high power DSTBs. We will cover the AC-DC section first, then the DC-DC section.

AC-DC Conversion

Figure 7–23 shows the entire conversion chain, from wall power to an intermediate DC-DC voltage (V_{OUT}) low enough to be safely distributed on the box motherboard. The AC line is rectified first, and then power factor corrected, and converted down to a manageable voltage V_{OUT} (12–28 V DC) for distribution.

The rectification is accomplished with a full bridge diode rectifier and converts the alternate line voltage into a continuous—but still poorly regulated—intermediate voltage. As best efficiency is obtained when voltage and current drawn from the line are "in phase", a PFC block forces the correct phasing by modulating the drawn current according to the shape of the input voltage. The switch Q1 (MOSFET) and the diode D1, controlled by half of FAN4803 in Figure 7–23, constitute the PFC section. The top portion of Figure 7–24 shows the PFC control loop with the multiplier block accomplishing the phase modulation. Finally this power-factor corrected voltage is converted down to a low voltage that is usable by the electronics on the motherboard by means of a "forward" converter (switches Q2 and Q3, diodes D1–D5, and the second half of FAN4803 in Figure 7–23). This last conversion requires electrical isolation between the high input and the low output voltages. This is accomplished via the utilization of a transformer (T) in the forward conversion path and an optocoupler in the feedback path.

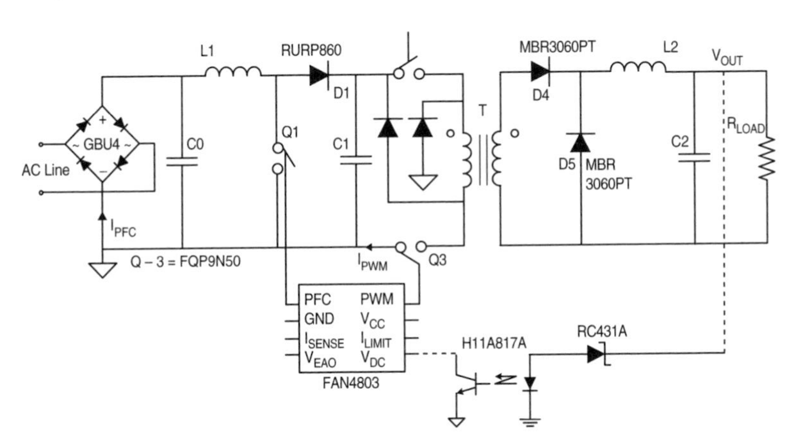

Figure 7–23 AC-DC power conversion with PFC.

DC-DC Conversion

With an appropriate DC voltage (12–24 V) delivered by the offline section, all the low voltage electronics on the motherboard can be safely powered. In Figure 7–24 the entire distribution of DC power on the motherboard is shown.

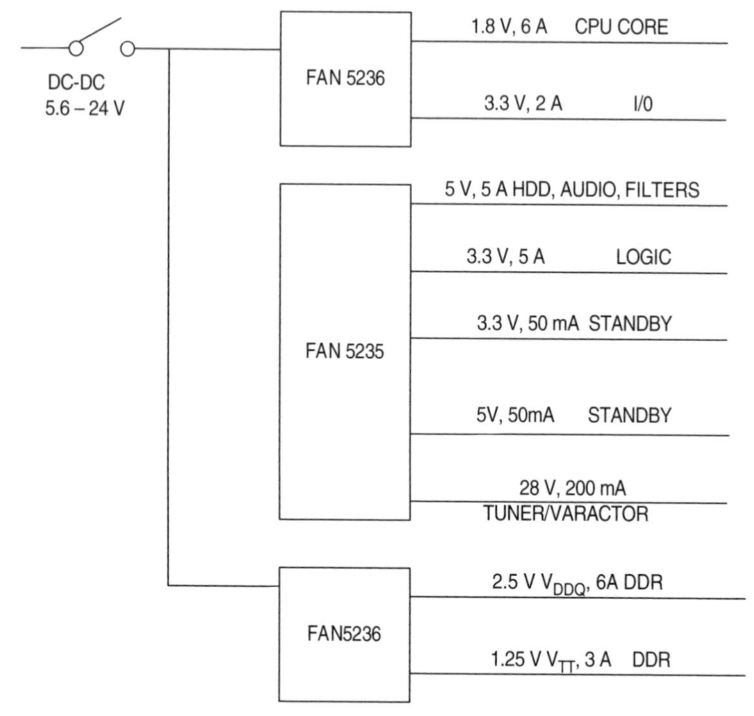

Figure 7–24 DC-DC regulation system for high power DSTB.

A total of nine different power lines are serviced, namely the nine output lines in Figure 7–24.

These power lines are described in more detail in the following text.

A dual PWM regulator, FAN5236, shown in Figure 7–25, powers the CPU core and I/O: these two regulators have adjustable voltages down to 0.9 V. This allows them to be easily set to power multiple generations of CPUs, from 0.18 μm lithography requiring 1.8 V, to 0.13 μm requiring 1.2 V, to future 0.1 μm lithography requiring sub band-gap voltage rails.

A highly integrated PWM controller (FAN5235) produces another five of the nine voltages: two buck regulators (3.3 and 5 V), one boost regulator (28 V) and two low power/low dropout regulators for standby operation. Figure 7–26 shows the typical application of this PWM controller.

A second dual PWM regulator provides DDR memory power V_{DDQ} (2.5 V, 6 A) and termination V_{TT} ($V_{DDQ}/2$ = 1.25 V, 3 A). The associated application diagram is similar to the one in Figure 7–25 so it is not repeated here.

Finally, Figure 7–27 shows a simplified internal functional diagram for one of the two PWM control loops of FAN5236. This controller is

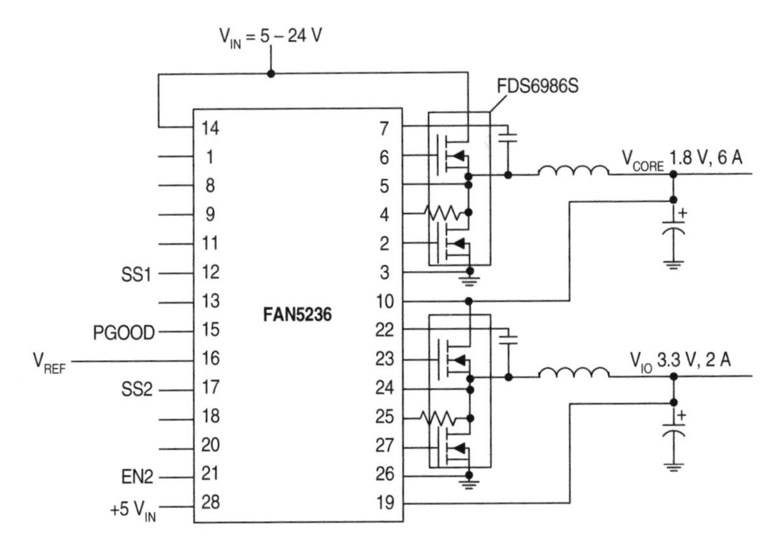

Figure 7–25 DC-DC regulation for CPU, I/O with FAN5236.

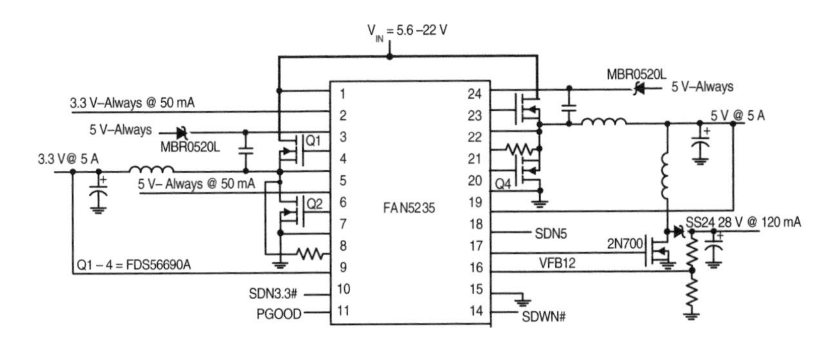

Figure 7–26 DC-DC regulation of five rails with FAN5235.

designed for very high efficiency: notice how the current sense (ISEN line) is done across the low side MOSFET R_{DSON} (drain to source "on" resistance of the MOSFET), avoiding the losses and the cost of a high power current sense resistor. Notice also the dual mode control loop, PWM for constant frequency operation at high currents, and Hysteretic (a technique leading to low frequency operation at light load, with constant ripple and low switching losses) for high efficiency at light load.

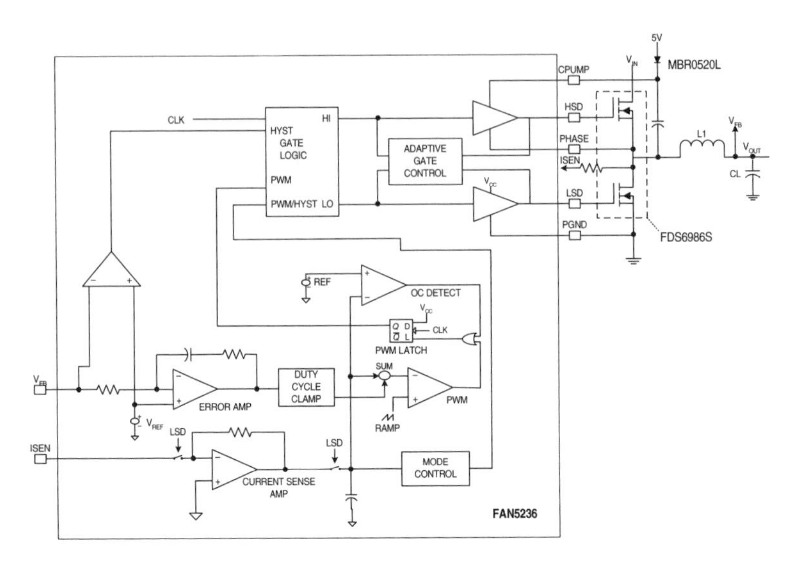

Figure 7–27 FAN5236 simplified diagram of one channel.

Low Power Set-Top Boxes

In this section we discuss a typical power management system for low power DSTB.

AC-DC Conversion

Below 50 W the architecture of the offline section becomes considerably more simple. The low level of power generally implies less sophisticated systems, for example those that lack HDDs and have less memory on board. Here the PFC section is no longer needed, and the lower power rating allows a simpler architecture. As shown in Figure 7–28, a diode bridge rectifier, in conjunction with a simple fly-back controller (KA5x03xx family) with a minimum number of external components, handles the entire offline section. The isolation requirements as per the high power offline discussed in the high power AC-DC conversion section still apply here.

The multi-chip approach to integration of the controller family allows such simplification (Figure 7–29). The SO8 package houses two dies, a controller die and a high voltage MOSFET die on board. Here again power-hungry discrete current sense resistors are avoided, in this case by means of a ratioed sense-fet technique on board the discrete element.

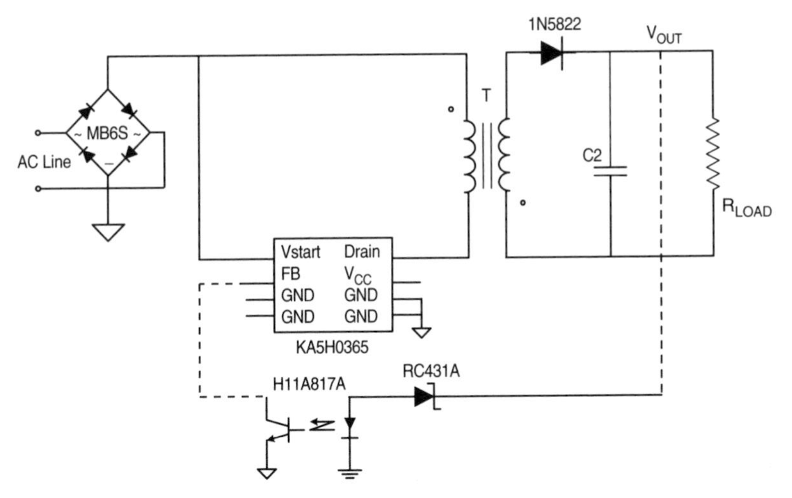

Figure 7–28 Low power AC-DC conversion.

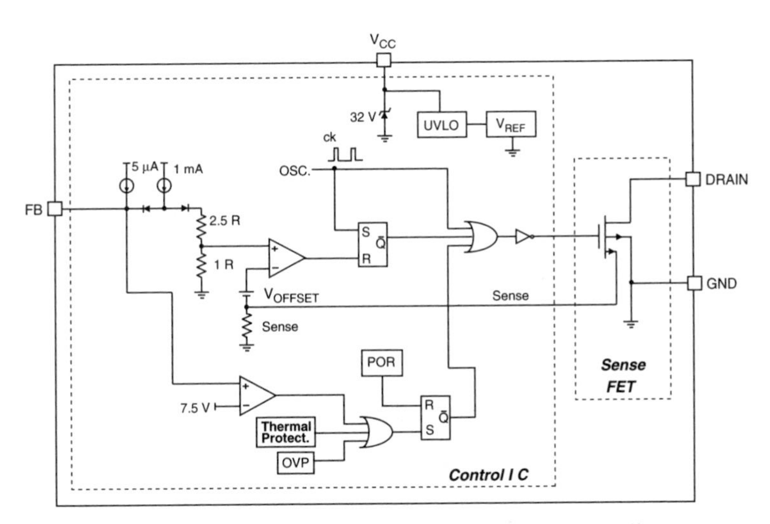

Figure 7–29 Offline controller KA5H0365 simplified block diagram.

DC-DC Conversion

Here the same type of controllers utilized in the previous section can be employed, although with smaller external discrete transistors and passive components, which leads to a much more compact set-top box. Figure 7–30 shows a system that needs only two controllers to power the entire DC-DC on the motherboard.

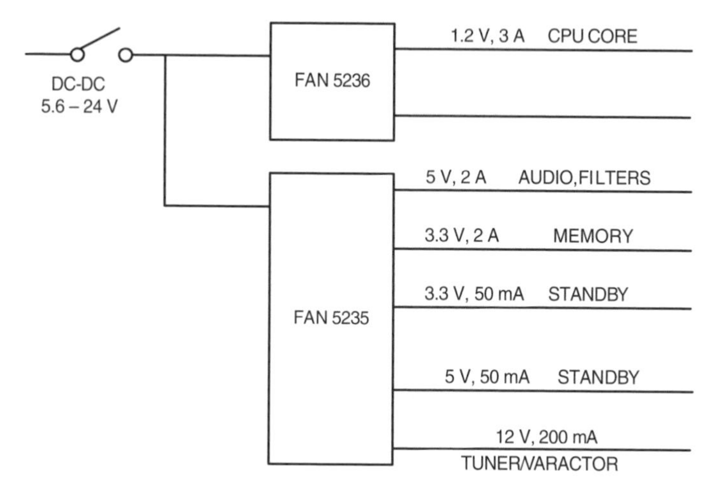

Figure 7–30 DC-DC regulation for low power systems.

Conclusion

We have discussed the power management needs of set-top boxes, covering two cases at opposite ends of the power spectrum.

The current generation of set-top boxes can be powered by a slew of ASSPs developed for the PC and handheld markets. As volumes increase and architectures solidify around a few leading core logic chipsets, dedicated ASSP ICs for set-top boxes will become necessary to allow increased performance at competitive cost.

7.5 Power Conversion for the Data Communications Market

This section discusses the transition from traditionally voice-centric telephony to converged voice and data over Internet Protocol (IP) and its implications for the power conversion of such systems. A few power conversion examples are provided complete with application schematics.

Introduction

The arm wrestling between voice and data has concluded in favor of the latter, with all the major data communications players now posturing for leadership of the migration from traditional voice to IP telephony. In the short term, the huge investments locked in the traditional telephony infrastructure and the new investments in data over IP necessitate that over the next few years we will have to provide power conversion for both types of systems as well as for the converged systems to come.

Current Environment with Separate Networks

Figure 7–31 shows the current telephony situation. Voice travels from traditional Private Branch Office (PBX) to Central Office, Switch, and finally to the Public Switch Telephone Network (PSTN). The data travels from routers to Wide Area Networks (WAN), and the video goes through a third independent path.

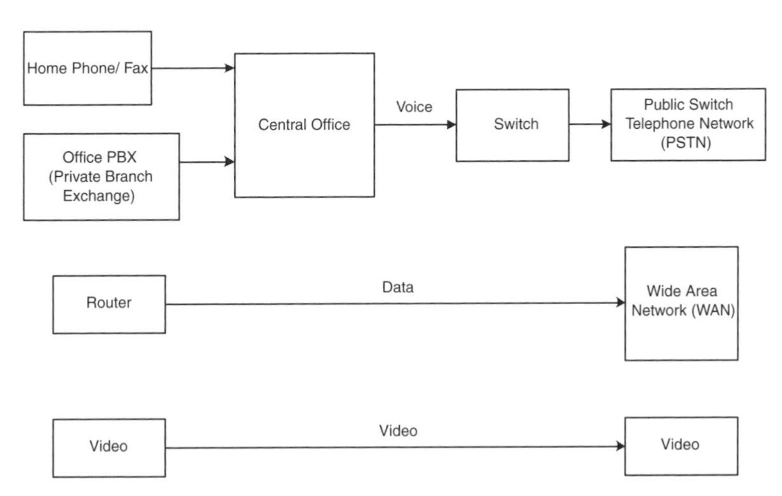

Figure 7–31 Separate networks for voice, data, and video.

Migration to Converged Voice/Data/Video IP

Figure 7–32 shows the envisioned converged Voice/Data/Video system over IP. At the center of this new universe is the Internet Protocol Wide Area Network, with all the services, including voice, data, video, and wireless communications gravitating around it.

Telecom –48 V DC Power Distribution

Usually telecom systems distribute a DC power (–48 V typically) obtained from a battery backup that is charged continually by a rectifier/charger from the AC line. Subsequently the –48 V is converted into various low positive DC voltages (Figure 7–33 shows 12 V only for simplicity) as well as back to AC voltages as necessary.

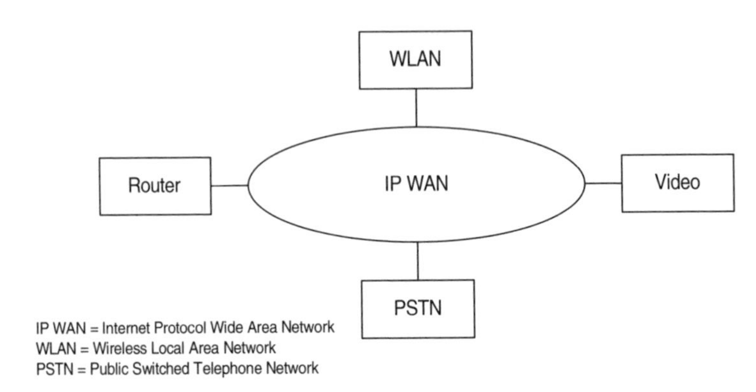

Figure 7–32 Voice/Data/Video over IP.

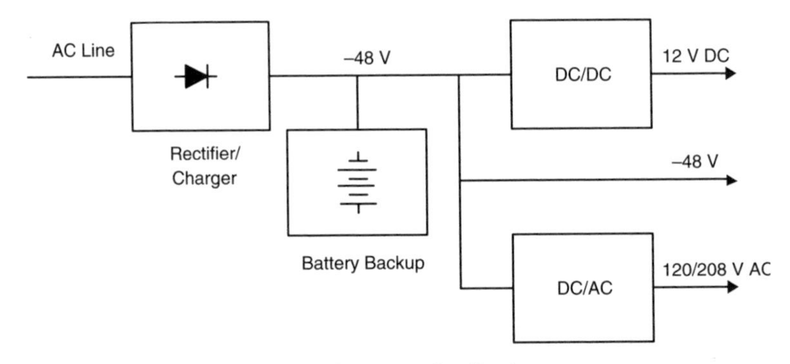

Figure 7–33 Telecom –48 V DC power distribution.

Datacom AC Power Distribution

Data centric systems tend to rely on an Uninterruptible Power Supply (AC UPS) front end for distributing AC power, which subsequently is converted into the basic constituents: –48 V, AC power, and low voltage DC (again, for simplicity Figure 7–34 only shows a 12 V DC).

With the advent of the converged systems, the telecom versus datacom separate approaches to power distribution will converge into new architectures. However, the bottom line is that at the board or backplane level the usual voltages will need to be delivered, namely 12 V and 5 V, as well as 0.9 V, 1.8 V, 2.5 V, and 3.3 V, with more to come.

The delivery of such low voltages starting from DC or AC power will be the focus of this document from here on.

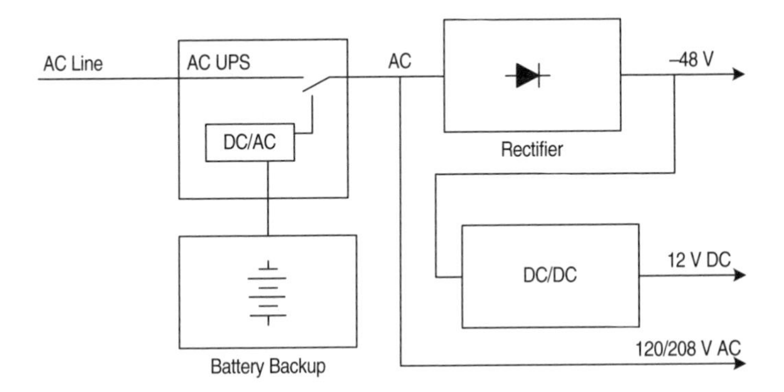

Figure 7–34 Datacom AC power distribution.

DC-DC Conversion

Figure 7–35 shows the –48 V to $+V_{OUT}$ (+5 V, +12 V etc.) with a forward converter architecture based on the ML4823 high frequency PWM controller.

Figure 7–36 shows the DC-DC conversion from 12 V and 5 V down to a variety of typical low voltages required by modern electronic loads.

The conversion down to heavy loads is done with synchronous rectification switching regulators of single or multiphase interleaved type, while for lighter loads linear regulators can be utilized.

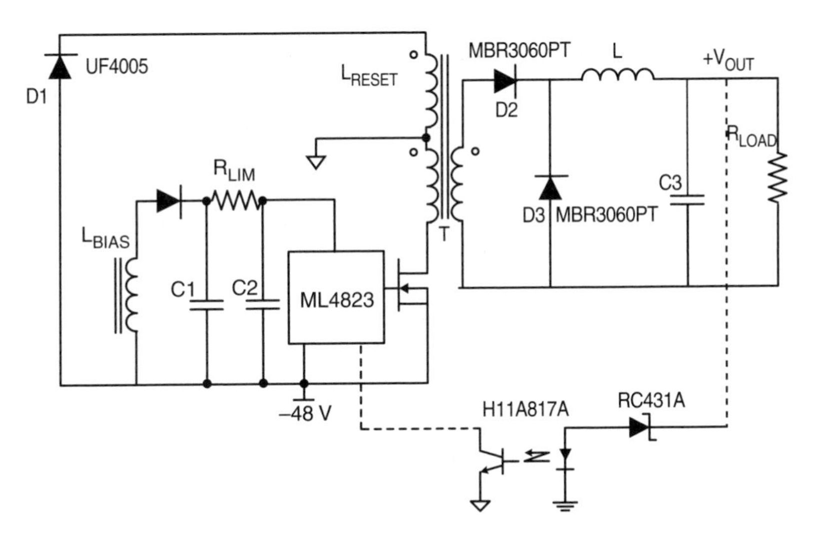

Figure 7–35 –48 V to $+V_{OUT}$ conversion.

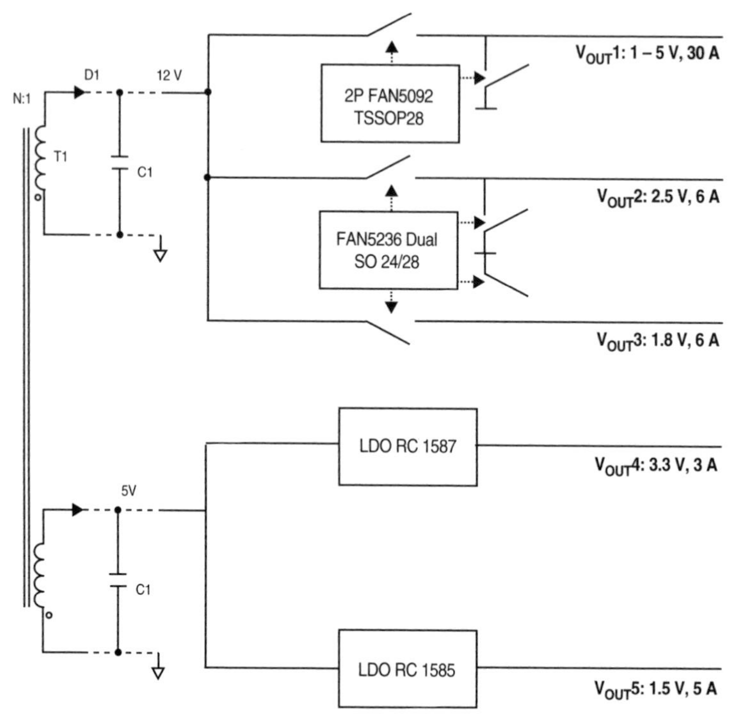

Figure 7–36 DC-DC conversion diagram.

FAN5092 Two-Phase Interleaved Buck Converter

The FAN5092 step-down (buck) converter (Figure 7–37) is ideal for data communications applications. This IC is a two-phase interleaved buck converter switching up to 1 MHz per phase. The application diagram illustrates conversion from 12 V down to 3.3 V in a 12 V-only input voltage source environment. The chip integrates the controller and the drivers on a single die. The high frequency of operation is enabled by:

- the monolithic approach of integrating controller and drivers on board
- a fast proprietary leading edge valley control architecture with 100 nanoseconds of response time
- the strongest drivers in the industry at 1 Ω of source and sink impedance for both high and low side driver of each phase

Such combination of features, together with loss-less current sensing via R_{DSON} sense, allows for a very efficient delivery of power with very small passive components, leading to record levels of power density.

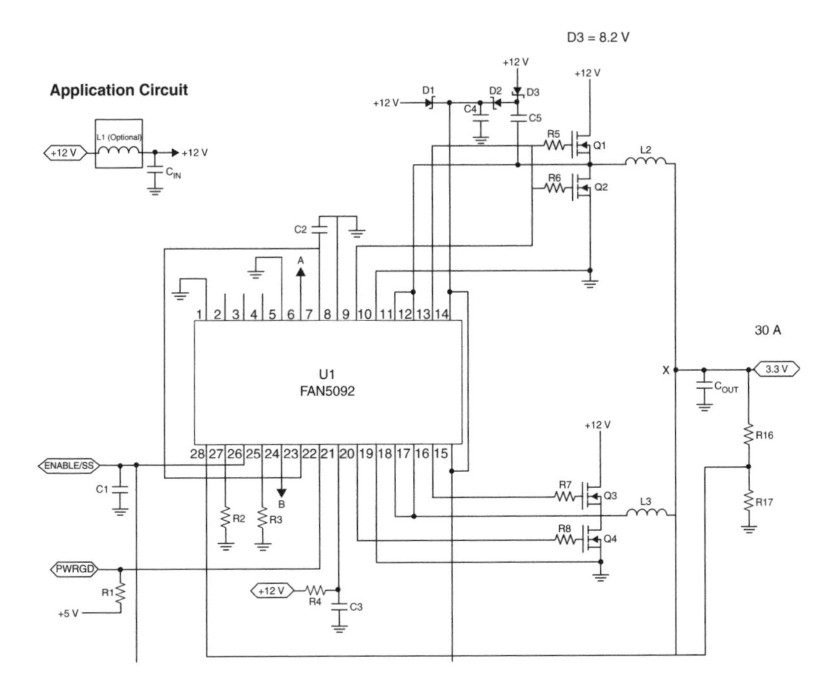

Figure 7–37 FAN5092 application circuit.

The application diagram of the IC is shown in Figure 7–37 for a 3.3 V, 30 A load. Optimum companions of the FAN5092 are the Fairchild discrete DMOS FDB6035AL for high side pass transistors Q1,2 and FDB6676S for low side synchronous rectification transistors Q2,4.

Two FAN5092 converters can be paralleled by means of doubling the above application and connecting together two pins (pin 26 and pin 15). This will allow handling of loads up to 120 A.

FAN5236 Dual Synchronous Buck Converter

The FAN5236 PWM controller (Figure 7–38) provides high efficiency and regulation for two output voltages adjustable in the range from 0.9 V to 5.5 V. Synchronous rectification and hysteretic operation at light loads contribute to a high efficiency over a wide range of loads. The hysteretic mode of operation can be disabled separately on each PWM converter if PWM mode is desired for all load levels. Again high efficiency is obtained by using MOSFET's R_{DSON} for current sensing. Out-of-phase operation with 180 degree phase shift reduces input current ripple.

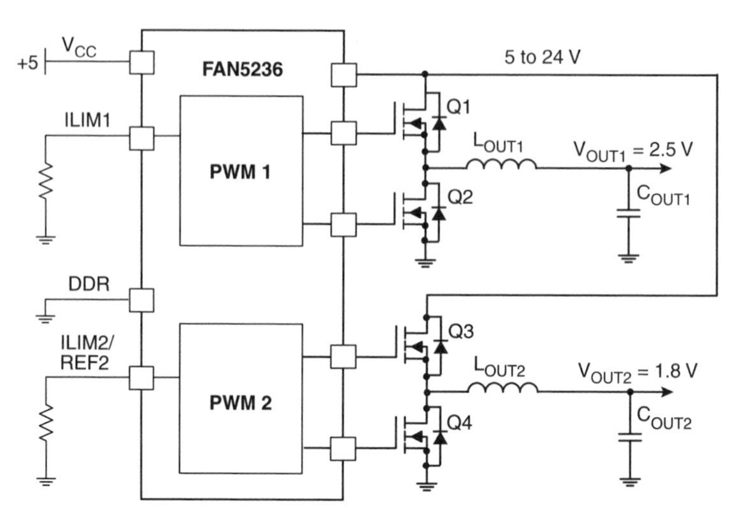

Figure 7–38 FAN5236 block diagram.

RC1585/7 Linear Regulators

In some cases, it makes sense to use linear regulators if the input to output voltage difference is sensibly less than the output voltage. Figure 7–36 showcases Fairchild's RC1587, 3 A and RC1585, 5 A linear regulators.

For more details and a complete bill of materials please refer to the FAN5092, FAN5236, RC1585, RC1587, and FOD2712 data sheets available on the Fairchild website *www.fairchildsemi.com.*

For KA5H0365, please refer to the data sheet as well as to Fairchild Power Switch (FPS) Application Notes for *Switch Mode Power Supply* (SMPS) design, also available on the Fairchild website.

Conclusion

The merging of data, voice, and video blurs the line between computing and communications. The smart loads of either application draw from the same advanced, high-density, sub-micron, low voltage CMOS technologies and require similar solutions for distributed power conversion. Fairchild expertise in power conversion for computing and communications offers proven solutions to the merging converged data communications market.

Chapter 8

Future Directions and Special Topics

8.1 Beyond Productivity and Toys: Designing ICs for the Health Care Market

As a veteran IC developer for the semiconductor industry, I have been, and still am, involved in efforts to design better consumer technologies. The exciting projects I have worked on range from making better electronic typewriters (late 1970s), to better hard disk drives (1980s), to better computers (1990s), and now, to designing better cell phone handsets and other portable electronics. Such technological advances have brought increased productivity to the industry and have enhanced peoples' lives, offering new forms of communication and expression, as well as creating new toys for entertainment.

All of these improvements, from the serious to the frivolous, are worthwhile, but they seem to lack the true nobility of "changing the world"; a catch-phrase worn out by almost daily use in our industry. However, within the fledgling fields of telemedicine and biosilicon opportunities are now presenting themselves which will enable us to focus our industry's aim on a truly substantive and meaningful purpose; namely enhancing lives by helping people fight against, or better cope with, diseases.

I will offer a personal example of how attention to health care technology could improve lives. In the last few years I have seen people close to me struggle with diabetes, a disease in which the body does not produce or properly use insulin, a hormone produced by the pancreas that is needed to convert sugar, starches, and other foods into energy needed for daily life. It is mind boggling to me that in the age of space traveling and the invention of the World Wide Web the high-tech industry has yet to succeed in putting together a glucose meter with an insulin pump that will deliver a viable artificial pancreas to diabetic patients.

At the same time it is heartening to see that our industry is beginning to focus its attention on health care products, with major players already foreseeing health care as the next market to turn into a silicon-based industry. It may not be coincidental that this market shift is happening as the leaders of our industry are aging and hence becoming more sensitive about health care issues. At the same time, given the huge potential numbers involved, the fact that attention to health care is good for people *and* good for business is certainly not lost on the industry.

This new growth is a welcome addition to our industry. In developing health care technologies, silicon design takes on a higher meaning and purpose, and indeed literally enhances our lives, by helping all of us live longer and better lives more free from disease.

8.2 Power Management Protocols Help Save Energy

Computing, communications, and consumer products fuel the race toward more integrated functions in smaller form factors, and consequently, escalate the rise in power density and power dissipation. Efficient power management inside an appliance long ago moved from a design afterthought to a principal concern, spurring a series of power management protocols and initiatives aimed at efficiently converting power from the source to the load. A new set of concerns has been prompted by the billions of such products sold each year. The number and rate of growth of these electronic appliances create a huge demand of power from the AC line, triggering concerns for power distribution and energy conservation and prompting a new set of protocols and initiatives.

A major phase transition in power management is happening before our eyes. *Power management*—often defined by the amount of heat safely disposable by the appliance—is evolving into *energy management*, driven by new concerns for energy conservation and environmental protection. This section reviews the main power management initiatives and protocols addressing power and energy management, progressing from the main board (DC-DC) to the wall (AC-DC) side of a system, and will point to challenges, opportunities, and limits associated with these techniques.

ACPI

At the highest level of power management techniques is Advanced Configuration and Power Interface (ACPI). ACPI power ICs take the available voltages from the silver box or AC adapter and, under specific operating system commands applied to the power chip via logic inputs, translates them into useful system voltages on the motherboard. This allows technologies to evolve independently while ensuring compatibility with operating systems and hardware.

Motherboard (DC-DC) Voltage Regulators

By far the most demanding load on the motherboard is the CPU. Efficient powering of a CPU—the core of modern electronic appliances—is done with special voltage regulators often described as voltage regulator modules. These regulators include power management techniques such as *Voltage Positioning* (VP), or dynamic voltage adjustment of the output (via D-A converter) to accommodate transitions to and from low power modes. Such techniques, first applied to desktop CPUs, have moved subsequently to notebooks and are now becoming popular in ultraportable devices.

The following is a list of a number of specifications, some of them proprietary, which addresses these challenges.

VRM Specifications

VRM specifications for desktop computing go into great detail about which architectures (interleaved buck converters), which external components (inductors and electrolytic and ceramic capacitors), and which protocols to apply in powering every new generation of CPU.

Notebook Power

Notebooks employ a set of aggressive power management techniques aimed at maximizing performance with the minimum expenditure of energy. Such techniques are similar to those discussed for VRMs and go well beyond. In addition to the previously-mentioned voltage positioning, alternate power management techniques for notebooks are:

Light Load Operation

At light load, voltage regulator switching losses become dominant over ohmic losses. For this reason, the switching regulator clock frequency of operation is scaled down at light load. This is done either automatically, commuting to light load mode below a set current threshold, or under micro control, via a digital input toggling between the two modes of operation.

Clock Speed on Demand

One of the most effective ways to contain power in notebooks is to manage the CPU clock speed and supply voltage as power dissipation goes with the square of the voltage and in proportion to the frequency (CV^2f). Different CPU manufacturers offer varying flavors of this technique. SpeedStep™ is Intel's recipe for mobile CPU power management while PowerNow™ is AMD's flavor. The bottom line is that for demanding applications—such as playing a movie from a hard disk drive—the CPU gets maximum clock speed and highest supply voltage, thereby yielding maximum power. On the other hand, for light tasks, such as typing a memo, the power is reduced considerably.

Offline (AC-DC) Voltage Regulators with Power Factor Correction (PFC)

In the past, the conversion and regulation of power from the wall has been concerned with the satisfaction of safety requirements. Recently, however, power management has become important in this area as well. PFC regulation is concerned with the efficient drawing of power from the wall, as opposed to minimization of power dissipation inside the gadget. Optimum conditions for power delivery from the AC line are achieved when the electric load, a PC, for example, draws current that is in phase with the input voltage (AC line) and when such a current is undistorted (sinusoidal). To this end, IEC 6100-2-3 is the European standard specifying the harmonic limits of various equipment classes. For example, all personal computers drawing more than 75 W must have harmonics at or below the profile demonstrated in Figure 8–1. Europe leads the world in compliance to these regulations, restricting all imported PCs. The rest of the world is following their example to varying degrees.

Figure 8–1 shows that the European allowance grows stricter for higher harmonics; however, these harmonics also have less energy content and are easier to filter. According to the specification, the allowed harmonic current does max out above 600 W, making it more challenging to achieve compliance at higher power.

Power factor is a global parameter speaking to the general quality of the power drawn from the line and it is related to the input current total harmonic distortion (THD) by Eq. 8–1.

$$PF = \frac{|\cos\varphi|}{(1 + THD^2)^{1/2}} \qquad \text{Eq. 8–1}$$

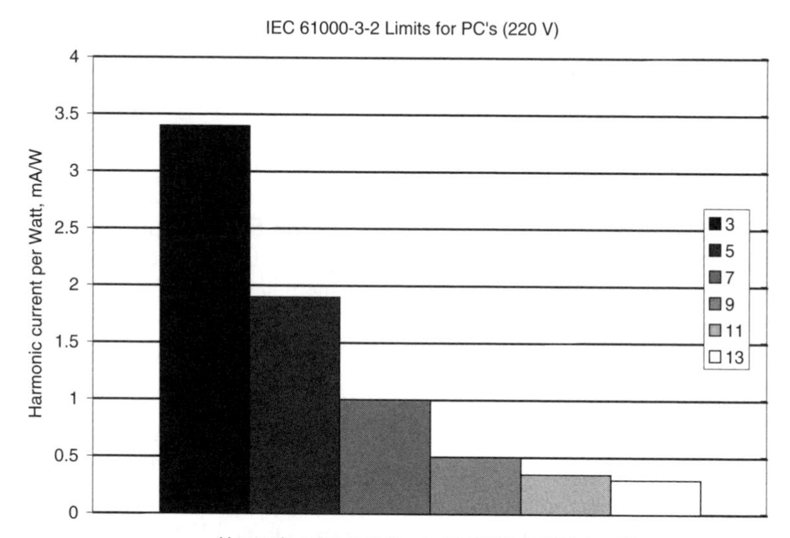

Figure 8–1 IEC 61000-3-2 harmonic current limits.

where φ is the phase shift between line voltage and drawn current. With no phase shift ($\varphi = 0$) and no distortion ($THD = 0$) it follows that $PF = 1$. Since the numerator $|\cos\varphi|$ is bounded between 0 and 1 and the denominator is always greater than or equal to one it follows that $PF \leq 1$.

Green Power (Energy Management)

Green power refers to sustainable energy systems that are based on renewable energy, such as power from the sun, wind, plants, or moving water. With respect to power conversion, green power loosely refers to a set of initiatives aimed at reducing power consumption of electrical appliances in standby and in the future, also in operation. Some major initiatives are briefly illustrated below:

Blue Angel

In 1977 Germany became the first country in the world to use an "eco-label" when the Federal Minister of the Interior and the Ministers of the Environment of the Federal States first introduced the Blue Angel label in order to promote environmentally compatible products.

Energy Star

The Energy Star label was developed by the United States Environmental Protection Agency and first appeared in 1993 on personal computer equipment. To bear the Energy Star label, a product must operate significantly more efficiently than its counterparts, while maintaining or improving performance. For example, a 300 W silver box in sleep mode should draw less than 20 W from the AC line to meet the Energy Star efficiency requirements (CFX12V power supply design guide). Products displaying the Energy Star logo now range from washing machines to commercial air conditioning systems to homes.

1-Watt Initiative

The International Energy Agency (IEA) created the 1-Watt Initiative aimed at reducing standby power losses to below 1 Watt. This initiative was launched in 1997 and adopted readily by Australia first. In July 2001 U.S. President George Bush issued Executive Order 13221, requiring the federal government to purchase products with standby power below 1 W, lending further weight to the IEA initiative. As an example, to meet the Blue Angel requirements (RAL-UZ 78), E.O. 13221, and other low power system demands, the PC 5 V standby efficiency should be greater than 50 percent with a load of 100 mA.

New Low Power System Requirements

Recently, the focus has shifted from standby to operating power savings. Intel, for example, is driving up the efficiency of the silver box (CFX12V Design Guide and others) as per Table 8–1. The efficiency targets recommended in Table 8–1 can largely be achieved today at moderate cost increases. Initiatives like Efficiency Challenge 2004, a power supply design competition sponsored by EPA Energy Star and the California Energy Commission, will likely push the limits even further.

Table 8–1 Loading Table from CFX12V Power Supply Design Guide

Loading	20% Load	50% Load	100% Load
2003 Intel required spec	50%	60%	70%
2004 Intel required spec	60%	70%	70%
2004 Intel recommended spec	67%	80%	75%

Conclusion

Miniaturization trends for modern electronic appliances and their market diffusion by the billions are fueling a keen interest in moving toward more efficient designs. It is becoming clear that an extra cost of a few dollars for an appliance is returned many times over in terms of energy savings and environmental protection—and this realization is strengthening the recommendation and even the mandate of new protocols and requirements. These requirements will push technology advancements beyond the traditional cost-oriented model of minimizing the appliance's bill of materials. These trends will lead to a more rational use of our energy resources and will stimulate the development of new power management technologies, injecting renewed energy inside the power semiconductor industry.

8.3 Heat Disposal in Electronics Applications

Active versus Passive Cooling

Introduction

Miniaturization and portability trends in combination with increasing performance are contributing to the well known problem of heat concentration and dissipation in modern electronic appliances. The electronics industry's answer has so far mostly consisted of trying to improve existing methods and technologies. The processor industry is moving to Silicon On Insulator (SOI) technology to reduce the heat dissipation per transistor, while the power supply industry is trying to squeeze every last percentage point of efficiency out of their regulators. And the two together are working closer than ever in an effort to devise efficient management schemes to consume as little power as possible.

Such measures are slowing down the speed of the rise in temperature, without actually taming it.

In portable electronics the issue of power dissipation is compounded by the lack of good energy sources. Eventually fuel cells will become viable, charging will yield to fueling and energy availability will no longer be a problem in portable systems. When that happens the heat will remain the lowest common denominator; the ultimate problem to solve—unless we do something about it sooner, that is.

Limits of Passive Cooling

The vast majority of heat management systems today relies on passive methods of cooling, typically based on a bulky mass of heat-conducting material shaped for maximum radiating surface (*heatsink*), attached to the heat source. The heatsink may be complemented as necessary by forced air circulation. In cases where space is at a premium heat pipes are utilized as means to transport the heat from the hot spot to peripheral areas where heat can be more easily disposed off. While heat pipes are state-of-the-art in modern notebook computers, such technology is less than desirable, as it is based on encapsulated fluids that may leak and damage the electronics. The fundamental limitation of passive cooling methods, including heat pipes, is that they rely on a negative temperature gradient to work. In other words the heat always has to flow from the higher temperature point to a lower temperature point. It follows that the device or load to be cooled will always be at higher temperature with respect to the heatsink and the ambient. With ambient temperature varying easily from 25 to 70°C and silicon failure rates proportional to the square of the silicon junction temperature, passive cooling resembles more a torture chamber for silicon rather than real refrigeration.

Active Cooling

Active cooling is a forced means of refrigeration in which heat can be made to flow from the lower to the higher temperature spot. This is obviously the principle on which common refrigeration is based. While active cooling overcomes the "negative temperature gradient" barrier, it pays a price in terms of additional heat generation. Can active cooling be the solution?

The theoretical limit for efficient heat transport is achieved by the reversible heat engine obeying the Carnot cycle. The transport of heat by a Carnot cycle is described by Eq. 8–2

$$P_{COOL} = \frac{P_{LOAD} \times T_C}{(T_H - T_C)} \qquad \text{Eq. 8–2}$$

where

P_{COOL} = Power expenditure to cool with Carnot engine (W)

P_{LOAD} = Power dissipated by the load to be cooled (W)

T_C = Temperature of the cooled side (°K)

T_H = Temperature of the hot side (°K)

Accordingly, in order to transport 100 W of heat from a cold surface (27°C) to a hot surface (say 300°C), an expenditure of power is theoretically necessary in absence of mechanical friction and other irreversibilites amounting to

$$P_{COOL} = \frac{100\ \mathrm{W} \times (27 + 273)}{(300 - 27)} = 109\ \mathrm{W} \qquad \text{Eq. 8–3}$$

In thermodynamic terms, this transport can be looked at as a refrigeration process or a heat pump process.

This can be described as a refrigeration process with the *Coefficient Of Performance* (COP), defined as the ratio of the work required to the energy transferred for cooling (COPC), equal to 109 W/100 W = 1.09. Or it can be seen as a heating process. In this case the cost of cooling, 109 W, is effectively "free" heat and hence the effective coefficient of performance (COPH) is 209 W/100 W = 2.09.

Moving from thermodynamic to electronic terminology, let us now assume that 100 W is the power generated by a chip powered by a voltage regulator (whose efficiency is 100 percent for simplicity) and cooled by Carnot.

We have

$$P_{LOAD} = 100\ \mathrm{W} \qquad \text{Eq. 8–4}$$

$$P_{COOL} = 109\ \mathrm{W} \qquad \text{Eq. 8–5}$$

$$\eta\% = 100 \times P_{LOAD}/(P_{LOAD} + P_{COOL}) = 100/209 = 48\% \qquad \text{Eq. 8–6}$$

where η is the efficiency, or ratio between useful power and overall power expenditure. Table 8–2 illustrates the relationships between these parameters and Figure 8–2 illustrates the elements at play and the power flow.

Notice that $\eta\%$ can also be calculated as $1/(1 + COPC)$, still 48 percent for Carnot.

Adding to this the inefficiency of the voltage regulators powering the load and the engine and mechanical frictions, we can conclude that active cooling at best will yield overall efficiencies in the range of 40 percent.

Active Cooling—Yes or No?

Can active cooling be viable at such levels of efficiency? Yes! Low efficiency is only a killer when it generates heat in the wrong places, namely at the junction of silicon transistors. Other than that, inefficiency is quite cheap.

Table 8–2 Watts Required to Transport 100 W of Power in a Carnot Cycle

Carnot Efficiency	Carnot Figures	Formulas
P_{COOL}	109 W	100 Watt × COPC
COPC (Cooling)	1.09	$T_C/(T_H - T_C)$ = COPC (T in °K)
COPH (Heating)	2.09	$T_H/(T_H - T_C)$ = COPH = 1 + COPC
η% Efficiency	48%	η = 1/(1 + COPH)

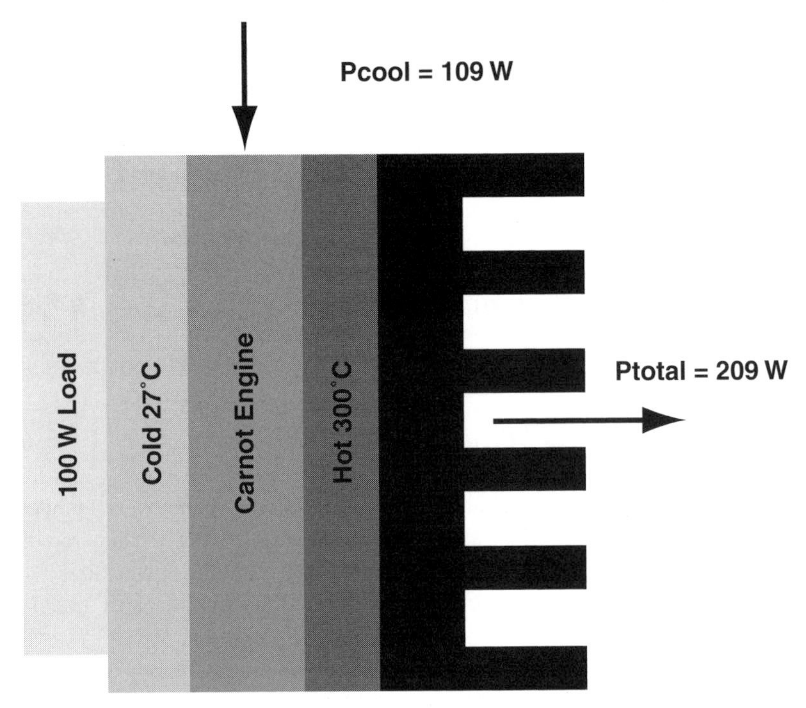

Figure 8–2 Schematic diagram of a Carnot engine cooling a 100 W load.

Watts are cheap; at 8 c/kWh a 100 W load consumes 0.8 c/h. Depending on usage patterns a CPU may not work at full speed for more than a few hours a day, making the daily cost of such features around a few cents per day (say three) and a few dollars a year (say ten). This is not an unacceptable cost.

We can do a similar calculation with respect to fuel cells and find that active cooling technology would burn energy (by burning methane or whatever fuel we end up using) at twice the rate or more of conventional technology. But then again is that really a problem? People care about long untethered operation—which they can get with fuel cells—not about the rate of energy burning. So in the long run we would not discount active cooling technology in portable computing either, once fuel cells become a viable technology.

The Hot Plate

Now that we are moving heat from the "cold" silicon junction to a "hot" plate for heat disposal, we have indeed turned the industry on its head; the hottest place is now the heat radiator, hotter than ambient, and the coldest place is the silicon junction. Doesn't this feel right? Would you want it the other way around ever again? Is a hot plate—perhaps as hot as 300°C or more—a problem? I don't think so. We deal routinely with hot surfaces at home (kitchen appliances, light bulbs) and on the road (motorcycles' tail pipes).

Active Cooling Implementation

Peltier

Examples of active cooling, like thermoelectric cooling based on a Peltier array, are found in satellite receivers where lowering the temperature of the LNA allows a lower noise figure, and in fiber optic network equipment where again, precision temperature control is required.

With thermoelectric cooling (Figure 8–3), a voltage is applied to an ohmic junction of two different conducting (thermocouple) or semi-conducting (P- and N-type) materials, and the ensuing current flow results in absorption or release of energy (heat) at the junction as the electrons cross a corresponding "uphill" or "downhill" potential. The intensity of heat flow is proportional to the current and the process is reversible, namely a heat source at the junction will produce a corresponding current flow.

In Figure 8–3 the mechanism of an electron acquiring energy in order to overcome the opposing electric field E_C—and hence cooling the "cold" plate—in crossing the NP junction, as well as releasing energy in the presence of a favorable electric field E_H—and hence heating the "hot" plate—in crossing the PN junction as illustrated.

The heat flow being proportional to the current means that any current controller in the semiconductor manufacturer product portfolio can be easily adapted to control current and hence, via a thermistor, temperature in a Peltier array.

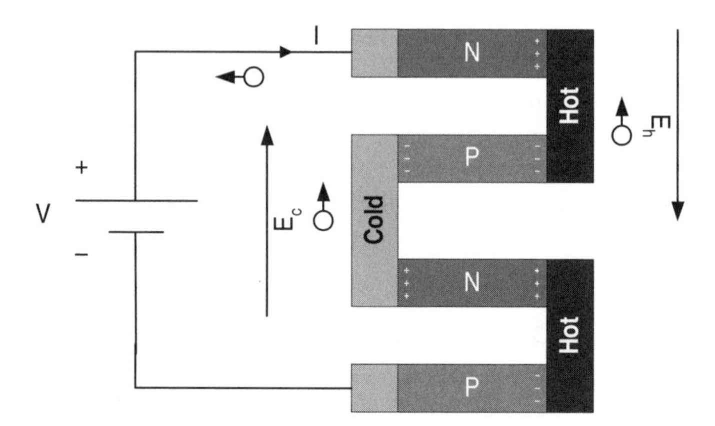

Figure 8–3 Illustration of Peltier effect with $V \times I = P_{COOL}$.

For example, Class D power amplifiers normally used to drive audio amplifiers are being applied successfully to drive Peltier arrays.

A few adventurous souls have applied Peltier cooling to their CPUs and managed to over-clock their PCs sensibly thanks to the lowered temperature. That Peltier could become mainstream in mass-market applications is dubious because of the extremely low efficiency, in the range of five percent, of Carnot. While we have made a case for cheap energy, an expenditure of 2 kW to cool a 100 W load seems to be a bit too much.

Stirling

Stirling refrigerators—and variations on the theme—are mechanical systems based on compression and expansion of an inert gas by a piston. These systems yield efficiency closest to the ideal Carnot cycle and are being seriously investigated for high end CPU applications like IA64. Because of their complexity, mechanical nature with moving parts, and cost it is unlikely that this will be the technology that will shrink heatsinks and displace heat pipes for high volume applications either.

What, Then?

What we need is an active cooling technology that has the efficiency of the Stirling and the solid state electronics makeup of Peltier. A solid state, electronically inherent solution to heat management must exist, if for no other reason than to mitigate the embarrassing dependence of the high tech solid state electronics industry on Iron Age passive heatsink technology.

A few pioneering companies have already advanced claims to this extent, but they still have little to show other than papers.

What we need now is a mindset change for our industry; enough with the single minded push of efficiency gains and on with creative ways to bring to the mainstream a viable active cooling technology.

8.4 Web Based Design Tools

The Tools on the Web

It is becoming increasingly popular to aid the sale of semiconductor products with web programs that facilitate the selection or the design of such products and in some cases even their design into the customer application.

There are essentially three primary classes of offerings on the web today:

Visual Basic or Excel Based Programs

Like POWER4-5-6, these programs are equation based and focus on predetermined topologies. As each "case" is represented by a set of equations, the flexibility is null: the simplest modification requires the creation of a new "case" or application. It is also doubtful that this approach can solve complex mathematical problems such as treating systems that don't rely on constant clock frequency of operation (hysteretic mode, constant on-, and constant off-time systems) or resolve transient response, start-up, and short circuit behavior with good accuracy.

Visual Basic/Excel programs are generally intended for distribution, servicing a large number of small customers in need of hand holding. These programs are inexpensive and some of them are provided free of charge by semiconductor companies.

Simulation Based Tools

Offered for free—by many semiconductor companies bundled with the company models (FSC-FETBench, NSC-Web-Bench)

Offered for sale—but immediately usable as DEMO download—by software companies like SPECTRUM-SOFT (MICRO-CAP simulator) bundled with available models and pointers to models that can be downloaded from semiconductor companies.
As with any simulator, this is potentially an "all crunching" tool with adequate flexibility to cover any possible case, say simulate a circuit with an extra input filter or without one.

The flexibility of a simulation tool will be more evident as the web tool begins to offer more complex functions (non-constant frequency systems, transient response, startup behavior, etc.) and becomes interactive, allowing the designer to modify (however slightly) on the fly the circuit they want to simulate. In schemes where the topology is fixed and the designer can only input a few values, it may be difficult to differentiate between the performances of the two tools (simulator versus equation based).

A simulation based system can be exploited internally by a design and apps community to a degree of flexibility (real system design) vastly superior to the one achievable on the web. Naturally these tools can do all that Visual Basic/Excel tools can do and more, satisfying any level of demand.

For system simulations, high level simulators like SIMPLIS from Transim can do the job in seconds, whereas the normal SPICE approach would take much longer, sometimes weeks. SIMPLIS requires a block diagram specific to the chip at hand and general parameters for the elements, like bandwidth, gain of the error amp, etc.

This comes from the data sheet and contacts with designers and apps engineers, so generally does not directly draw on the SPICE models accumulated in the design community.

Models

The semiconductor company puts out models that the electrical engineer can use with his preferred SPICE simulator. This tool is most appropriate for components (discretes) and simple devices (opamps, comparator). Typically the system designer is interested in in-depth simulation of subsets of his total system, as attempting to simulate the entire system with SPICE might take weeks.

The intended audience for models on the web is the system designers of both big and small companies.

Publishing models on the web is a good investment for semiconductor companies; once a company puts its models on the web, dozens of software companies like SPECTRUMSOFT will propagate the use of these models via their own simulators.

8.5 Motor Drivers for Portable Electronic Appliances

Introduction

The consumer segment is the biggest and fastest growing of the electronics markets for ICs. This market is fueled by the success of gadgets like digital still cameras, MP3 and DVD players, set-top boxes, video game consoles, cell phones—all electronic "toys" that require motors. These motors are for features such as moving lenses in cameras, spinning disks, and moving read/write heads. In addition, auxiliary elements inside electronics, such as cooling fans and CD-ROM/RW/DVD-ROM drives, also employ motors. More recently, camera phones have even incorporated motors inside the cellular handset. Each one of the motors in these applications requires a motor driver, making motor driver ICs a very fast-growing and exciting market.

This section will discuss the motion control requirements in modern digital still cameras, possibly the gadgets housing the highest density of motors in a small volume. It will also discuss a motor driver solution based on a highly integrated motor driver IC (circled in Figure 8–4).

Figure 8–4 DSC mainboard with a motor driver IC on board.

Camera Basics

Most digital cameras have an optical zoom, a digital zoom, or both. An optical zoom lens actually moves outward toward the subject to take sharp close-up photographs; this is the same kind of zoom lens found in traditional cameras. Digital zoom is a function of software inside the camera that crops the edges from a photograph and electronically enlarges the center portion of the image to fill the frame, resulting in a photograph with less detail. The opening (or aperture), through which light enters the camera is controlled by an iris diaphragm (mimicking the human eye) inside the lens. A shutter controls the time during which light is permitted to enter the camera. Adjusting the shutter speed in conjunction with the width of aperture allows the proper amount of light in to expose the film.

Motors and Motor Drivers

Figure 8–5 illustrates the three types of motors in a DSC. A DC actuator is typically used to drive the shutter. This actuator is controlled in Constant Current Drive (CCD) mode for high-speed shutter operation (higher than 1/2000th of a second in modern digital cameras). A step motor will assure precise position control for auto focusing. The optical zoom can be implemented with a DC motor or a step motor. For fast zooming operation, especially when DSC power is turned on, the DC motor is preferable because of its higher torque compared to a step motor. On the other hand, for smooth and precise zooming during normal operation the step motor can be preferable. Iris operation is done with DC actuators for single iris applications and with step motors when precise position control is required (multi iris).

Fairchild Semiconductor offers a motor driver IC, the FAN8702, with six channel drivers (CH1 through 6, each driving a single winding) on board that are capable of powering the entire camera set.
Figure 8–6 shows a full set of motors in a DSC. The six channels are specialized as follows:

DC actuator for shutter (Channel 5)

Step motor for auto focus (Channels 1 and 2—two windings)

Step motor for iris (Channels 3 and 4—two windings)

DC motor for optical zoom (Channel 6—this channel has the brake function necessary in a DC motor because of its moment of inertia)

Driving Implementation

The main motor driving methods in DSCs are Constant Voltage Drive (CVD) and Constant Current Drive (CCD). Voltage PWM is sometimes used to supply proper voltages to different motors from a fixed power supply. For example, a lens barrel maker may release a new DC zoom motor

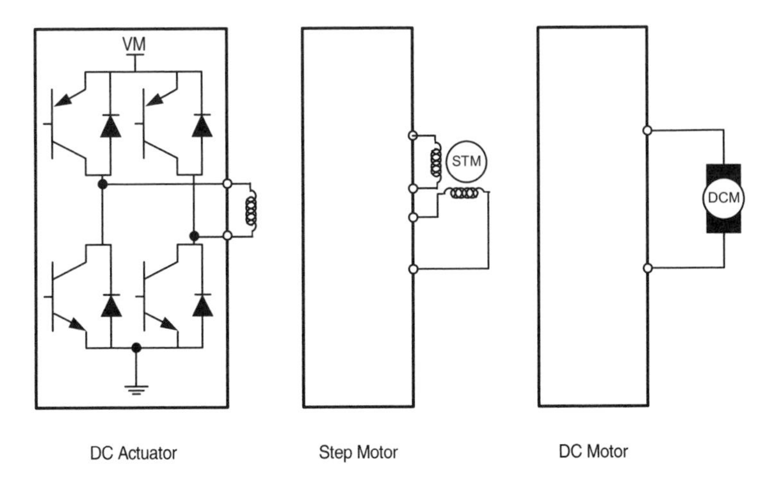

Auto Focus	Shutter	Zoom	Iris
Step Motor	DC Actuator	Step or DC Motor	Step or DC Motor

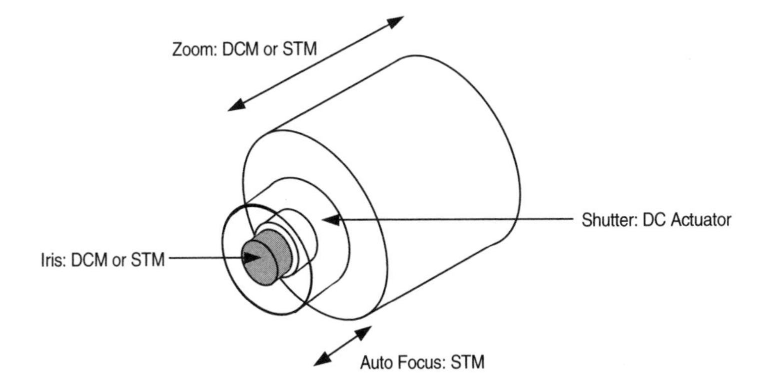

Figure 8–5 Lens motor assembly in DSC.

and DC actuator iris specified at 3 V and 3.6 V respectively, against an available VM (motor supply voltage) fixed at 4 V. In this case the DSP or Servo IC may have to generate a PWM signal with a 0.75 and 0.9 duty cycle to take the 4 V down to 3 V and 3.6 V. Such PWM signal frequency will have to be set above 100 kHz due to the low inductance value in small DSC motors.

Figure 8–6 (1) DC actuator shutter, (2) Step motor auto focus, (3) Step motor iris, (4) DC zoom.

Efficiency

The FAN8702 combo motor driver has four command inputs for adjusting output voltage and current levels, set respectively by pins VC1, VC2, IAE, and ISH. The proper choice of output voltages and currents leads to minimum power dissipation. Such high levels of flexibility make simple control methods like CVD and CCD the most popular. These methods, in addition to being simple, compact, and cost effective, are also inherently low noise thanks to the absence of switching transitions typical of PWM architectures. Appendix G provides the data sheet of FAN8702 for more technical details.

DSC Power Consumption

Digital still cameras can be powered by one or more Lithium-Ion cells. As a reference, a DSC in the class of the Samsung Digimax V50 is powered by one Li^+ cell (3.7 V, 1440 mAh), storing roughly 5 Wh of energy. The Canon EOS300D Digital Rebel is powered by two Li^+ cells (7.4 V, 1100 Ah) storing roughly 8 Wh of energy. This camera typically consumes 1 W in idle mode with peaks of 8 W when photographing with flash. Hence, the untethered operation of this DSC will vary, ranging from one to eight hours depending on its frequency and mode of use.

Conclusion

Amazingly, motors are able to adapt to the most demanding miniaturization trends and in fact, now we are starting to find "micro" (1 inch, 1.5 Gbyte) hard disk drives in smart phones. HDDs need a spindle and a voice coil motor to operate. The new camera phones coming to market may soon have on board all the motors mentioned so far, including optics and mass storage motors—adding up to six of them—all in a pretty cramped space. Along with all these motors comes the need for efficient and up-integrated motor drivers, such as the one discussed earlier in this chapter. This type of motor driver IC can meet the increasing demand for low power dissipation, control, and diminishing space of modern portable appliances.

Appendix A

Fairchild Specifications for FAN5093

www.fairchildsemi.com

FAN5093

Two Phase Interleaved Synchronous Buck Converter for VRM 9.x Applications

Features

- Programmable output from 1.10V to 1.85V in 25mV steps using an integrated 5-bit DAC
- Two interleaved synchronous phases for maximum performance
- 100nsec transient response time
- Built-in current sharing between phases
- Remote sense
- Programmable Active Droop™ (Voltage Positioning)
- Programmable switching frequency from 100KHz to 1MHz per phase
- Adaptive delay gate switching
- Integrated high-current gate drivers
- Integrated Power Good, OV, UV, Enable/Soft Start functions
- Drives N-channel MOSFETs
- Operation optimized for 12V operation
- High efficiency mode (E*) at light load
- Overcurrent protection using MOSFET sensing
- 24 pin TSSOP package

Applications

- Power supply for Pentium® IV
- Power supply for Athlon®
- VRM for Pentium IV processor
- Programmable step-down power supply

Description

The FAN5093 is a synchronous two-phase DC-DC controller IC which provides a highly accurate, programmable output voltage for VRM 9.x processors. Two interleaved synchronous buck regulator phases with built-in current sharing operate 180° out of phase to provide the fast transient response needed to satisfy high current applications while minimizing external components.

The FAN5093 features Programmable Active Droop™ for transient response with minimum output capacitance. It has integrated high-current gate drivers, with adaptive delay gate switching, eliminating the need for external drive devices.

The FAN5093 uses a 5-bit D/A converter to program the output voltage from 1.10V to 1.85V in 25mV steps with an accuracy of 1%. The FAN5093 uses a high level of integration to deliver load currents in excess of 50A from a 12V source with minimal external circuitry.

The FAN5093 also offers integrated functions including Power Good, Output Enable/Soft Start, under-voltage lockout, over-voltage protection, and adjustable current limiting with independent current sense on each phase. It is available in a 24 pin TSSOP package.

Block Diagram

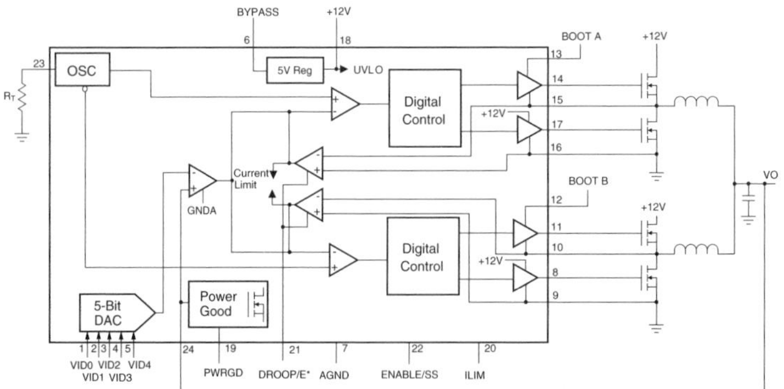

REV. 1.1.0 4/20/05

Pin Assignments

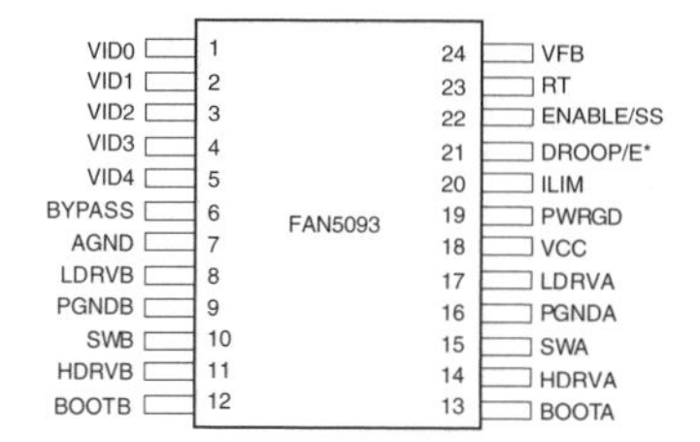

Pin Definitions

Pin Number	Pin Name	Pin Function Description
1-5	VID0-4	**Voltage Identification Code Inputs.** Open collector/TTL compatible inputs will program the output voltage over the ranges specified in Table 1. Internally Pulled-Up.
6	BYPASS	**5V Rail.** Bypass this pin with a 0.1µF ceramic capacitor to AGND.
7	AGND	**Analog Ground.** Return path for low power analog circuitry. This pin should be connected to a low impedance system ground plane to minimize ground loops.
8	LDRVB	**Low Side FET Driver for B.** Connect this pin to the gate of an N-channel MOSFET for synchronous operation. The trace from this pin to the MOSFET gate should optimally be <0.5".
9	PGNDB	**Power Ground B.** Return pin for high currents flowing in low-side MOSFET. Connect directly to low-side MOSFET source.
10	SWB	**High side driver source and low side driver drain switching node B.** Gate drive return for high side MOSFET, and negative input for low-side MOSFET current sense.
11	HDRVB	**High Side FET Driver B.** Connect this pin to the gate of an N-channel MOSFET. The trace from this pin to the MOSFET gate should optimally be <0.5".
12	BOOTB	**Bootstrap B.** Input supply for high-side MOSFET.
13	BOOTA	**Bootstrap A.** Input supply for high-side MOSFET.
14	HDRVA	**High Side FET Driver A.** Connect this pin to the gate of an N-channel MOSFET. The trace from this pin to the MOSFET gate should optimally be <0.5".
15	SWA	**High side driver source and low side driver drain switching node A.** Gate drive return for high side MOSFET, and negative input for low-side MOSFET current sense.
16	PGNDA	**Power Ground A.** Return pin for high currents flowing in low-side MOSFET. Connect directly to low-side MOSFET source.
17	LDRVA	**Low Side FET Driver for A.** Connect this pin to the gate of an N-channel MOSFET for synchronous operation. The trace from this pin to the MOSFET gate should optimally be <0.5".
18	VCC	**VCC.** Internal IC supply. Connect to system 12V supply, and decouple with a 10¾ resistor and 1µF ceramic capacitor.
19	PWRGD	**Power Good Flag.** An open collector output that will be logic LOW if the output voltage is less than 350mV less than the nominal output voltage setpoint. Power Good is prevented from going low until the output voltage is out of spec for 500µsec.

Pin Number	Pin Name	Pin Function Description
20	ILIM	**Current Limit.** A resistor from this pin to ground sets the over current trip level.
21	DROOP/E*	**Droop Control/Energy Star Mode Control.** A resistor from this pin to ground sets the amount of droop by controlling the gain of the current sense amplifier. When this pin is pulled high to BYPASS, the phase A drivers are turned off for Energy-star operation.
22	ENABLE/SS	**Output Enable/Softstart.** A logic LOW on this pin will disable the output. An 10μA internal current source allows for open collector control. This pin also doubles as soft start.
23	RT	**Frequency Set.** A resistor from this pin to ground sets the switching frequency.
24	VFB	**Voltage Feedback.** Connect to the desired regulation point at the output of the converter.

Absolute Maximum Ratings

(Absolute Maximum Ratings are the values beyond which the device may be damaged or have it's useful life impaired. Functional operation under these conditions is not implied.)

Parameter	Min.	Max.	Unit
Supply Voltage VCC		15	V
Supply Voltages BOOT to PGND		24	V
BOOT to SW		24	V
Voltage Identification Code Inputs, VID0-VID4		6	V
VFB, ENABLE/SS, PWRGD, DROOP/E*		6	V
SWA, SWB to AGND (<1μs)	-3	15	V
PGNDA, PGNDB to AGND	-0.5	0.5	V
Gate Drive Current, peak pulse		3	A
Junction Temperature, T_J	-55	150	°C
Storage Temperature	-65	150	°C

Thermal Ratings

Parameter	Min.	Typ.	Max.	Unit
Lead Soldering Temperature, 10 seconds			300	°C
Power Dissipation, P_D			650	mW
Thermal Resistance Junction-to-Case, Θ_{JC}		16		°C/W
Thremal Resistance Junction-to-Ambient, Θ_{JA}		84		°C/W

Recommended Operating Conditions (See Figure 2)

Parameter	Conditions	Min.	Max.	Units
Output Driver Supply, BOOTA, B		16	22	V
Ambient Operating Temperature		0	70	°C
Supply Voltage V_{CC}		10.8	13.2	V

Electrical Specifications

(V_{CC} = 12V, VID = [01111] = 1.475V, and T_A = +25°C using circuit in Figure 2, unless otherwise noted.)
The • denotes specifications which apply over the full operating temperature range.

Parameter	Conditions		Min.	Typ.	Max.	Units
Input Supply						
UVLO Hysteresis				0.5		V
12V UVLO	Rising Edge	•	8.5	9.5	10.3	V
12V Supply Current	PWM Output Open			20		mA
Internal Voltage Regulator						
BYPASS Voltage			4.75	5	5.25	V
BYPASS Capacitor			100			nF
VREF and DAC						
Output Voltage	See Table 1	•	1.100		1.850	V
Initial Voltage Setpoint[1]	I_{LOAD} = 0A, VID = [01111]		1.460	1.475	1.490	V
Output Temperature Drift	T_A = 0 to 70°C			5		mV
Line Regulation	V_{CC} = 11.4V to 12.6V	•		130		μV
Droop[2]	I_{LOAD} = 69A, R_{DROOP} = 13.3k¾			56		mV
Programmable Droop Range			0		1.25	m¾
Response Time	ΔV_{out} = 10mV			100		nsec
Current Mismatch	$R_{DS,on}$ (A) = $R_{DS,on}$ (B), I_{LOAD} = 69A, Droop = 1m¾				5	%
VID Inputs						
Input LOW current, VID pins	V_{VID} = 0.4V		-60			μA
VID V_{IH}			2.0			V
VID V_{IL}					0.8	V
Oscillator						
Oscillator Frequency	RT = 54.9k¾	•	440	500	560	kHz
Oscillator Range	RT = 137.5k¾ to 13.75 k¾		200		2000	kHz
Maximum Duty Cycle	RT = 137.5k¾			90		%
Minimum LDRV on-time	RT = 13.75k¾			330		nsec
Gate Drive						
Gate Drive On-Resistance	Sink & Source			1.0		¾
Output Driver Rise & Fall Time	See Figure 1, C_L = 3000pF			20		nsec
Enable/Soft Start						
Soft Start Current				10		μA
Enable Threshold	ON OFF		1.0		 0.4	V
Power Good						
PWRGD Threshold	Logic LOW, $V_{VID} - V_{PWRGD}$	•	85	88	92	%V_{OUT}
PWRGD Output Voltage	I_{sink} = 4mA				0.4	V
PWRGD Delay	High → Low			500		μsec
OVP and OTP						
Output Overvoltage Detect		•	2.1	2.2	2.3	V
Over Temperature Shutdown			130	140	150	°C
Over Temperature Hysteresis				40		°C

Notes:

1. As measured at the converter's VFB sense point. For motherboard applications, the PCB layout should exhibit no more than 0.5mΩ trace resistance between the converter's output capacitors and the CPU. Remote sensing should be used for optimal performance.
2. Using the VFB pin for remote sensing of the converter's output at the load, the converter will be in compliance with VRM 9.x specification.

Gate Drive Test Circuit

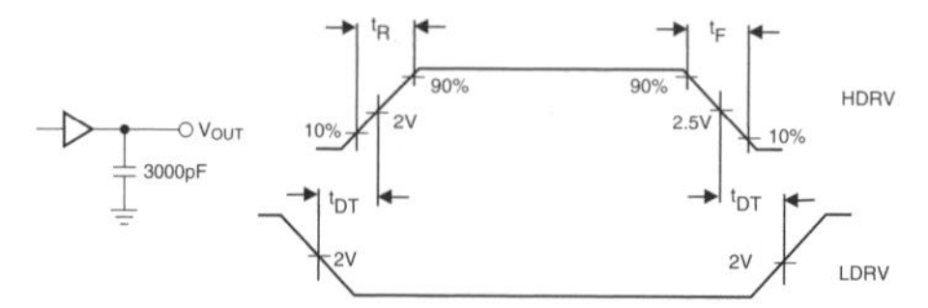

Figure 1. Output Drive Timing Diagram

Table 1. Output Voltage Programming Codes

VID4	VID3	VID2	VID1	VID0	V_{OUT} to CPU
1	1	1	1	1	OFF
1	1	1	1	0	1.100V
1	1	1	0	1	1.125V
1	1	1	0	0	1.150V
1	1	0	1	1	1.175V
1	1	0	1	0	1.200V
1	1	0	0	1	1.225V
1	1	0	0	0	1.250V
1	0	1	1	1	1.275V
1	0	1	1	0	1.300V
1	0	1	0	1	1.325V
1	0	1	0	0	1.350V
1	0	0	1	1	1.375V
1	0	0	1	0	1.400V
1	0	0	0	1	1.425V
1	0	0	0	0	1.450V
0	1	1	1	1	1.475V
0	1	1	1	0	1.500V
0	1	1	0	1	1.525V
0	1	1	0	0	1.550V
0	1	0	1	1	1.575V
0	1	0	1	0	1.600V
0	1	0	0	1	1.625V
0	1	0	0	0	1.650V
0	0	1	1	1	1.675V
0	0	1	1	0	1.700V
0	0	1	0	1	1.725V
0	0	1	0	0	1.750V
0	0	0	1	1	1.775V
0	0	0	1	0	1.800V
0	0	0	0	1	1.825V
0	0	0	0	0	1.850V

Note:
1. 0 = VID pin is tied to GND.
   1 = VID pin is pulled up to 5V.

Typical Operating Characteristics

(V_{CC} = 12V, V_{OUT} = 1.475V, and T_A = +25°C using circuit in Figure 2, unless otherwise noted.)

EFFICIENCY VS. OUTPUT CURRENT

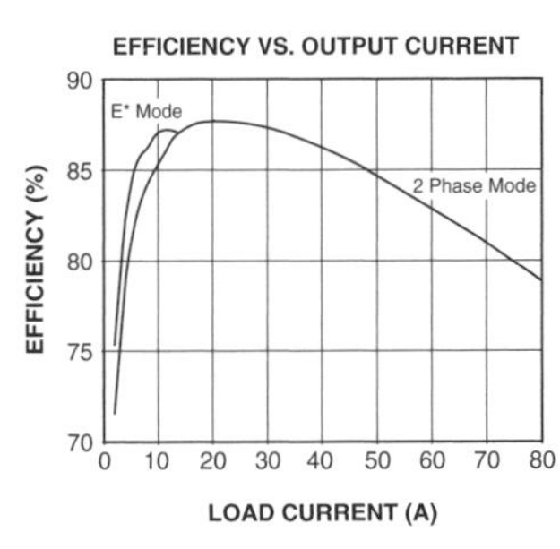

ADAPTIVE GATE DELAY

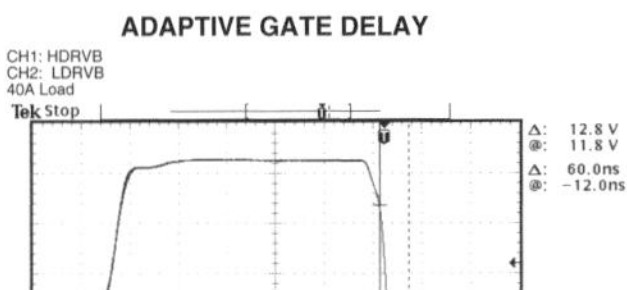

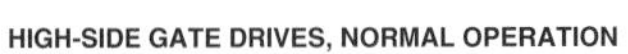

HIGH-SIDE GATE DRIVES, NORMAL OPERATION

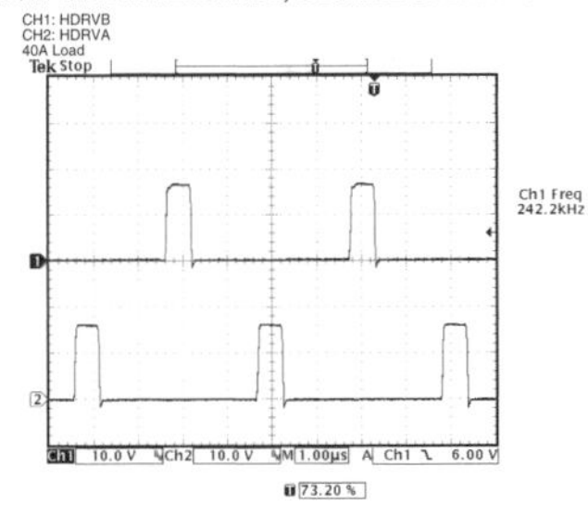

HIGH-SIDE GATE DRIVES, E*-MODE

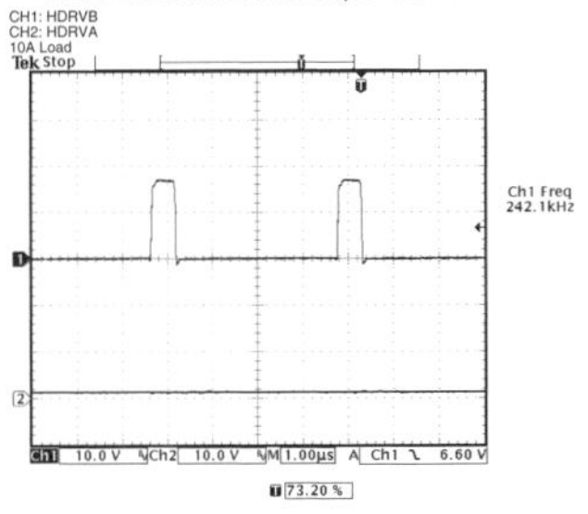

HIGH-SIDE GATE DRIVES, RISE / FALL TIME

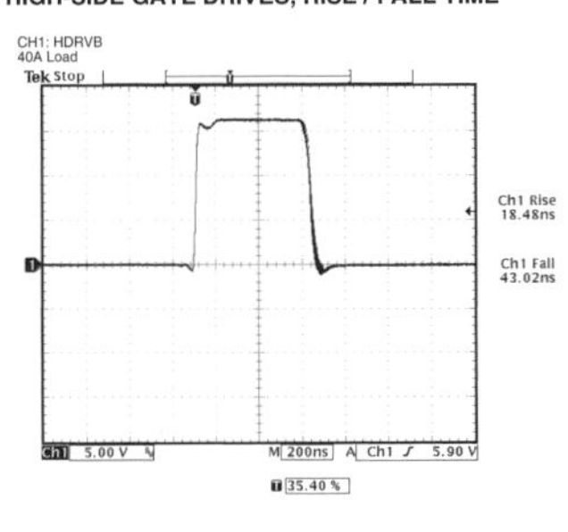

LOW-SIDE GATE DRIVES, RISE / FALL TIME

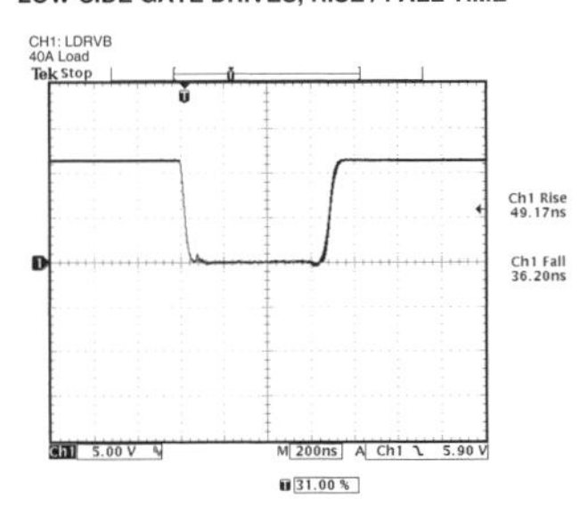

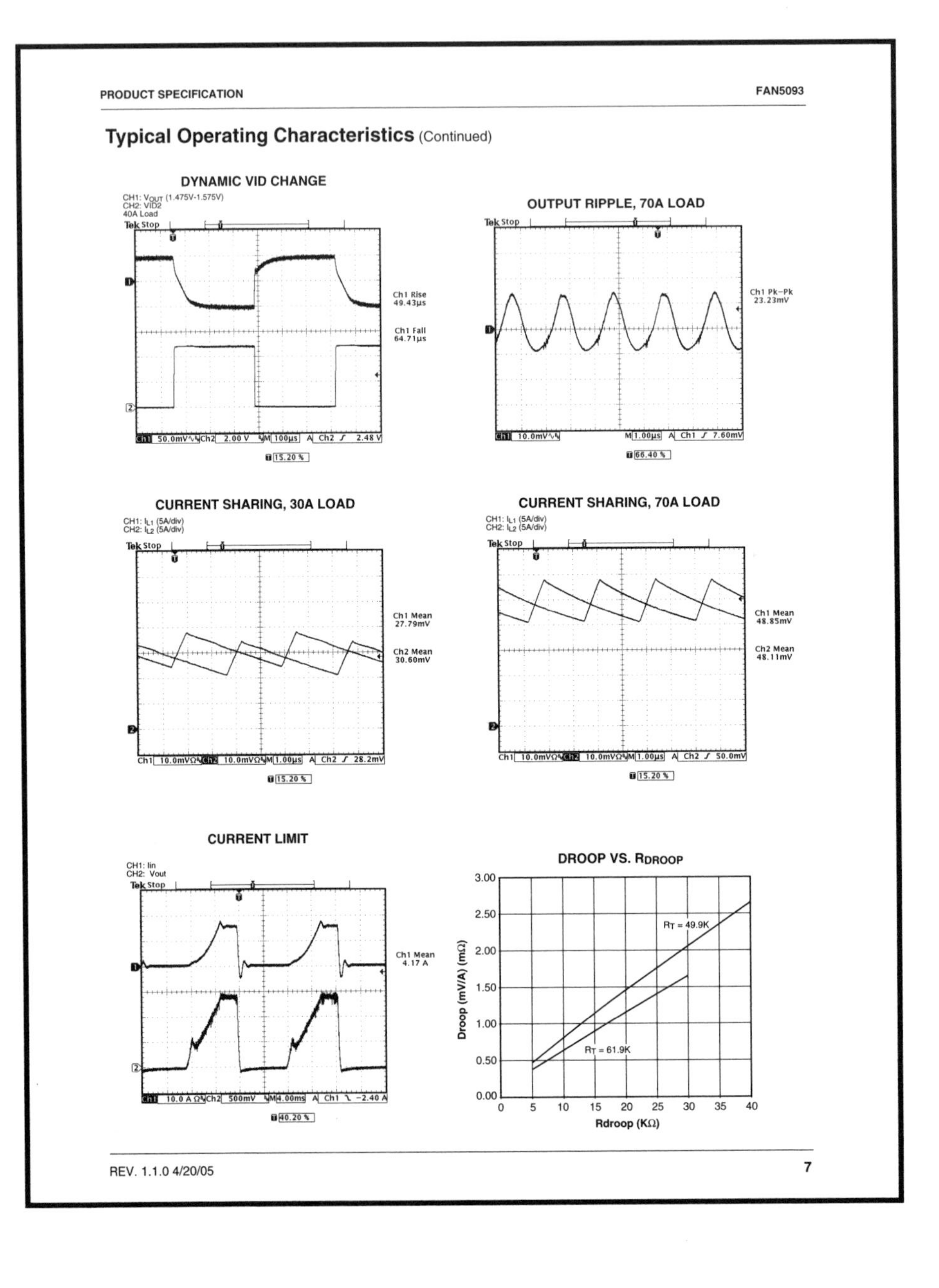
PRODUCT SPECIFICATION
FAN5093
Typical Operating Characteristics (Continued)
DYNAMIC VID CHANGE
CH1: VOUT (1.475V-1.575V)
CH2: VID2
40A Load
Ch1 Rise 49.43µs
Ch1 Fall 64.71µs
OUTPUT RIPPLE, 70A LOAD
Ch1 Pk-Pk 23.23mV
CURRENT SHARING, 30A LOAD
CH1: IL1 (5A/div)
CH2: IL2 (5A/div)
Ch1 Mean 27.79mV
Ch2 Mean 30.60mV
CURRENT SHARING, 70A LOAD
CH1: IL1 (5A/div)
CH2: IL2 (5A/div)
Ch1 Mean 48.85mV
Ch2 Mean 48.11mV
CURRENT LIMIT
CH1: Iin
CH2: Vout
Ch1 Mean 4.17 A
DROOP VS. RDROOP
Droop (mV/A) (mΩ)
Rdroop (KΩ)
RT = 49.9K
RT = 61.9K
REV. 1.1.0 4/20/05
7

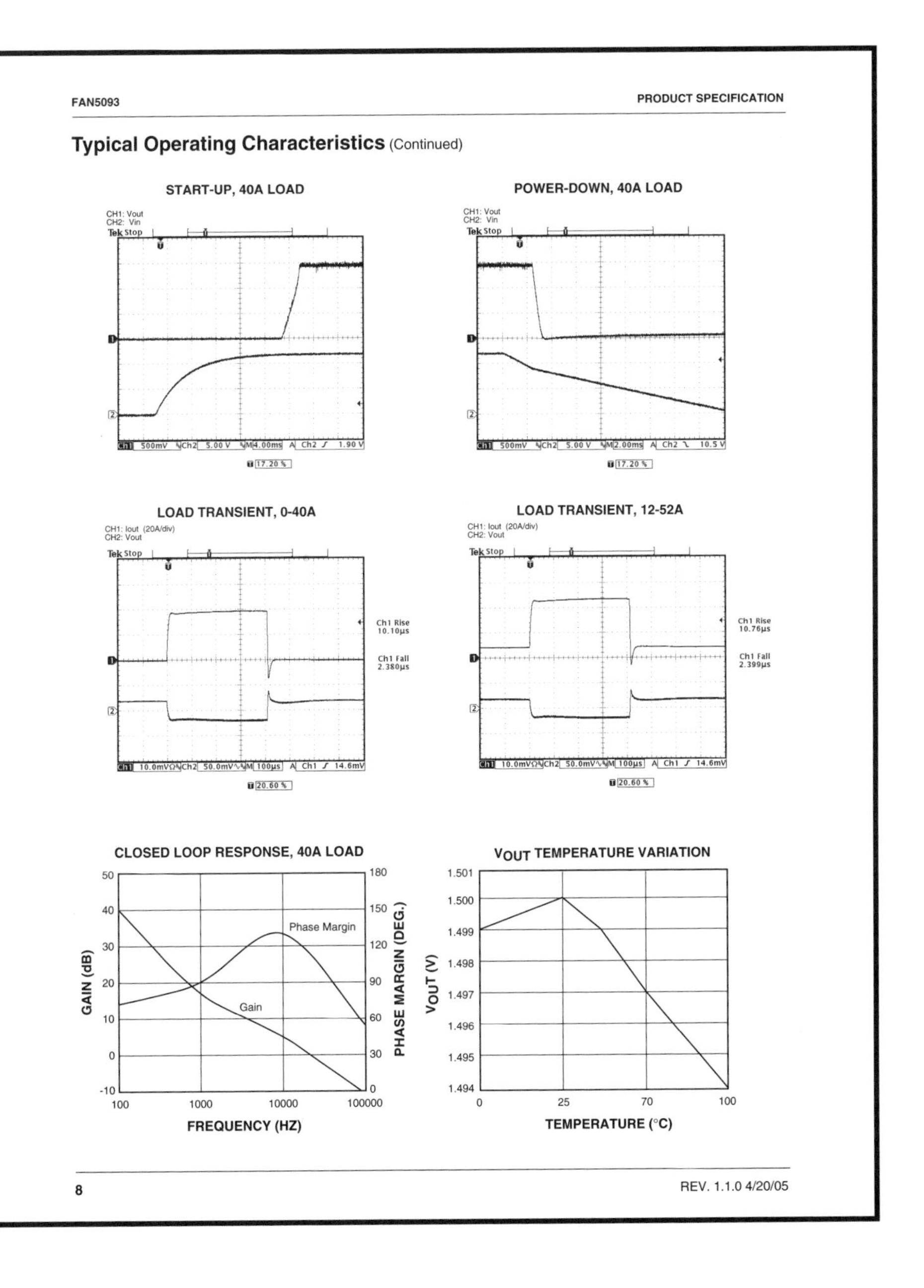

FAN5093 PRODUCT SPECIFICATION

Typical Operating Characteristics (Continued)

8 REV. 1.1.0 4/20/05

Application Circuit

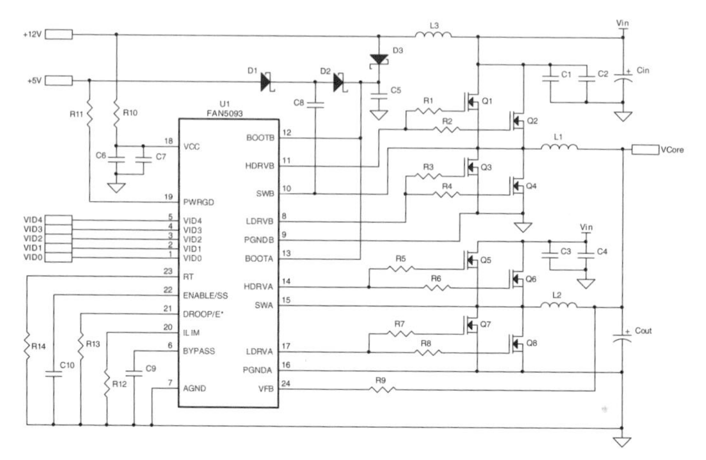

Figure 2. Application Circuit for 70A VRM 9.x Desktop Application

Table 2. FAN5093 Application Bill of Materials for Figure 2

Reference	QTY	Description	Manufacturer / Number
U1	1	IC, PWM, FAN5093	Fairchild FAN5093
Q1-8	8	NFET, 30V, 50A, 9m¾	Fairchild FDD6696
D1, 2, 3	3	DIOS, 40V, 500mA	Fairchild MBR0540
L1, 2	2	IND, 850nH, 30A, 0.9m¾	Inter-Technical SCTA5022A-R85M
L3	opt	IND, 750nH, 20A, 3.5m¾	Inter-Technical SC4015-R75M
R1-4, 9	5	4.7¾, 5%	
R5-8	4	2.2¾, 5%	
R10	1	10¾, 5%	
R11	1	10K, 5%	
R12	1	75.0K, 1%	
R13	1	13.3K, 1%	
R14	1	56.2K, 1%	
C1-6	6	1.0µf, 25V, 10%, X7R	
C7-10	4	0.1µf, 16V, 10%, X7R	
Cin	4	1500µf, 16V, 20%,12m¾, Aluminum Electrolytic	Rubycon 16MBZ1500M
Cout	8	2200µf, 6.3V, 20%, 12m¾, Aluminum Electrolytic	Rubycon 6.3MBZ2200M

Application Information

Operation

The FAN5093 Controller

The FAN5093 is a programmable synchronous two-phase DC-DC controller IC. When designed with the appropriate external components, the FAN5093 can be configured to deliver more than 50A of output current, for VRM 9.x applications. The FAN5093 functions as a fixed frequency PWM step down regulator, with a high efficiency mode (E*) at light load.

Main Control Loop

Refer to the FAN5093 Block Diagram on page 1. The FAN5093 consists of two interleaved synchronous buck converters, implemented with summing-mode control. Each phase has its own current feedback, and there is a common voltage feedback.

The two buck converters controlled by the FAN5093 are interleaved, that is, they run 180° out of phase. This minimizes the RMS input ripple current, minimizing the number of input capacitors required. It also doubles the effective switching frequency, improving transient response.

The FAN5093 implements "summing mode control", which is different from both classical voltage-mode and current-mode control. It provides superior performance to either by allowing a large converter bandwidth over a wide range of output loads and external components. No external compensation is required.

The control loop of the regulator contains two main sections: the analog control block and the digital control block. The analog section consists of signal conditioning amplifiers feeding into a comparator which provides the input to the digital control block. The signal conditioning section accepts inputs from a current sensor and a voltage sensor, with the voltage sensor being common to both phases, and the current sensor separate for each. The voltage sensor amplifies the difference between the VFB signal and the reference voltage from the DAC and presents the output to each of the two comparators. The current control path for each phase takes the difference between its PGND and SW pins when the low-side MOSFET is on, reproducing the voltage across the MOSFET and thus the input current; it presents the resulting signal to the same input of its summing amplifier, adding its signal to the voltage amplifier's with a certain gain. These two signals are thus summed together. This sum is then presented to a comparator looking at the oscillator ramp, which provides the main PWM control signal to the digital control block. The oscillator ramps are 180° out of phase with each other, so that the two phases are on alternately.

The digital control block takes the analog comparator input to provide the appropriate pulses to the HDRV and LDRV output pins for each phase. These outputs control the external power MOSFETs.

Response Time

The FAN5093 utilizes leading-edge, not trailing-edge control. Conventional trailing-edge control turns on the high-side MOSFET at a clock signal, and then turns it off when the error amplifier output voltage is equal to the ramp voltage. As a result, the response time of a trailing-edge converter can be as long as the off-time of the high-side driver, nearly an entire switching period. The FAN5093's leading-edge control turns the high-side MOSFET on when the error amplifier output voltage is equal to the ramp voltage, and turns it off at the clock signal. As a result, when a transient occurs, the FAN5093 responds immediately by turning on the high-side MOSFET. Response time is set by the internal propagation delays, typically 100nsec. In worst case, the response time is set by the minimum on-time of the low-side MOSFET, 330nsec.

Oscillator

The FAN5093 oscillator section runs at a frequency determined by a resistor from the RT pin to ground according to the formula

$$R_T(\Omega) = \frac{27.5E9}{f(Hz)}$$

The oscillator generates two internal sawtooth ramps, each at one-half the oscillator frequency, and running 180° out of phase with each other. These ramps cause the turn-on time of the two phases to be phased apart. The oscillator frequency of the FAN5093 can be programmed from 200KHz to 2MHz with each phase running at 100KHz to 1MHz, respectively. Selection of a frequency will depend on various system performance criteria, with higher frequency resulting in smaller components but typically lower efficiency.

Remote Voltage Sense

The FAN5093 has true remote voltage sense capability, eliminating errors due to trace resistance. To utilize remote sense, the VFB and AGND pins should be connected as a Kelvin trace pair to the point of regulation, such as the processor pins. The converter will maintain the voltage in regulation at that point. Care is required in layout of these grounds; see the layout guidelines in this datasheet.

High Current Output Drivers

The FAN5093 contains four high current output drivers that utilize MOSFETs in a push-pull configuration. The drivers for the high-side MOSFETs use the BOOT pin for input power and the SW pin for return. The drivers for the low-side MOSFETs use the VCC pin for input power and the PGND pin for return. Typically, the BOOT pin will use a charge pump as shown in Figure 2. Note that the BOOT and VCC pins are separated from the chip's internal power and ground, BYPASS and AGND, for switching noise immunity.

Adaptive Delay Gate Drive

The FAN5093 embodies an advanced design that ensures minimum MOSFET transition times while eliminating shoot-through current. It senses the state of the MOSFETs and adjusts the gate drive adaptively to ensure that they are never on simultaneously. When the high-side MOSFET turns off, the voltage on its source begins to fall. When the voltage there reaches approximately 2.5V, the low-side MOSFETs gate drive is applied. When the low-side MOSFET turns off, the voltage at the LDRV pin is sensed. When it drops below approximately 2V, the high-side MOSFET's gate drive is applied.

Maximum Duty Cycle

In order to ensure that the current-sensing and charge-pumping work, the FAN5093 guarantees that the low-side MOSFET will be on a certain portion of each period. For low frequencies, this occurs as a maximum duty cycle of approximately 90%. Thus at 250KHz, with a period of 4μsec, the low-side will be on at least 4μsec • 10% = 400nsec. At higher frequencies, this time might fall so low as to be ineffective. The FAN5093 guarantees a minimum low-side on-time of approximately 330nsec, regardless of duty cycle.

Current Sensing

The FAN5093 has two independent current sensors, one for each phase. Current sensing is accomplished by measuring the source-to-drain voltage of the low-side MOSFET during its on-time. Each phase has its own power ground pin, to permit the phases to be placed in different locations without affecting measurement accuracy. For best results, it is important to connect the PGND and SW pins for each phase as a Kelvin trace pair directly to the source and drain, respectively, of the appropriate low-side MOSFET. Care is required in the layout of these grounds; see the layout guidelines in this datasheet.

Current Sharing

The two independent current sensors of the FAN5093 operate with their independent current control loops to guarantee that the two phases each deliver half of the total output current. The only mismatch between the two phases occurs if there is a mismatch between the $R_{DS,on}$ of the low-side MOSFETs.

Light Load Efficiency

At light load, the FAN5093 uses a number of techniques to improve efficiency. Because a synchronous buck converter is two quadrant, able to both source and sink current, during light load the inductor current will flow away from the output and towards the input during a portion of the switching cycle. This reverse current flow is detected by the FAN5093 as a positive voltage appearing on the low-side MOSFET during its on-time. When reverse current flow is detected, the low-side MOSFET is turned off for the rest of the cycle, and the current instead flows through the body diode of the high-side MOSFET, returning the power to the source. This technique substantially enhances light load efficiency.

Short Circuit Current Characteristics (ILIM Pin)

The FAN5093 short circuit current characteristic includes a function that protects the DC-DC converter from damage in the event of a short circuit. The short circuit limit is set with the R_S resistor, as given by the formula

$$R_S(\Omega) = I_{SC} \bullet R_{DS,on} \bullet R_T \bullet 3.33$$

with I_{SC} the desired output current limit, RT the oscillator resistor and $R_{DS,on}$ one phase's low-side MOSFET's on resistance. Remember to make the R_S large enough to include the effects of initial tolerance and temperature variation on the MOSFETs' $R_{DS,on}$.

Important Note! The oscillator frequency must be selected before selecting the current limit resistor, because the value of RT is used in the calculation of R_S.

When an overcurrent is detected, the high-side MOSFETs are turned off, and the low-side MOSFETs are turned on, and they remain in this state until the measured current through the low-side MOSFET has returned to zero amps. After reaching zero, the FAN5093 re-soft-starts, ensuring that it can also safely turn on into a short.

A limitation on the current sense circuit is that $I_{SC} \bullet R_{DS,on}$ must be less that 375mV. To ensure correct operation, use $I_{SC} \bullet R_{DS,on}$ ð 300mV; between 300mV and 375mV, there will be some non-linearity in the short-circuit current not accounted for in the equation.

As an example, consider the typical characteristic of the DC-DC converter circuit with two FDP6670AL low-side MOSFETs (R_{DS} = 6.5mΩ maximum at 25°C • 1.2 at 75°C = 7.8mΩ each, or 3.9mΩ total) in each phase, RT = 42.1K¾ (600KHz oscillator) and a 50K¾ R_S.

The converter exhibits a normal load regulation characteristic until the voltage across the MOSFETs exceeds the internal short circuit threshold of 50K¾/(3.9m¾ • 41.2K¾ • 6.66) = 47A. [Note that this current limit level can be as high as 50K¾/(3.5m¾ • 41.2K¾ • 6.66) = 52A, if the MOSFETs have typical $R_{DS,on}$ rather than maximum, and are at 25°C.] At this point, the internal comparator trips and signals the controller to leave on the low-side MOSFETs and keep off the high-side MOSFETs. The inductor current decreases, and power is not applied again until the inductor current reaches 0A and the converter attempts to re-softstart.

E*-mode

In addition, further enhancement in efficiency can be obtained by putting the FAN5093 into E*-mode. When the Droop pin is pulled to the 5V BYPASS voltage, the "A" phase of the FAN5093 is completly turned off, reducing in half the amount of gate charge power being consumed. E*-mode can be implemented with the circuit shown in Figure 3.

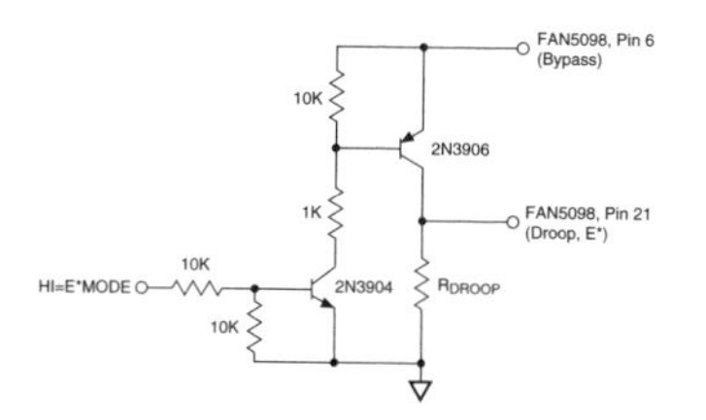

Figure 3. Implementing E*-mode Control

Note: The charge pump for the HIDRVs should be based on the "B" phase of the FAN5093, since the "A" phase is off in E*-mode.

Internal Voltage Reference

The reference included in the FAN5093 is a precision band-gap voltage reference. Its internal resistors are precisely trimmed to provide a near zero temperature coefficient (TC). Based on the reference is the output from an integrated 5-bit DAC. The DAC monitors the 5 voltage identification pins, VID0-4, and scales the reference voltage from 1.100V to 1.850V in 25mV steps.

BYPASS Reference

The internal logic of the FAN5093 runs on 5V. To permit the IC to run with 12V only, it produces 5V internally with a linear regulator, whose output is present on the BYPASS pin. This pin should be bypassed with a 100nF capacitor for noise suppression. The BYPASS pin should not have any external load attached to it.

Dynamic Voltage Adjustment

The FAN5093 can have its output voltage dynamically adjusted to accommodate low power modes. The designer must ensure that the transitions on the VID lines all occur simultaneously (within less than 500nsec) to avoid false codes generating undesired output voltages. The Power Good flag tracks the VID codes, but has a 500μsec delay transitioning from high to low; this is long enough to ensure that there will not be any glitches during dynamic voltage adjustment.

Power Good (PWRGD)

The FAN5093 Power Good function is designed in accordance with the Pentium IV DC-DC converter specifications and provides a continuous voltage monitor on the VFB pin. The circuit compares the VFB signal to the VREF voltage and outputs an active-low interrupt signal to the CPU should the power supply voltage deviate more than -12% of its nominal setpoint. The Power Good flag provides no control functions to the FAN5093.

Output Enable/Soft Start (ENABLE/SS)

The FAN5093 will accept an open collector/TTL signal for controlling the output voltage. The low state disables the output voltage. When disabled, the PWRGD output is in the low state.

Even if an enable is not required in the circuit, this pin should have attached a capacitor (typically 100nF) to soft-start the switching. A softstart capacitor may be approximately chosen by the formula:

$$t_D = \frac{C_{SS}}{10\mu A} \bullet \frac{(1.7 + 0.9074 \bullet V_{OUT})}{2.5}$$

where: t_D is the delay time before the output starts to ramp

$$t_R = \frac{C_{SS}}{10\mu A} \bullet \frac{V_{OUT} \bullet 0.9}{V_{IN}}$$

t_R is the ramp time of the output
C_{SS} = softstart cap
V_{OUT} = nominal output voltage

However, C must be Š 100nF.

Programmable Active Droop™

The FAN5093 features Programmable Active Droop™: as the output current increases, the output voltage drops proportionately an amount that can be programmed with an external resistor. This feature is offered in order to allow maximum headroom for transient response of the converter. The current is sensed losslessly by measuring the voltage across the low-side MOSFET during its on time. Consult the section on current sensing for details. The droop is adjusted by the droop resistor changing the gain of the current loop. Note that this method makes the droop dependent on the temperature and initial tolerance of the MOSFET, and the droop must be calculated taking account of these tolerances. Given a maximum output current, the amount of droop can be programmed with a resistor to ground on the droop pin, according to the formula

$$R_{Droop}(\Omega) = \frac{V_{Droop} \bullet R_T}{I_{max} \bullet R_{DS,on}}$$

with V_{Droop} the desired droop voltage, RT the oscillator resistor, I_{max} the output current at which the droop is desired, and $R_{DS,on}$ the on-state resistance of one phase's low-side MOSFET.

Important Note! The oscillator frequency must be selected before selecting the droop resistor, because the value of R_T is used in the calculation of R_{Droop}.

Over-Voltage Protection

The FAN5093 constantly monitors the output voltage for protection against over-voltage conditions. If the voltage at the VFB pin exceeds 2.2V, an over-voltage condition is assumed and the FAN5093 latches on the external low-side MOSFET and latches off the high-side MOSFET. The DC-DC converter returns to normal operation only after V_{CC} has been recycled.

Over Temperature Protection

If the FAN5093 die temperature exceeds approximately 150°C, the IC shuts itself off. It remains off until the temperature has dropped approximately 25°C, at which time it resumes normal operation.

Component Selection

MOSFET Selection

This application requires N-channel Enhancement Mode Field Effect Transistors. Desired characteristics are as follows:

- Low Drain-Source On-Resistance,
- $R_{DS,ON} < 10m\Omega$ (lower is better);
- Power package with low Thermal Resistance;
- Drain-Source voltage rating > 15V;
- Low gate charge, especially for higher frequency operation.

For the low-side MOSFET, the on-resistance ($R_{DS,ON}$) is the primary parameter for selection. Because of the small duty cycle of the high-side, the on-resistance determines the power dissipation in the low-side MOSFET and therefore significantly affects the efficiency of the DC-DC converter. For high current applications, it may be necessary to use two MOSFETs in parallel for the low-side for each phase.

For the high-side MOSFET, the gate charge is as important as the on-resistance, especially with a 12V input and with higher switching frequencies. This is because the speed of the transition greatly affects the power dissipation. It may be a good trade-off to select a MOSFET with a somewhat higher $R_{DS,on}$, if by so doing a much smaller gate charge is available. For high current applications, it may be necessary to use two MOSFETs in parallel for the high-side for each phase.

At the FAN5093's highest operating frequencies, it may be necessary to limit the total gate charge of both the high-side and low-side MOSFETs together, to avert excess power dissipation in the IC.

For details and a spreadsheet on MOSFET selection, refer to Applications Bulletin AB-8.

Gate Resistors

Use of a gate resistor on every MOSFET is mandatory. The gate resistor prevents high-frequency oscillations caused by the trace inductance ringing with the MOSFET gate capacitance. The gate resistors should be located physically as close to the MOSFET gate as possible.

The gate resistor also limits the power dissipation inside the IC, which could otherwise be a limiting factor on the switching frequency. It may thus carry significant power, especially at higher frequencies. As an example: The FDB7045L has a maximum gate charge of 70nC at 5V, and an input capacitance of 5.4nF. The total energy used in powering the gate during one cycle is the energy needed to get it up to 5V, plus the energy to get it up to 12V:

$$E = QV + \frac{1}{2}C \bullet \Delta V^2 = 70nC \bullet 5V + \frac{1}{2}5.4nF \bullet (12V - 5V)^2$$
$$= 482nJ$$

This power is dissipated every cycle, and is divided between the internal resistance of the FAN5093 gate driver and the gate resistor. Thus,

$$P_{Rgate} = \frac{E \bullet f \bullet R_{gate}}{(R_{gate} + R_{internal})} = 482nJ \bullet 300KHz \bullet \frac{4.7\Omega}{4.7\Omega + 0.5\Omega} = 131mW$$

and each gate resistor thus requires a 1/4W resistor to ensure worst case power dissipation.

Inductor Selection

Choosing the value of the inductor is a tradeoff between allowable ripple voltage and required transient response. A smaller inductor produces greater ripple while producing better transient response. In any case, the minimum inductance is determined by the allowable ripple. The first order equation (close approximation) for minimum inductance for a two-phase converter is:

$$L_{min} = \frac{V_{in} - 2 \bullet V_{out}}{f} \bullet \frac{V_{out}}{V_{in}} \bullet \frac{ESR}{V_{ripple}}$$

where:
Vin = Input Power Supply
Vout = Output Voltage
f = DC/DC converter switching frequency
ESR = Equivalent series resistance of all output capacitors in parallel
Vripple = Maximum peak to peak output ripple voltage budget.

Schottky Diode Selection

The application circuit of Figure 2 shows a Schottky diode, D1 (D2 respectively), one in each phase. They are used as free-wheeling diodes to ensure that the body-diodes in the low-side MOSFETs do not conduct when the upper MOSFET is turning off and the lower MOSFETs are turning on. It is undesirable for this diode to conduct because its high forward voltage drop and long reverse recovery time degrades efficiency, and so the Schottky provides a shunt path for the current. Since this time duration is extremely short, being minimized by the adaptive gate delay, the selection criterion for the diode is that the forward voltage of the Schottky at the output current should be less than the forward voltage of the MOSFET's body diode. Power capability is not a criterion for this device, as its dissipation is very small.

Output Filter Capacitors

The output bulk capacitors of a converter help determine its output ripple voltage and its transient response. It has already been seen in the section on selecting an inductor that the ESR helps set the minimum inductance. For most converters, the number of capacitors required is determined by the transient response and the output ripple voltage, and these are determined by the ESR and not the capacitance value. That is, in order to achieve the necessary ESR to meet the transient and ripple requirements, the capacitance value required is already very large.

The most commonly used choice for output bulk capacitors is aluminum electrolytics, because of their low cost and low ESR. The only type of aluminum capacitor used should be those that have an ESR rated at 100kHz. Consult Application Bulletin AB-14 for detailed information on output capacitor selection.

For higher frequency applications, particularly those running the FAN5093 oscillator at >1MHz, Oscon or ceramic capacitors may be considered. They have much smaller ESR than comparable electrolytics, but also much smaller capacitance.

The output capacitance should also include a number of small value ceramic capacitors placed as close as possible to the processor; 0.1μF and 0.01μF are recommended values.

Input Filter

The DC-DC converter design may include an input inductor between the system main supply and the converter input as shown in Figure 2. This inductor serves to isolate the main supply from the noise in the switching portion of the DC-DC converter, and to limit the inrush current into the input capacitors during power up. A value of 1.3μH is recommended.

It is necessary to have some low ESR capacitors at the input to the converter. These capacitors deliver current when the high side MOSFET switches on. Because of the interleaving, the number of such capacitors required is greatly reduced from that required for a single-phase buck converter. Figure 2 shows 3 x 1500μF, but the exact number required will vary with the output voltage and current, according to the formula

$$I_{rms} = \frac{I_{out}}{2}\sqrt{2DC - 4DC^2}$$

for the two phase FAN5093, where DC is the duty cycle, DC = Vout / Vin. Capacitor ripple current rating is a function of temperature, and so the manufacturer should be contacted to find out the ripple current rating at the expected operational temperature. For details on the design of an input filter, refer to Applications Bulletin AB-16.

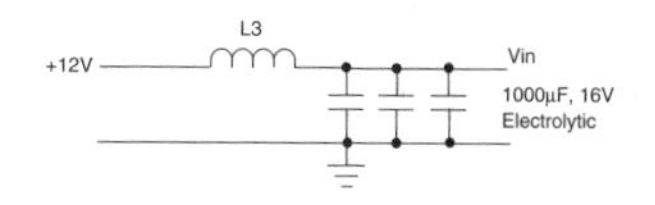

Figure 4. Input Filter

Design Considerations and Component Selection

Additional information on design and component selection may be found in Fairchild's Application Note 59.

PCB Layout Guidelines

- Placement of the MOSFETs relative to the FAN5093 is critical. Place the MOSFETs such that the trace length of the HIDRV and LODRV pins of the FAN5093 to the FET gates is minimized. A long lead length on these pins will cause high amounts of ringing due to the inductance of the trace and the gate capacitance of the FET. This noise radiates throughout the board, and, because it is switching at such a high voltage and frequency, it is very difficult to suppress.
- In general, all of the noisy switching lines should be kept away from the quiet analog section of the FAN5093. That is, traces that connect to pins 8-17 (LODRV, HIDRV, PGND and BOOT) should be kept far away from the traces that connect to pins 1 through 7, and pins 18-24.
- Place the 0.1μF decoupling capacitors as close to the FAN5093 pins as possible. Extra lead length on these reduces their ability to suppress noise.
- Each power and ground pin should have its own via to the appropriate plane. This helps provide isolation between pins.
- Place the MOSFETs, inductor, and Schottky of a given phase as close together as possible for the same reasons as in the first bullet above. Place the input bulk capacitors as close to the drains of the high side MOSFETs as possible. In addition, placement of a 0.1μF decoupling cap right on the drain of each high side MOSFET helps to suppress some of the high frequency switching noise on the input of the DC-DC converter.
- Place the output bulk capacitors as close to the CPU as possible to optimize their ability to supply instantaneous current to the load in the event of a current transient. Additional space between the output capacitors and the CPU will allow the parasitic resistance of the board traces to degrade the DC-DC converter's performance under severe load transient conditions, causing higher voltage deviation. For more detailed information regarding capacitor placement, refer to Application Bulletin AB-5.
- A PC Board Layout Checklist is available from Fairchild Applications. Ask for Application Bulletin AB-11.

PC Motherboard Sample Layout and Gerber File

A reference design for motherboard implementation of the FAN5093 along with the PCAD layout Gerber file and silk screen can be obtained through your local Fairchild representative.

FAN5093 Evaluation Board

Fairchild provides an evaluation board to verify the system level performance of the FAN5093. It serves as a guide to performance expectations when using the supplied external components and PCB layout. Please contact your local Fairchild representative for an evaluation board.

Additional Information

For additional information contact your local Fairchild representative.

Mechanical Dimensions – 24 Lead TSSOP

Symbol	Inches		Millimeters		Notes
	Min.	Max.	Min.	Max.	
A	—	.047	—	1.20	
A1	.002	.006	0.05	0.15	
B	.007	.012	0.19	0.30	
C	.004	.008	0.09	0.20	
D	.303	.316	7.70	7.90	2
E	.169	.177	4.30	4.50	2
e	.026 BSC		0.65 BSC		
H	.252 BSC		6.40 BSC		
L	.018	.030	0.45	0.75	3
N	24		24		5
α	0°	8°	0°	8°	
ccc	—	.004	—	0.10	

Notes:

1. Dimensioning and tolerancing per ANSI Y14.5M-1982.
2. "D" and "E" do not include mold flash. Mold flash or protrusions shall not exceed .006 inch (0.15mm).
3. "L" is the length of terminal for soldering to a substrate.
4. Terminal numbers are shown for reference only.
5. Symbol "N" is the maximum number of terminals.

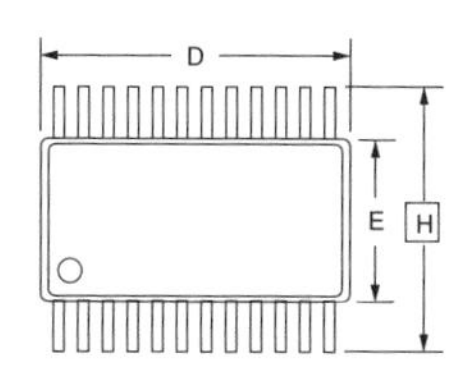

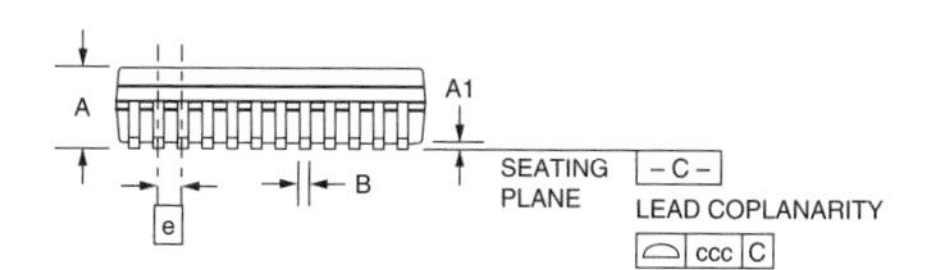

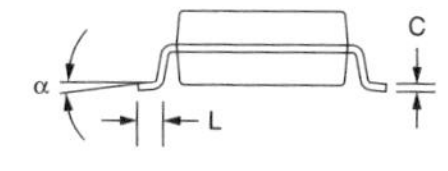

Ordering Information

Product Number	Description	Package
FAN5093MTC	VRM 9.x DC-DC Controller	24 pin TSSOP
FAN5093MTCX	VRM 9.x DC-DC Controller	24 pin TSSOP in Tape and Reel

4/20/05 0.0m 005
Stock#DS30005093

Appendix B

Fairchild Specifications for FAN4803

www.fairchildsemi.com

FAN4803

8-Pin PFC and PWM Controller Combo

Features

- Internally synchronized PFC and PWM in one 8-pin IC
- Patented one-pin voltage error amplifier with advanced input current shaping technique
- Peak or average current, continuous boost, leading edge PFC (Input Current Shaping Technology)
- High efficiency trailing-edge current mode PWM
- Low supply currents; start-up: 150µA typ., operating: 2mA typ.
- Synchronized leading and trailing edge modulation
- Reduces ripple current in the storage capacitor between the PFC and PWM sections
- Overvoltage, UVLO, and brownout protection
- PFC V_{CC}OVP with PFC Soft Start

General Description

The FAN4803 is a space-saving controller for power factor corrected, switched mode power supplies that offers very low start-up and operating currents.

Power Factor Correction (PFC) offers the use of smaller, lower cost bulk capacitors, reduces power line loading and stress on the switching FETs, and results in a power supply fully compliant to IEC1000-3-2 specifications. The FAN4803 includes circuits for the implementation of a leading edge, average current "boost" type PFC and a trailing edge, PWM.

The FAN4803-1's PFC and PWM operate at the same frequency, 67kHz. The PFC frequency of the FAN4803-2 is automatically set at half that of the 134kHz PWM. This higher frequency allows the user to design with smaller PWM components while maintaining the optimum operating frequency for the PFC. An overvoltage comparator shuts down the PFC section in the event of a sudden decrease in load. The PFC section also includes peak current limiting for enhanced system reliability.

Block Diagram

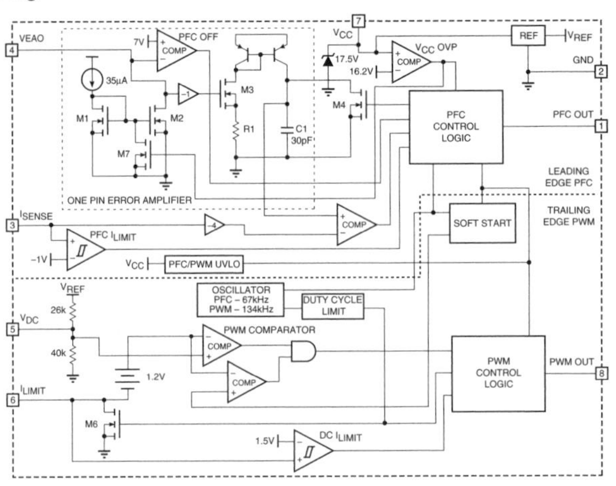

REV. 1.2.3 4/20/05

Pin Configuration

FAN4803
8-Pin PDIP (P08)
8-Pin SOIC (S08)

Pin	Name		Pin	Name
1	PFC OUT		8	PWM OUT
2	GND		7	V_{CC}
3	I_{SENSE}		6	I_{LIMIT}
4	VEAO		5	V_{DC}

TOP VIEW

Pin Description

Pin	Name	Function
1	PFC OUT	PFC driver output
2	GND	Ground
3	I_{SENSE}	Current sense input to the PFC current limit comparator
4	VEAO	PFC one-pin error amplifier input
5	V_{DC}	PWM voltage feedback input
6	I_{LIMIT}	PWM current limit comparator input
7	V_{CC}	Positive supply (may require an external shunt regulator)
8	PWM OUT	PWM driver output

Absolute Maximum Ratings

Absolute maximum ratings are those values beyond which the device could be permanently damaged. Absolute maximum ratings are stress ratings only and functional device operation is not implied.

Parameter	Min	Max	Unit
I_{CC} Current (average)		40	mA
V_{CC} MAX		18.3	V
I_{SENSE} Voltage	-5	1	V
Voltage on Any Other Pin	GND – 0.3	V_{CC} + 0.3	V
Peak PFC OUT Current, Source or Sink		1	A
Peak PWM OUT Current, Source or Sink		1	A
PFC OUT, PWM OUT Energy Per Cycle		1.5	μJ
Junction Temperature		150	°C
Storage Temperature Range	-65°	150	°C
Lead Temperature (Soldering, 10 sec)		260	°C
Thermal Resistance (θ_{JA})			
Plastic DIP		110	°C/W
Plastic SOIC		160	°C/W

Operating Conditions

Temperature Range	
FAN4803CS-X	0°C to 70°C
FAN4803CP-X	0°C to 70°C

Electrical Characteristics

Unless otherwise specified, V_{CC} = 15V, T_A = Operating Temperature Range (Note 1)

Symbol	Parameter	Conditions	Min	TYP	MAX	UNITS
One-pin Error Amplifier						
	V_{EAO} Output Current	T_A = 25°C, V_{EAO} = 6V	34.0	36.5	39.0	μA
	Line Regulation	10V < V_{CC} < 15V, V_{EAO} = 6V		0.1	0.3	μA
V_{CC} OVP Comparator						
	Threshold Voltage		15.5	16.3	16.8	V
PFC I_{LIMIT} Comparator						
	Threshold Voltage		-0.9	-1	-1.15	V
	Delay to Output			150	300	ns
DC I_{LIMIT} Comparator						
	Threshold Voltage		1.4	1.5	1.6	V
	Delay to Output			150	300	ns
Oscillator						
	Initial Accuracy	T_A = 25°C	60	67	74	kHz
	Voltage Stability	10V < V_{CC} < 15V		1		%
	Temperature Stability			2		%
	Total Variation	Over Line and Temp	60	67	74.5	kHz
	Dead Time	PFC Only	0.3	0.45	0.65	μs
PFC						
	Minimum Duty Cycle	V_{EAO} > 7.0V, I_{SENSE} = -0.2V			0	%
	Maximum Duty Cycle	V_{EAO} < 4.0V, I_{SENSE} = 0V	90	95		%
	Output Low Impedance			8	15	¾
	Output Low Voltage	I_{OUT} = –100mA		0.8	1.5	V
		I_{OUT} = –10mA, V_{CC} = 8V		0.7	1.5	V
	Output High Impedance			8	15	¾
	Output High Voltage	I_{OUT} = 100mA, V_{CC} = 15V	13.5	14.2		V
	Rise/Fall Time	C_L = 1000pF		50		ns
PWM						
	Duty Cycle Range	FAN4803-2	0-41	0-47	0-50	%
		FAN4803-1	0-49.5		0-50	%
	Output Low Impedance			8	15	¾
	Output Low Voltage	I_{OUT} = –100mA		0.8	1.5	V
		I_{OUT} = –10mA, V_{CC} = 8V		0.7	1.5	V
	Output High Impedance			8	15	¾
	Output High Voltage	I_{OUT} = 100mA, V_{CC} = 15V	13.5	14.2		V
	Rise/Fall Time	C_L = 1000pF		50		ns
Supply						
	V_{CC} Clamp Voltage (V_{CCZ})	I_{CC} = 10mA	16.7	17.5	18.3	V
	Start-up Current	V_{CC} = 11V, C_L = 0		0.2	0.4	mA
	Operating Current	V_{CC} = 15V, C_L = 0		2.5	4	mA
	Undervoltage Lockout Threshold		11.5	12	12.5	V
	Undervoltage Lockout Hysteresis		2.4	2.9	3.4	V

Note:

1. Limits are guaranteed by 100% testing, sampling, or correlation with worst case test conditions.

Functional Description

The FAN4803 consists of an average current mode boost Power Factor Corrector (PFC) front end followed by a synchronized Pulse Width Modulation (PWM) controller. It is distinguished from earlier combo controllers by its low pin count, innovative input current shaping technique, and very low start-up and operating currents. The PWM section is dedicated to peak current mode operation. It uses conventional trailing-edge modulation, while the PFC uses leading-edge modulation. This patented Leading Edge/Trailing Edge (LETE) modulation technique helps to minimize ripple current in the PFC DC buss capacitor.

The FAN4803 is offered in two versions. The FAN4803-1 operates both PFC and PWM sections at 67kHz, while the FAN4803-2 operates the PWM section at twice the frequency (134kHz) of the PFC. This allows the use of smaller PWM magnetics and output filter components, while minimizing switching losses in the PFC stage.

In addition to power factor correction, several protection features have been built into the FAN4803. These include soft start, redundant PFC over-voltage protection, peak current limiting, duty cycle limit, and under voltage lockout (UVLO). See Figure 12 for a typical application.

Detailed Pin Descriptions

VEAO

This pin provides the feedback path which forces the PFC output to regulate at the programmed value. It connects to programming resistors tied to the PFC output voltage and is shunted by the feedback compensation network.

ISENSE

This pin ties to a resistor or current sense transformer which senses the PFC input current. This signal should be negative with respect to the IC ground. It internally feeds the pulse-by-pulse current limit comparator and the current sense feedback signal. The I_{LIMIT} trip level is –1V. The I_{SENSE} feedback is internally multiplied by a gain of four and compared against the internal programmed ramp to set the PFC duty cycle. The intersection of the boost inductor current downslope with the internal programming ramp determines the boost off-time.

VDC

This pin is typically tied to the feedback opto-collector. It is tied to the internal 5V reference through a 26k¾ resistor and to GND through a 40k¾ resistor.

ILIMIT

This pin is tied to the primary side PWM current sense resistor or transformer. It provides the internal pulse-by-pulse current limit for the PWM stage (which occurs at 1.5V) and the peak current mode feedback path for the current mode control of the PWM stage. The current ramp is offset internally by 1.2V and then compared against the opto feedback voltage to set the PWM duty cycle.

PFC OUT and PWM OUT

PFC OUT and PWM OUT are the high-current power drivers capable of directly driving the gate of a power MOSFET with peak currents up to ±1A. Both outputs are actively held low when V_{CC} is below the UVLO threshold level.

VCC

V_{CC} is the power input connection to the IC. The V_{CC} start-up current is 150µA . The no-load I_{CC} current is 2mA. V_{CC} quiescent current will include both the IC biasing currents and the PFC and PWM output currents. Given the operating frequency and the MOSFET gate charge (Qg), average PFC and PWM output currents can be calculated as I_{OUT} = Qg x F. The average magnetizing current required for any gate drive transformers must also be included. The V_{CC} pin is also assumed to be proportional to the PFC output voltage. Internally it is tied to the V_{CC}OVP comparator (16.2V) providing redundant high-speed over-voltage protection (OVP) of the PFC stage. V_{CC} also ties internally to the UVLO circuitry, enabling the IC at 12V and disabling it at 9.1V. V_{CC} must be bypassed with a high quality ceramic bypass capacitor placed as close as possible to the IC. Good bypassing is critical to the proper operation of the FAN4803.

V_{CC} is typically produced by an additional winding off the boost inductor or PFC Choke, providing a voltage that is proportional to the PFC output voltage. Since the V_{CC}OVP max voltage is 16.2V, an internal shunt limits V_{CC} overvoltage to an acceptable value. An external clamp, such as shown in Figure 1, is desirable but not necessary.

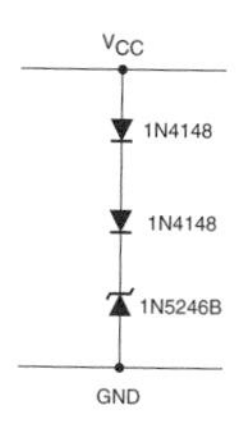

Figure 1. Optional V_{CC} Clamp

V_{CC} is internally clamped to 16.7V minimum, 18.3V maximum. This limits the maximum V_{CC} that can be applied to the IC while allowing a V_{CC} which is high enough to trip the V_{CC}OVP. The max current through this zener is 10mA. External series resistance is required in order to limit the current through this Zener in the case where the V_{CC} voltage exceeds the zener clamp level.

GND

GND is the return point for all circuits associated with this part. Note: a high-quality, low impedance ground is critical to the proper operation of the IC. High frequency grounding techniques should be used.

Power Factor Correction

Power factor correction makes a nonlinear load look like a resistive load to the AC line. For a resistor, the current drawn from the line is in phase with, and proportional to, the line voltage. This is defined as a unity power factor is (one). A common class of nonlinear load is the input of a most power supplies, which use a bridge rectifier and capacitive input filter fed from the line. Peak-charging effect, which occurs on the input filter capacitor in such a supply, causes brief high-amplitude pulses of current to flow from the power line, rather than a sinusoidal current in phase with the line voltage. Such a supply presents a power factor to the line of less than one (another way to state this is that it causes significant current harmonics to appear at its input). If the input current drawn by such a supply (or any other nonlinear load) can be made to follow the input voltage in instantaneous amplitude, it will appear resistive to the AC line and a unity power factor will be achieved.

To hold the input current draw of a device drawing power from the AC line in phase with, and proportional to, the input voltage, a way must be found to prevent that device from loading the line except in proportion to the instantaneous line voltage. The PFC section of the FAN4803 uses a boost-mode DC-DC converter to accomplish this. The input to the converter is the full wave rectified AC line voltage. No filtering is applied following the bridge rectifier, so the input voltage to the boost converter ranges, at twice line frequency, from zero volts to the peak value of the AC input and back to zero. By forcing the boost converter to meet two simultaneous conditions, it is possible to ensure that the current that the converter draws from the power line matches the instantaneous line voltage. One of these conditions is that the output voltage of the boost converter must be set higher than the peak value of the line voltage. A commonly used value is 385VDC, to allow for a high line of $270VAC_{RMS}$. The other condition is that the current that the converter is allowed to draw from the line at any given instant must be proportional to the line voltage.

Since the boost converter topology in the FAN4803 PFC is of the current-averaging type, no slope compensation is required.

Leading/Trailing Modulation

Conventional Pulse Width Modulation (PWM) techniques employ trailing edge modulation in which the switch will turn ON right after the trailing edge of the system clock. The error amplifier output voltage is then compared with the modulating ramp. When the modulating ramp reaches the level of the error amplifier output voltage, the switch will be turned OFF. When the switch is ON, the inductor current will ramp up. The effective duty cycle of the trailing edge modulation is determined during the ON time of the switch. Figure 2 shows a typical trailing edge control scheme.

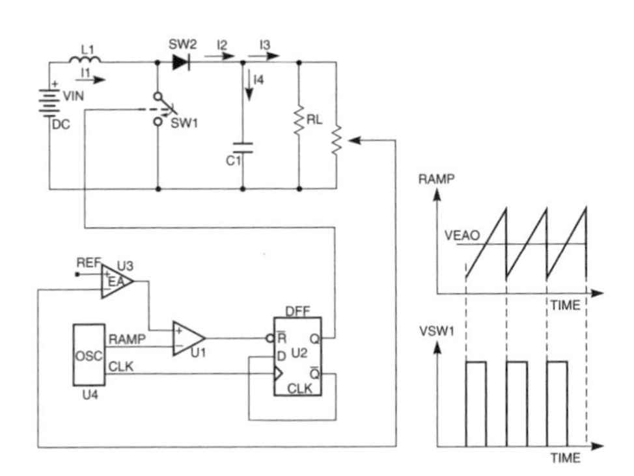

Figure 2. Typical Trailing Edge Control Scheme.

In the case of leading edge modulation, the switch is turned OFF right at the leading edge of the system clock. When the modulating ramp reaches the level of the error amplifier output voltage, the switch will be turned ON. The effective duty-cycle of the leading edge modulation is determined during the OFF time of the switch. Figure 3 shows a leading edge control scheme.

One of the advantages of this control technique is that it requires only one system clock. Switch 1 (SW1) turns OFF and Switch 2 (SW2) turns ON at the same instant to minimize the momentary "no-load" period, thus lowering ripple voltage generated by the switching action. With such synchronized switching, the ripple voltage of the first stage is reduced. Calculation and evaluation have shown that the 120Hz component of the PFC's output ripple voltage can be reduced by as much as 30% using this method, substantially reducing dissipation in the high-voltage PFC capacitor.

Typical Applications

One Pin Error Amp

The FAN4803 utilizes a one pin voltage error amplifier in the PFC section (VEAO). The error amplifier is in reality a current sink which forces 35μA through the output programming resistor. The nominal voltage at the VEAO pin is 5V. The VEAO voltage range is 4 to 6V. For a 11.3M¾ resistor chain to the boost output voltage and 5V steady state at the VEAO, the boost output voltage would be 400V.

Programming Resistor Value

Equation 1 calculates the required programming resistor value.

$$Rp = \frac{V_{BOOST} - V_{EAO}}{I_{PGM}} = \frac{400V - 5.0V}{35\mu A} = 11.3M\Omega \quad (1)$$

PFC Voltage Loop Compensation

The voltage-loop bandwidth must be set to less than 120Hz to limit the amount of line current harmonic distortion. A typical crossover frequency is 30Hz. Equation 1, for simplicity, assumes that the pole capacitor dominates the error amplifier gain at the loop unity-gain frequency. Equation 2 places a pole at the crossover frequency, providing 45 degrees of phase margin. Equation 3 places a zero one decade prior to the pole. Bode plots showing the overall gain and phase are shown in Figures 5 and 6. Figure 4 displays a simplified model of the voltage loop.

$$C_{COMP} = \frac{Pin}{R_P \times V_{BOOST} \times \Delta VEAO \times C_{OUT} \times (2 \times \pi \times f)^2} \quad (2)$$

$$C_{COMP} = \frac{300W}{11.3M\Omega \times 400V \times 0.5V \times 220\mu F \times (2 \times \pi \times 30Hz)^2}$$

$$C_{COMP} = 16nF$$

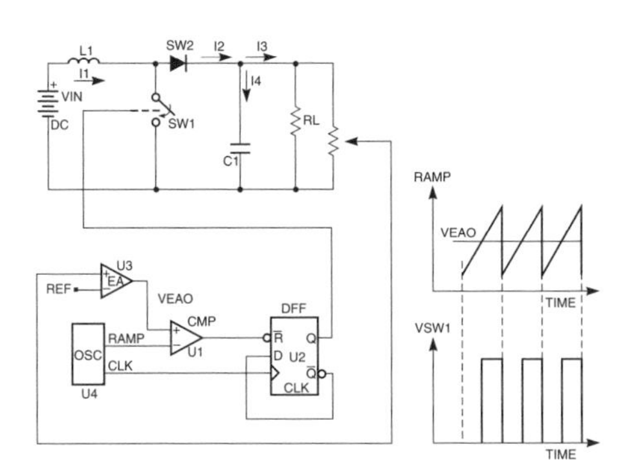

Figure 3. Typical Leading Edge Control Scheme.

$$R_{COMP} = \frac{1}{2 \times \pi \times f \times C_{COMP}} \quad (3)$$

$$R_{COMP} = \frac{1}{6.28 \times 30Hz \times 16nF} = 330k\Omega$$

$$C_{ZERO} = \frac{1}{2 \times \pi \times \frac{f}{10} \times R_{COMP}} \quad (4)$$

$$C_{ZERO} = \frac{1}{6.28 \times 3Hz \times 330k\Omega} = 0.16\mu F$$

Internal Voltage Ramp

The internal ramp current source is programmed by way of the VEAO pin voltage. Figure 7 displays the internal ramp current vs. the VEAO voltage. This current source is used to develop the internal ramp by charging the internal 30pF +12/ –10% capacitor. See Figures 10 and 11. The frequency of the internal programming ramp is set internally to 67kHz.

PFC Current Sense Filtering

In DCM, the input current wave shaping technique used by the FAN4803 could cause the input current to run away. In order for this technique to be able to operate properly under DCM, the programming ramp must meet the boost inductor current down-slope at zero amps. Assuming the programming ramp is zero under light load, the OFF-time will be terminated once the inductor current reaches zero.

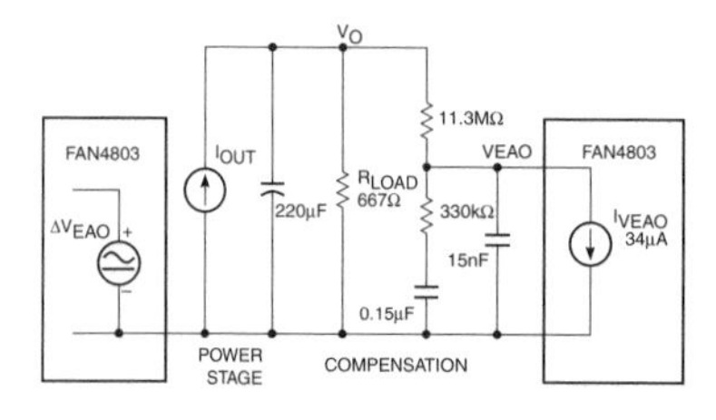

Figure 4. Voltage Control Loop

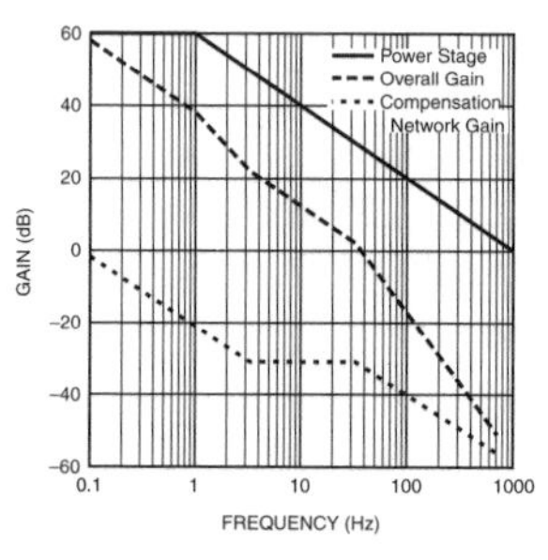

Figure 5. Voltage Loop Gain

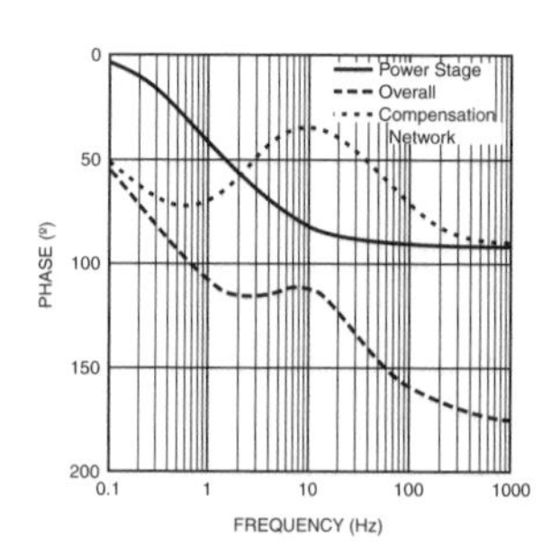

Figure 6. Voltage Loop Phase

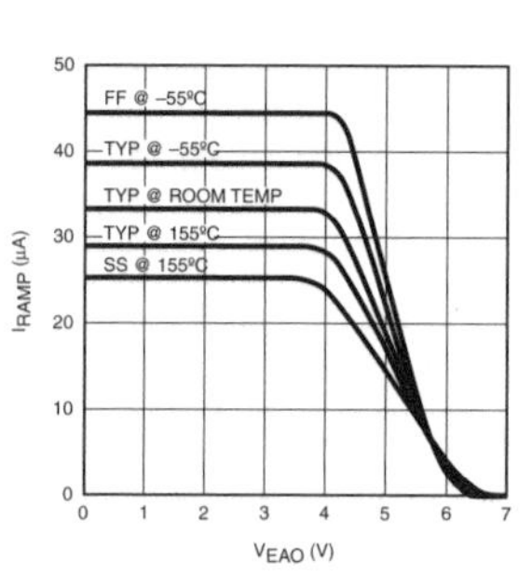

Figure 7. Internal Ramp Current vs. VEAO

Subsequently the PFC gate drive is initiated, eliminating the necessary dead time needed for the DCM mode. This forces the output to run away until the V_{CC} OVP shuts down the PFC. This situation is corrected by adding an offset voltage to the current sense signal, which forces the duty cycle to zero at light loads. This offset prevents the PFC from operating in the DCM and forces pulse-skipping from CCM to no-duty, avoiding DMC operation. External filtering to the current sense signal helps to smooth out the sense signal, expanding the operating range slightly into the DCM range, but this should be done carefully, as this filtering also reduces the bandwidth of the signal feeding the pulse-by-pulse current limit signal. Figure 9 displays a typical circuit for adding offset to I_{SENSE} at light loads.

PFC Start-Up and Soft Start

During steady state operation VEAO draws 35μA. At start-up the internal current mirror which sinks this current is defeated until V_{CC} reaches 12V. This forces the PFC error voltage to V_{CC} at the time that the IC is enabled. With leading edge modulation V_{CC} on the VEAO pin forces zero duty on the PFC output. When selecting external compensation components and V_{CC} supply circuits VEAO must not be prevented from reaching 6V prior to V_{CC} reaching 12V in the turn-on sequence. This will guarantee that the PFC stage will enter soft-start. Once V_{CC} reaches 12V the 35μA VEAO current sink is enabled. VEAO compensation components are then discharged by way of the 35μA current sink until the steady state operating point is reached. See Figure 8.

PFC Soft Recovery Following V_{CC} OVP

The FAN4803 assumes that V_{CC} is generated from a source that is proportional to the PFC output voltage. Once that source reaches 16.2V the internal current sink tied to the VEAO pin is disabled just as in the soft start turn-on sequence. Once disabled, the VEAO pin charges HIGH by way of the external components until the PFC duty cycle goes to zero, disabling the PFC. The V_{CC} OVP resets once the V_{CC} discharges below 16.2V, enabling the VEAO current sink and discharging the VEAO compensation components until the steady state operating point is reached. It should be noted that, as shown in Figure 8, once the VEAO pin exceeds 6.5V, the internal ramp is defeated. Because of this, an external Zener can be installed to reduce the maximum voltage to which the VEAO pin may rise in a shutdown condition. Clamping the VEAO pin externally to 7.4V will reduce the time required for the VEAO pin to recover to its steady state value.

UVLO

Once V_{CC} reaches 12V both the PFC and PWM are enabled. The UVLO threshold is 9.1V providing 2.9V of hysteresis.

Generating V_{CC}

An internal clamp limits overvoltage to V_{CC}. This clamp circuit ensures that the V_{CC} OVP circuitry of the FAN4803 will function properly over tolerance and temperature while protecting the part from voltage transients. This circuit allows the FAN4803 to deliver 15V nominal gate drive at PWM OUT and PFC OUT, sufficient to drive low-cost IGBTs.

It is important to limit the current through the Zener to avoid overheating or destroying it. This can be done with a single resistor in series with the V_{CC} pin, returned to a bias supply of typically 14V to 18V. The resistor value must be chosen to meet the operating current requirement of the FAN4803 itself (4.0mA max) plus the current required by the two gate driver outputs.

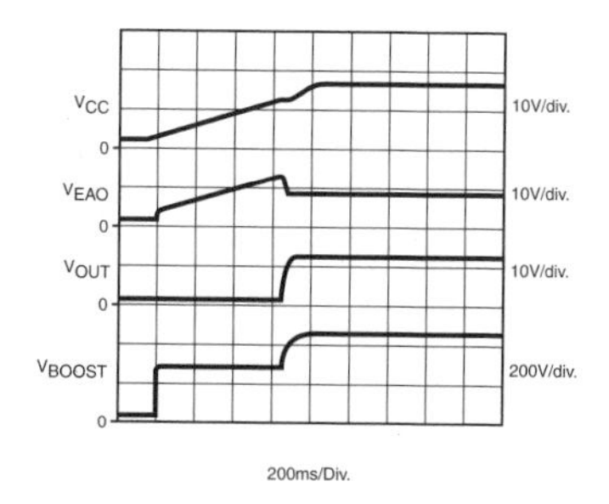

Figure 8. PFC Soft Start

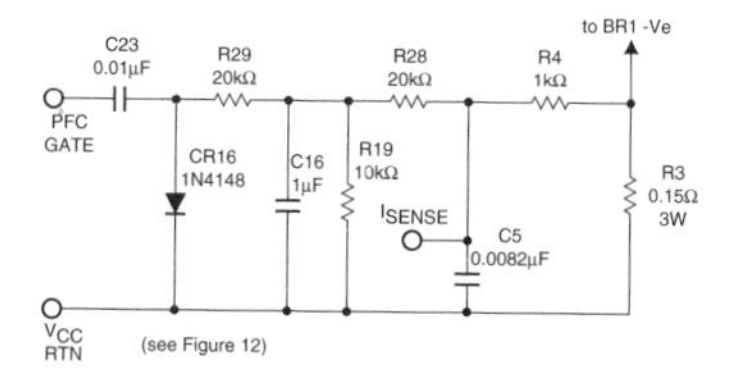

Figure 9. I_{SENSE} Offset for Light Load Conditions

V_{CC} OVP

V_{CC} is assumed to be a voltage proportional to the PFC output voltage, typically a bootstrap winding off the boost inductor. The V_{CC} OVP comparator senses when this voltage exceeds 16V, and terminates the PFC output drive while disabling the VEAO current sink. Once the VEAO current sink is disabled, the VEAO voltage will charge unabated, except for a diode clamp to V_{CC}, reducing the PFC pulse width. Once the V_{CC} rail has decreased to below 16.2V the VEAO sink will be enabled, discharging external VEAO compensation components until the steady state voltage is reached. Given that 15V on V_{CC} corresponds to 400V on the PFC output, 16V on V_{CC} corresponds to an OVP level of 426V.

Component Reduction

Components associated with the V_{RMS} and I_{RMS} pins of a typical PFC controller such as the ML4824 have been eliminated. The PFC power limit and bandwidth does vary with line voltage. Double the power can be delivered from a 220 V AC line versus a 110 V AC line. Since this is a combination PFC/PWM, the power to the load is limited by the PWM stage.

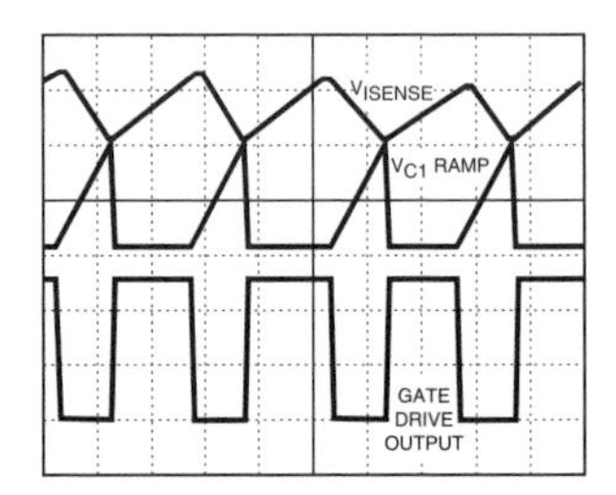

Figure 10. Typical Peak Current Mode Waveforms

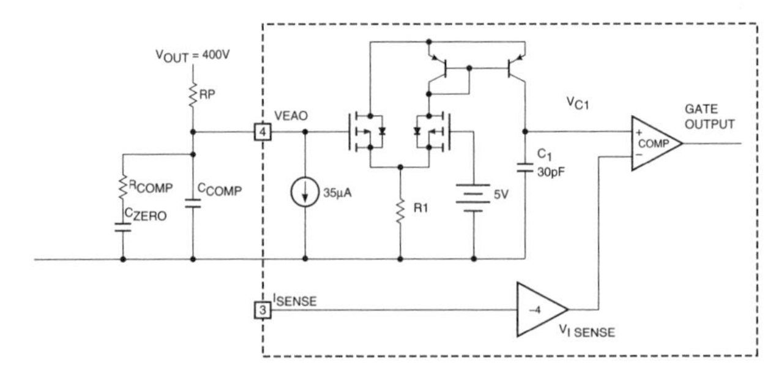

Figure 11. FAN4803 PFC Control

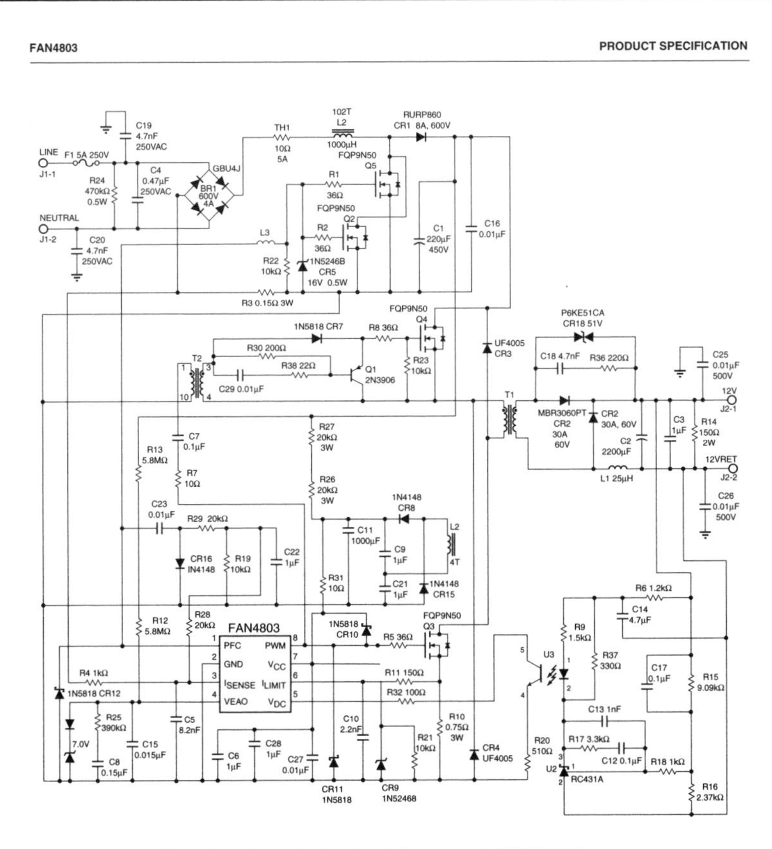

Figure 12. Typical Application Circuit. Universal Input 240W 12V DC Output

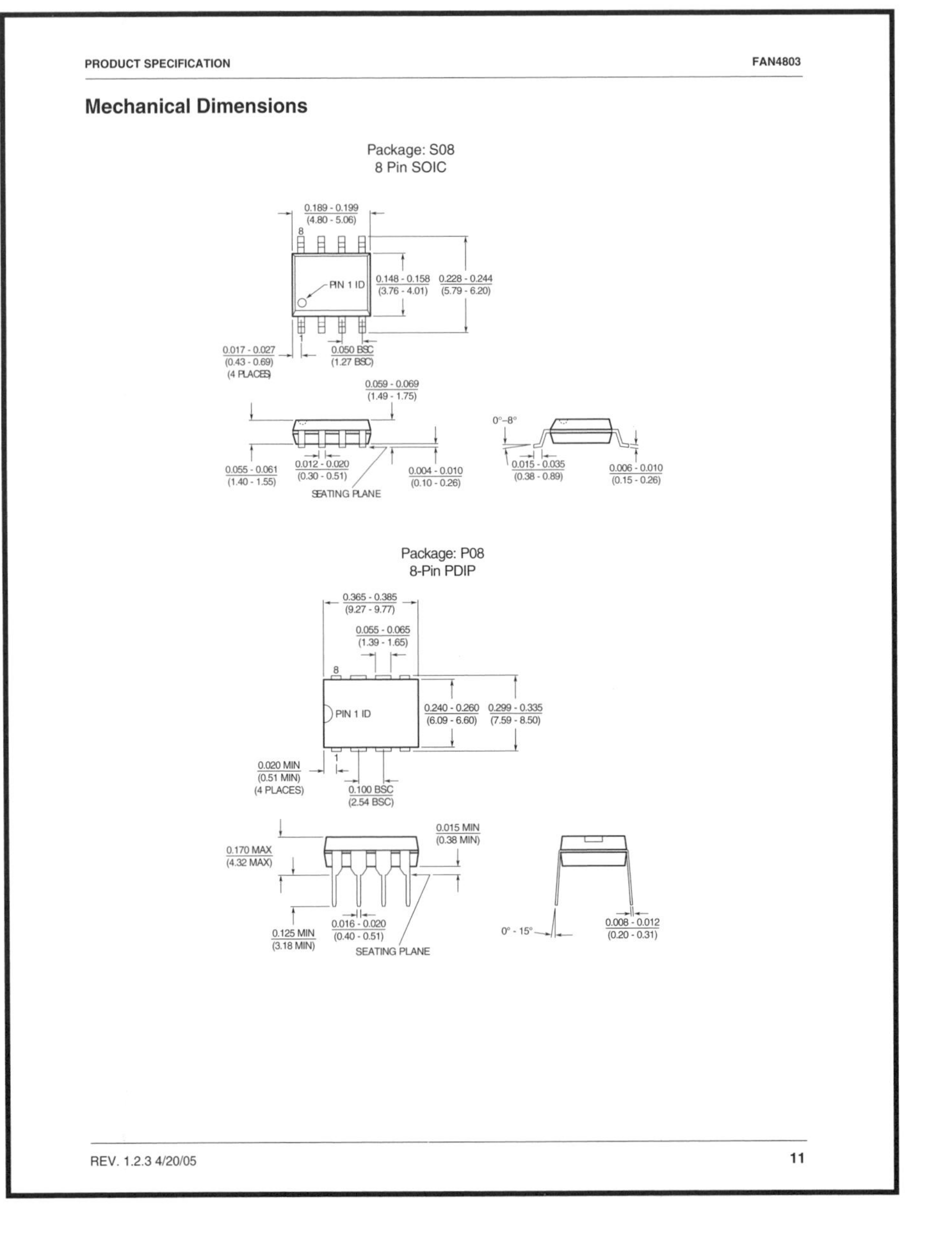

Mechanical Dimensions

Package: S08
8 Pin SOIC

Package: P08
8-Pin PDIP

Ordering Information

Part Number	PFC/PWM Frequency	Temperature Range	Package
FAN4803CS-1	67kHz / 67kHz	0°C to 70°C	8-Pin SOIC (S08)
FAN4803CS-2	67kHz / 134kHz	0°C to 70°C	8-Pin SOIC (S08)
FAN4803CP-1	67kHz / 67kHz	0°C to 70°C	8-Pin PDIP (P08)
FAN4803CP-2	67kHz / 134kHz	0°C to 70°C	8-Pin PDIP (P08)

4/20/05 0.0m 003
Stock#DS30004803

Appendix C

Fairchild Specifications for FSD210 and FSD200

www.fairchildsemi.com

FSD210, FSD200

Green Mode Fairchild Power Switch (FPS™)

Features

- Single Chip 700V Sense FET Power Switch
- Precision Fixed Operating Frequency (134kHz)
- Advanced Burst-Mode operation Consumes under 0.1W at 265Vac and no load (FSD210 only)
- Internal Start-up Switch and Soft Start
- Under Voltage Lock Out (UVLO) with Hysteresis
- Pulse by Pulse Current Limit
- Over Load Protection (OLP)
- Internal Thermal Shutdown Function (TSD)
- Auto-Restart Mode
- Frequency Modulation for EMI
- FSD200 does not require an auxiliary bias winding

Applications

- Charger & Adaptor for Mobile Phone, PDA & MP3
- Auxiliary Power for White Goods, PC, C-TV & Monitor

Description

The FSD200 and FSD210 are integrated Pulse Width Modulators (PWM) and Sense FETs specially designed for high performance off-line Switch Mode Power Supplies (SMPS) with minimal external components. Both devices are monolithic high voltage power switching regulators which combine an LDMOS Sense FET with a voltage mode PWM control block. The integrated PWM controller features include: a fixed oscillator with frequency modulation for reduced EMI, Under Voltage Lock Out (UVLO) protection, Leading Edge Blanking (LEB), optimized gate turn-on/turn-off driver, thermal shut down protection (TSD), temperature compensated precision current sources for loop compensation and fault protection circuitry. When compared to a discrete MOSFET and controller or RCC switching converter solution, the FSD200 and FSD210 reduce total component count, design size, weight and at the same time increase efficiency, productivity, and system reliability. The FSD200 eliminates the need for an auxiliary bias winding at a small cost of increased supply power. Both devices are a basic platform well suited for cost effective designs of flyback converters.

OUTPUT POWER TABLE				
PRODUCT	230VAC ±15%(3)		85-265VAC	
	Adapter(1)	Open Frame(2)	Adapter(1)	Open Frame(2)
FSD210	5W	7W	4W	5W
FSD200	5W	7W	4W	5W
FSD210M	5W	7W	4W	5W
FSD200M	5W	7W	4W	5W

Table 1. Notes: 1. Typical continuous power in a non-ventilated enclosed adapter measured at 50°C ambient. 2. Maximum practical continuous power in an open frame design at 50°C ambient. 3. 230 VAC or 100/115 VAC with doubler.

Typical Circuit

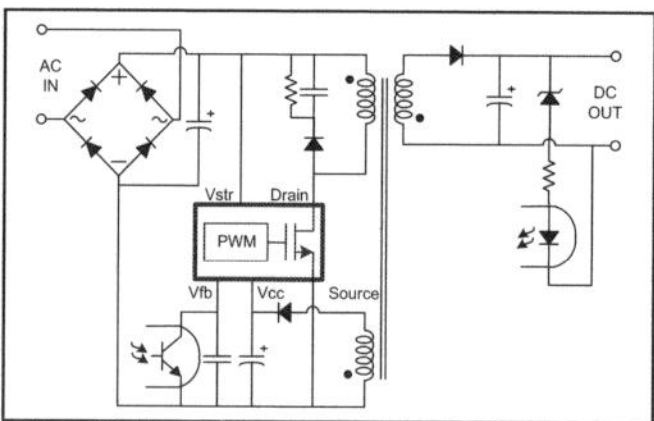

Figure 1. Typical Flyback Application using FSD210

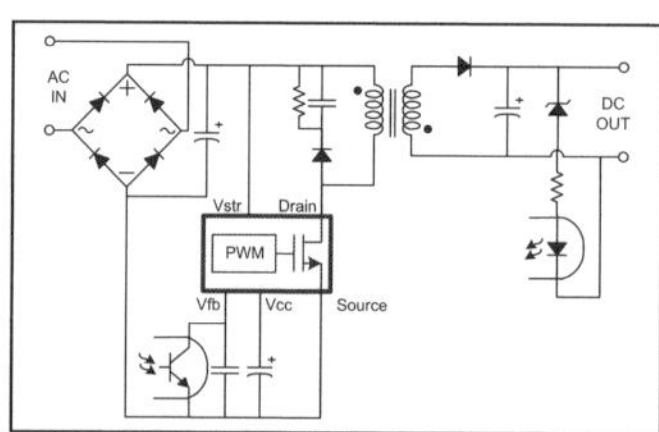

Figure 2. Typical Flyback Application using FSD200

Rev.1.0.2

Internal Block Diagram

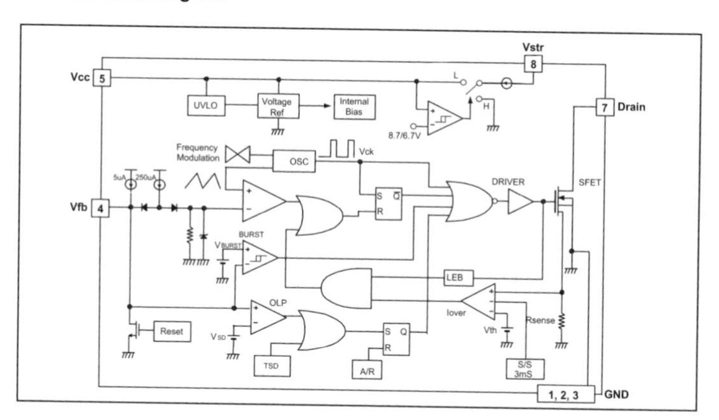

Figure 3. Functional Block Diagram of FSD210

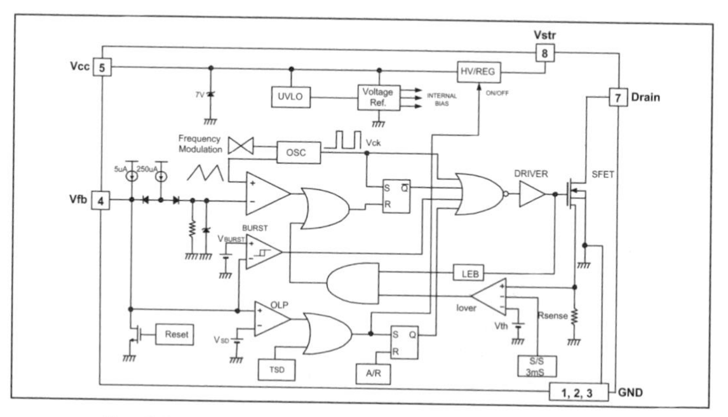

Figure 4. Functional Block Diagram of FSD200 showing internal high voltage regulator

Pin Definitions

Pin Number	Pin Name	Pin Function Description
1, 2, 3	GND	Sense FET source terminal on primary side and internal control ground.
4	Vfb	The feedback voltage pin is the inverting input to the PWM comparator with nominal input levels between 0.5Vand 2.5V. It has a 0.25mA current source connected internally while a capacitor and opto coupler are typically connected externally. A feedback voltage of 4V triggers overload protection (OLP). There is a time delay while charging between 3V and 4V using an internal 5uA current source, which prevents false triggering under transient conditions but still allows the protection mechanism to operate under true overload conditions.
5	Vcc	**FSD210** Positive supply voltage input. Although connected to an auxiliary transformer winding, current is supplied from pin 8 (Vstr) via an internal switch during startup (see Internal Block Diagram section). It is not until Vcc reaches the UVLO upper threshold (8.7V) that the internal start-up switch opens and device power is supplied via the auxiliary transformer winding. **FSD200** This pin is connected to a storage capacitor. A high voltage regulator connected between pin 8 (Vstr) and this pin, provides the supply voltage to the FSD200 at startup and when switching during normal operation. The FSD200 eliminates the need for auxiliary bias winding and associated external components.
7	Drain	The Drain pin is designed to connect directly to the primary lead of the transformer and is capable of switching a maximum of 700V. Minimizing the length of the trace connecting this pin to the transformer will decrease leakage inductance.
8	Vstr	The startup pin connects directly to the rectified AC line voltage source for both the FSD200 and FSD210. For the FSD210, at start up the internal switch supplies internal bias and charges an external storage capacitor placed between the Vcc pin and ground. Once this reaches 8.7V, the internal current source is disabled. For the FSD200, an internal high voltage regulator provides a constant supply voltage.

Pin Configuration

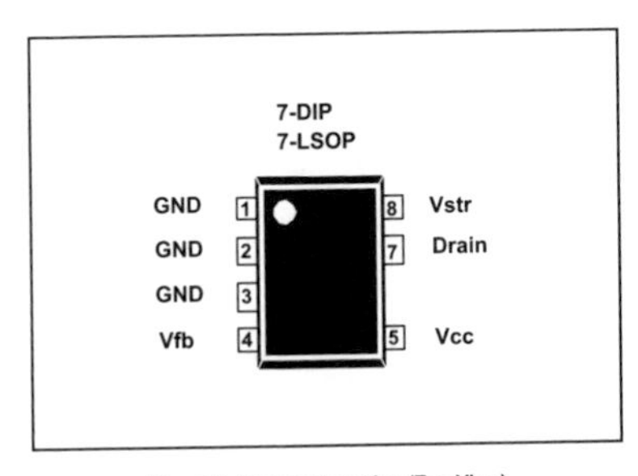

Figure 5. Pin Configuration (Top View)

Absolute Maximum Ratings

(Ta=25°C unless otherwise specified)

Parameter	Symbol	Value	Unit
Maximum Supply Voltage (FSD200)	$V_{CC,MAX}$	10	V
Maximum Supply Voltage (FSD210)	$V_{CC,MAX}$	20	V
Input Voltage Range	V_{FB}	–0.3 to V_{STOP}	V
Operating Junction Temperature.	T_J	+150	°C
Operating Ambient Temperature	T_A	–25 to +85	°C
Storage Temperature Range	T_{STG}	–55 to +150	°C

Thermal Impedance

Parameter	Symbol	Value	Unit
7DIP			
Junction-to-Ambient Thermal	θ_{JA}(1)	74.07(3)	°C/W
	θ_{JA}(1)	60.44(4)	°C/W
Junction-to-Case Thermal	θ_{JC}(2)	22.00	°C/W
7LSOP			
Junction-to-Ambient Thermal	θ_{JA}(1)	-	°C/W
	θ_{JA}(1)	-	°C/W
Junction-to-Case Thermal	θ_{JC}(2)	-	°C/W

Note:

1. Free standing without heat sink.
2. Measured on the GND pin close to plastic interface.
3. Soldered to 100mm^2 copper clad.
4. Soldered to 300mm^2 copper clad.

Electrical Characteristics

(Ta=25°C unless otherwise specified)

Parameter	Symbol	Condition	Min.	Typ.	Max.	Unit
Sense FET SECTION						
Drain-Source Breakdown Voltage	BV_{DSS}	V_{CC} = 0V, I_D = 100μA	700	-	-	V
Startup Voltage (Vstr) Breakdown	BV_{STR}		700	-	-	V
Off-State Current	I_{DSS}	V_{DS} = 560V	-	-	100	μA
On-State Resistance	$R_{DS(ON)}$	Tj = 25°C, I_D = 25mA	-	28	32	Ω
		Tj = 100°C, I_D = 25mA	-	42	48	Ω
Rise Time	T_R	V_{DS} = 325V, I_D = 50mA	-	100	-	ns
Fall Time	T_F	V_{DS} = 325V, I_D = 25mA	-	50	-	ns
CONTROL SECTION						
Output Frequency	F_{OSC}	Tj = 25°C	126	134	142	kHz
Output Frequency Modulation	F_{MOD}	Tj = 25°C	-	±4	-	kHz
Feedback Source Current	I_{FB}	Vfb = 0V	0.22	0.25	0.28	mA
Maximum Duty Cycle	D_{MAX}	Vfb = 3.5V	60	65	70	%
Minimum Duty Cycle	D_{MIN}	Vfb = 0V	0	0	0	%
UVLO Threshold Voltage (FSD200)	V_{START}		6.3	7	7.7	V
	V_{STOP}	After turn on	5.3	6	6.7	V
UVLO Threshold Voltage (FSD210)	V_{START}		8.0	8.7	9.4	V
	V_{STOP}	After turn on	6.0	6.7	7.4	V
Supply Shunt Regulator (FSD200)	V_{CCREG}	-	-	7	-	V
Internal Soft Start Time	$T_{S/S}$		-	3	-	ms
BURST MODE SECTION						
Burst Mode Voltage	V_{BURH}	Tj = 25°C	0.58	0.64	0.7	V
	V_{BURL}		0.5	0.58	0.64	V
	Hysteresis		-	60	-	mV
PROTECTION SECTION						
Drain to Source Peak Current Limit	I_{OVER}		0.275	0.320	0.365	A
Current Limit Delay[1]	T_{CLD}	Tj = 25°C	-	220	-	ns
Thermal Shutdown Temperature (Tj)[1]	T_{SD}		125	145	160	°C
Shutdown Feedback Voltage	V_{SD}	-	3.5	4.0	4.5	V
Feedback Shutdown Delay Current	I_{DELAY}	Vfb = 4.0V	3	5	7	μA
Leading Edge Blanking Time[2]	T_{LEB}		200	-	-	ns
TOTAL DEVICE SECTION						
Operating Supply Current (FSD200)	I_{OP}	Vcc = 7V	-	600	-	μA
Operating Supply Current (FSD210)	I_{OP}	Vcc = 11V	-	700	-	μA
Start Up Current (FSD200)	I_{START}	Vcc = 0V	-	1	1.2	mA
Start Up Current (FSD210)	I_{START}	Vcc = 0V	-	700	900	μA
Vstr Supply Voltage		Vcc = 0V	20	-	-	V

Note:
1. These parameters, although guaranteed, are not 100% tested in production
2. This parameter is derived from characterization

Comparison Between FSDH565 and FSD210

Function	FSDH0565	FSD210	FSD210 Advantages
Soft-Start	not applicable	3mS	• Gradually increasing current limit during soft-start further reduces peak current and voltage component stresses • Eliminates external components used for soft-start in most applications • Reduces or eliminates output overshoot
Switching Frequency	100kHz	134kHz	• Smaller transformer
Frequency Modulation	not applicable	±4kHz	• Reduced conducted EMI
Burst Mode Operation	not applicable	Yes-built into controller	• Improve light load efficiency • Reduces no-load consumption • Transformer audible noise reduction
Drain Creepage at Package	1.02mm	3.56mm DIP 3.56mm LSOP	• Greater immunity to acting as a result of build-up of dust, debris and other contaminants

Typical Performance Characteristics

(These characteristic graphs are normalized at Ta=25℃)

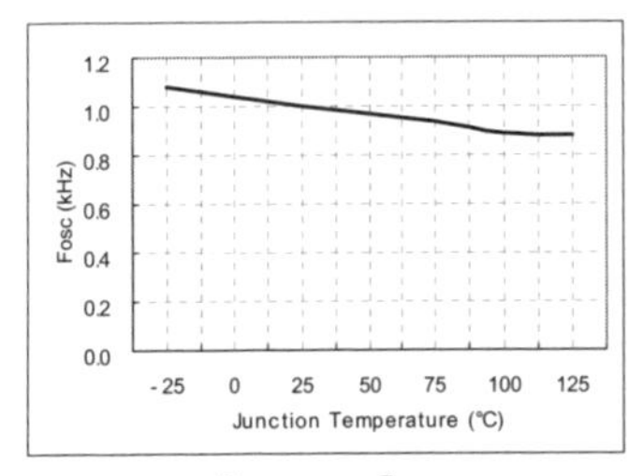

Frequency vs. Temp

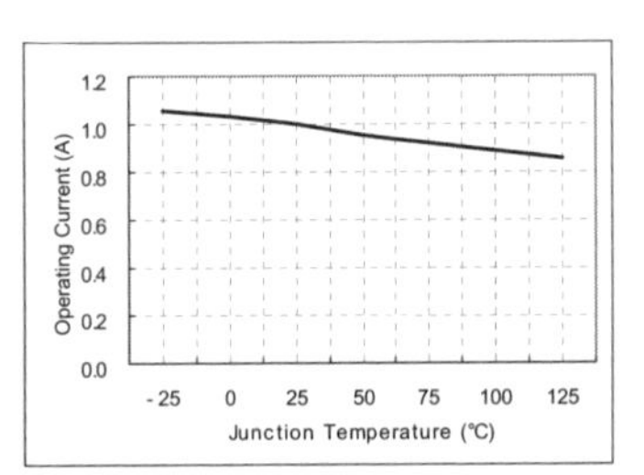

Operating Current vs. Temp

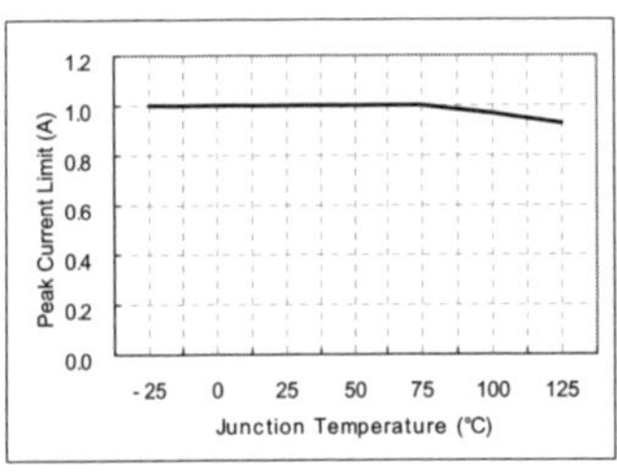

Peak Current Limit vs. Temp

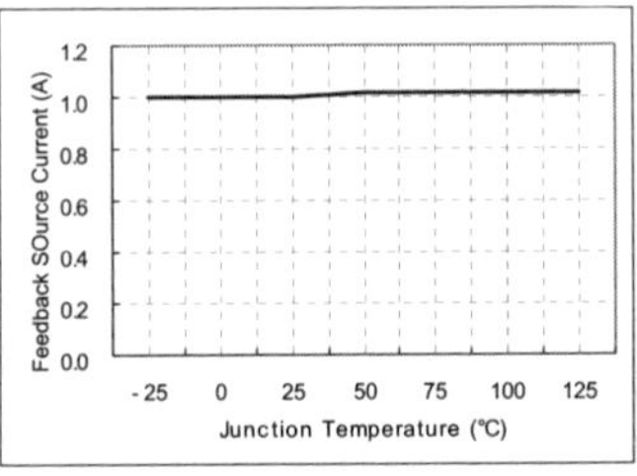

Feedback Source Current vs. Temp

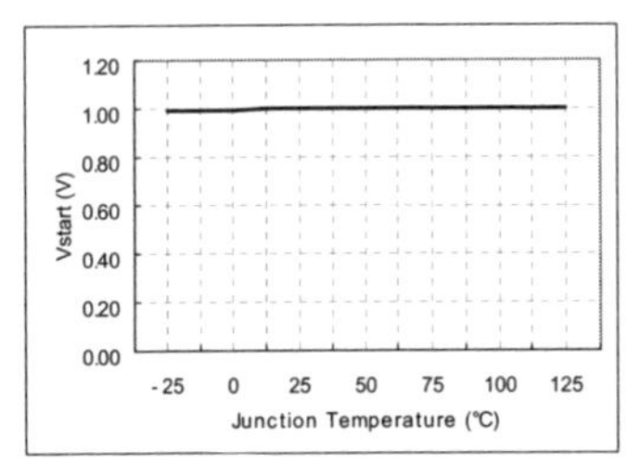

Vstart Voltage vs. Temp

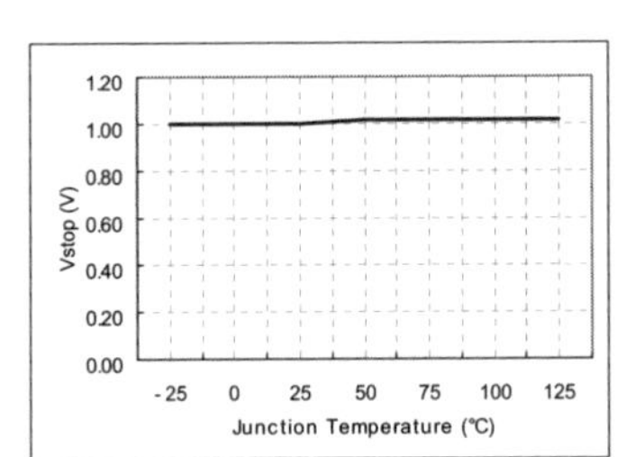

Vstop Voltage vs. Temp

Typical Performance Characteristics (Continued)

(These characteristic graphs are normalized at Ta=25℃)

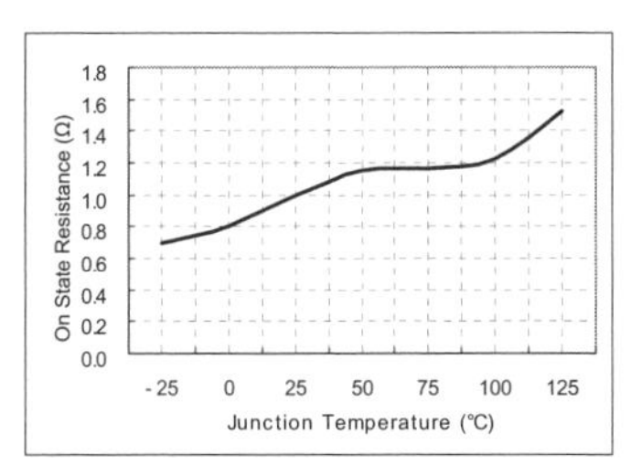

On State Resistance vs. Temp

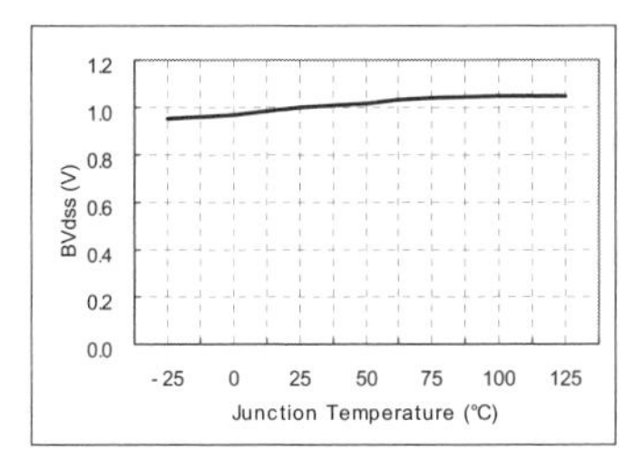

Breakdown Voltage vs. Temp

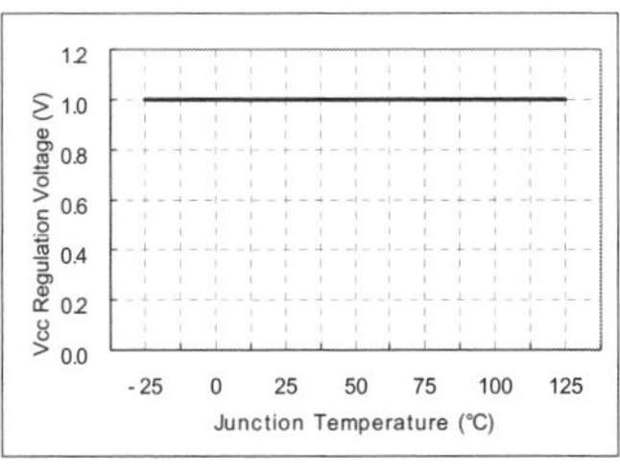

Vcc Regulation Voltage vs. Temp (for FSD200)

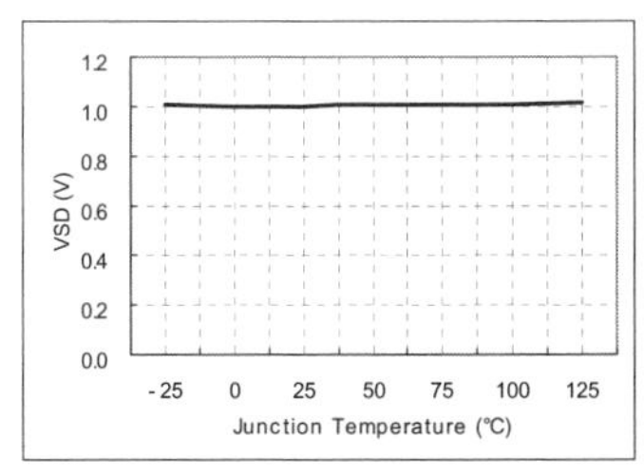

Shutdown Feedback Voltage vs. Temp

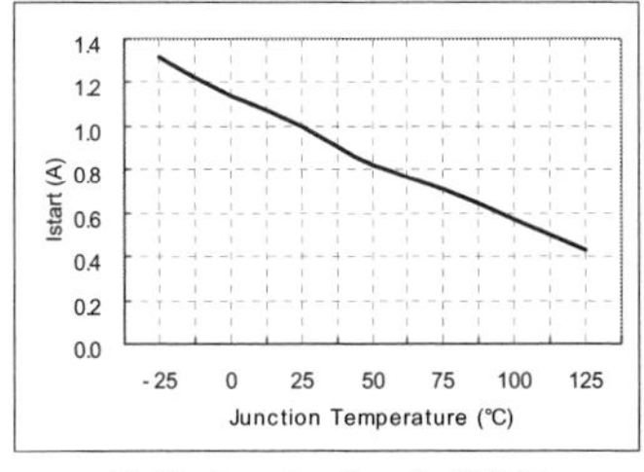

Start Up Current vs. Temp (for FSD210)

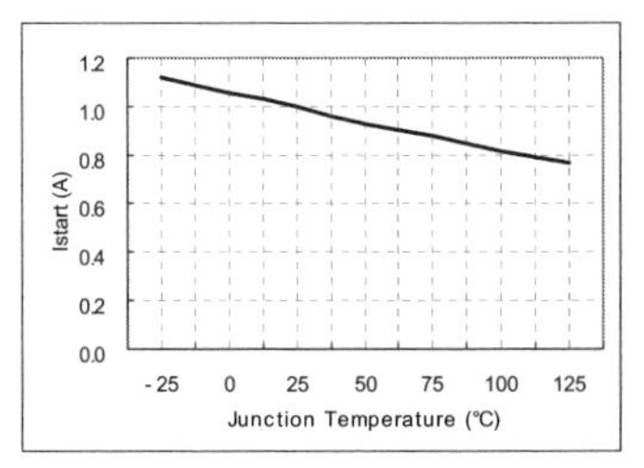

Start Up Current vs. Temp (for FSD200)

Functional Description

1. Startup : At startup, the internal high voltage current source supplies the internal bias and charges the external Vcc capacitor as shown in figure 7. In the case of the FSD210, when Vcc reaches 8.7V the device starts switching and the internal high voltage current source is disabled (see figure 1). The device continues to switch provided that Vcc does not drop below 6.7V. For FSD210, after startup, the bias is supplied from the auxiliary transformer winding. In the case of FSD200, Vcc is continuously supplied from the external high voltage source and Vcc is regulated to 7V by an internal high voltage regulator (HVReg), thus eliminating the need for an auxiliary winding (see figure 2).

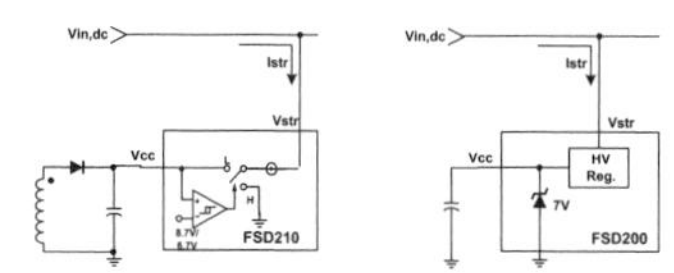

Figure 6. Internal startup circuit

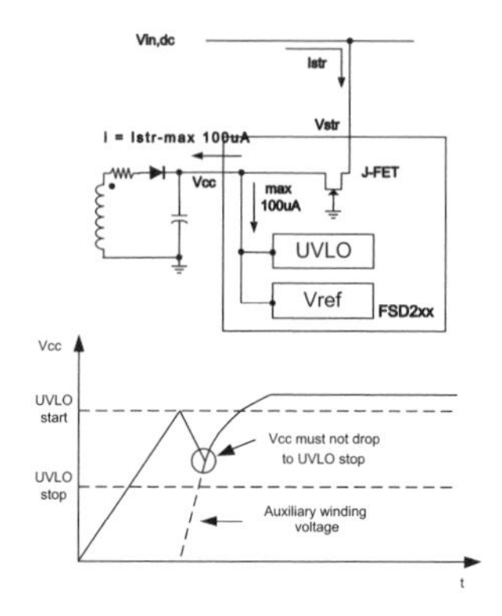

Figure 7. Charging the Vcc capacitor through Vstr

Calculating the Vcc capacitor is an important step to designing in the FSD200/210. At initial start-up in both the FSD200/210, the stand-by maximum current is 100uA, supplying current to UVLO and Vref Block. The charging current (i) of the Vcc capacitor is equal to Istr - 100uA. After Vcc reaches the UVLO start voltage only the bias winding supplies Vcc current to device. When the bias winding voltage is not sufficient, the Vcc level decreases to the UVLO stop voltage. At this time Vcc oscillates. In order to prevent this ripple it is recommended that the Vcc capacitor be sized between 10uF and 47uF.

2. Feedback Control : The FSD200/210 are both voltage mode devices as shown in Figure 8. Usually, a H11A817 optocoupler and KA431 voltage reference (or a FOD2741 integrated optocoupler and voltage reference) are used to implement the isolated secondary feedback network. The feedback voltage is compared with an internally generated sawtooth waveform, directly controlling the duty cycle. When the KA431 reference pin voltage exceeds the internal reference voltage of 2.5V, the optocoupler LED current increases pulling down the feedback voltage and reducing the duty cycle. This event will occur when either the input voltage increases or the output load decreases.

3. Leading edge blanking (LEB) : At the instant the internal Sense FET is turned on, there usually exists a high current spike through the Sense FET, caused by the primary side capacitance and secondary side rectifier diode reverse recovery. Exceeding the pulse-by-pulse current limit could cause premature termination of the switching pulse (see Protection Section). To counter this effect, the FPS employs a leading edge blanking (LEB) circuit. This circuit inhibits the over current comparator for a short time (TLEB) after the Sense FET is turned on.

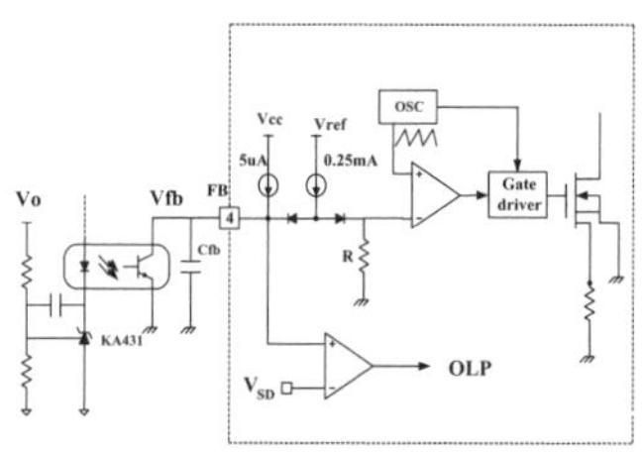

Figure 8. PWM and feedback circuit

4. Protection Circuit : The FSD200/210 has 2 self protection functions: over load protection (OLP) and thermal shutdown (TSD). Because these protection circuits are fully integrated into the IC with no external components, system reliability is improved without a cost increase. If either of these thresholds are triggered, the FPS starts an auto-restart cycle. Once the fault condition occurs, switching is terminated and the Sense FET remains off. This causes Vcc to fall. When Vcc reaches the UVLO stop voltage (6.7V:FSD210, 6V:FSD200), the protection is reset and the internal high voltage current source charges the Vcc capacitor. When Vcc reaches the UVLO start voltage (8.7V:FSD210,7V:FSD200), the device attempts to resume normal operation. If the fault condition is no longer present start up will be successful. If it is still present the cycle is repeated (see figure 10).

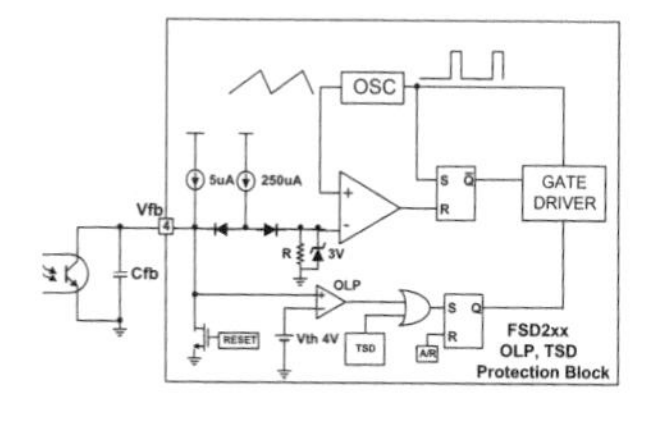

Figure 9. Protection block

4.1 Over Load Protection (OLP) : Over load protection occurs when the load current exceeds a pre-set level due to an abnormal situation. If this occurs, the protection circuit should be triggered to protect the SMPS. It is possible that a short term load transient can occur under normal operation. In order to avoid false shutdowns, the over load protection circuit is designed to trigger after a delay. Therefore the device can differentiate between transient over loads and true fault conditions. The maximum input power is limited using the pulse-by-pulse current limit feature. If the load tries to

draw more than this, the output voltage will drop below its set value. This reduces the optocoupler LED current which in turn reduces the photo-transistor current (see figure 9). Therefore, the 250uA current source will charge the feedback pin capacitor, Cfb, and the feedback voltage, Vfb, will increase. The input to the feedback comparator is clamped at 3V. Once Vfb reaches 3V, the device switches at maximum power, the 250uA current source is blocked and the 5uA source continues to charge Cfb. Once Vfb reaches 4V, switching stops.and overload protection is triggered. The resultant shutdown delay time is set by the time required to charge Cfb from 3Vto 4Vwith 5uA as shown in Fig. 10.

4.2 Thermal Shutdown (TSD) : The Sense FET and the control IC are integrated, making it easier for the control IC to detect the temperature of the Sense FET. When the temperature exceeds approximately 145°C, thermal shutdown is activated.

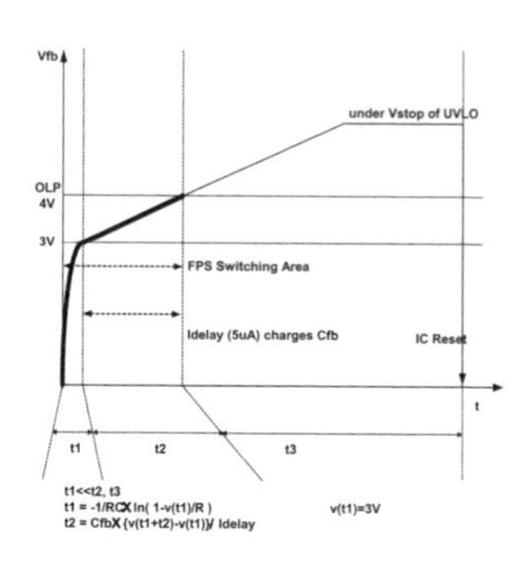

Figure 10. Over load protection delay

5. Soft Start : FSD200/210 has an internal soft start circuit that gradually increases current through the Sense FET as shown in figure 11. The soft start time is 3msec in FSD200/210.

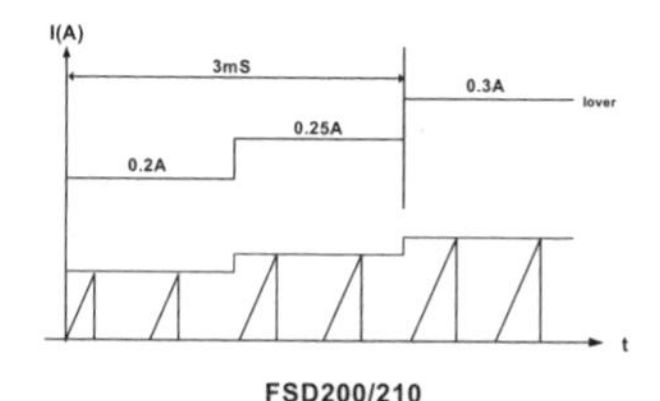

FSD200/210

Figure 11. Internal Soft Start

6. Burst operation : In order to minimize the power dissipation in standby mode, the FSD200/210 implements burst mode functionality (see figure 12). As the load decreases, the feedback voltage decreases. As shown in figure 13, the device automatically enters burst mode when the feedback voltage drops below V_{BURL}(0.58V). At this point switching stops and the output voltages start to drop at a rate dependant on standby current load. This causes the feedback voltage to rise. Once it passes V_{BURH}(0.64V) switching starts again. The feedback voltage falls and the process repeats. Burst mode operation alternately enables and disables switching of the power Sense FET thereby reducing switching loss in standby mode.

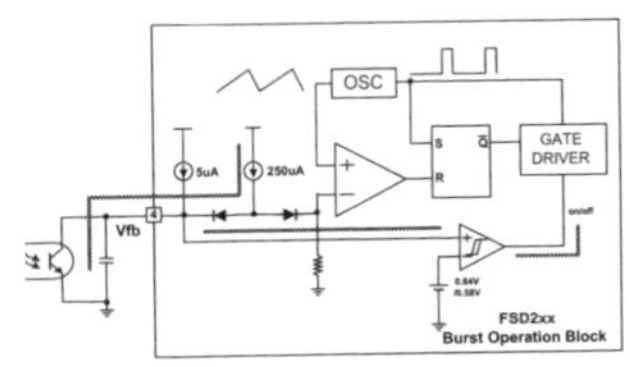

Figure 12. Circuit for burst operation

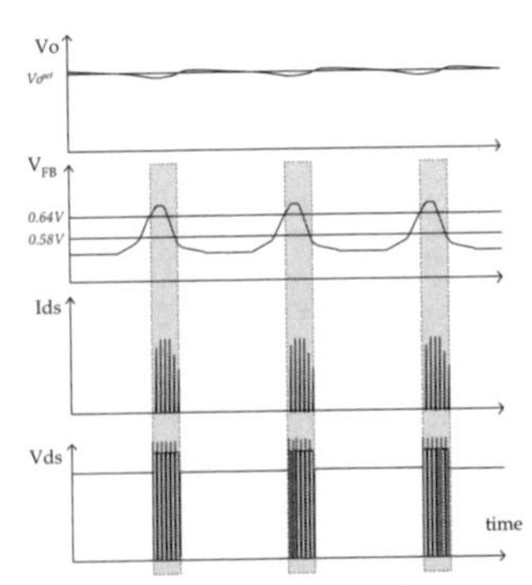

Figure 13. Burst mode operation

7. Frequency Modulation : EMI reduction can be accomplished by modulating the switching frequency of a SMPS. Frequency modulation can reduce EMI by spreading the energy over a wider frequency range. The amount of EMI reduction is directly related to the level of modulation (Fmod) and the rate of modulation. As can be seen in Figure 14, the frequency changes from 130kHz to 138kHz in 4mS for the FSD200/FSD210. Frequency modulation allows the use of a cost effective inductor instead of an AC input mode choke to satisfy the requirements of world wide EMI limits.

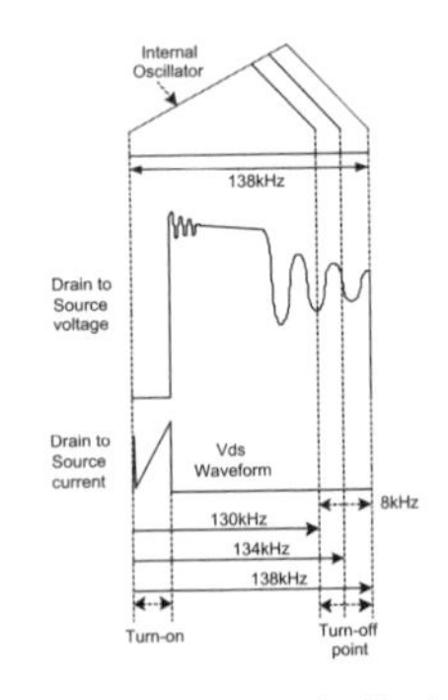

Figure 14. Frequency Modulation Waveforms

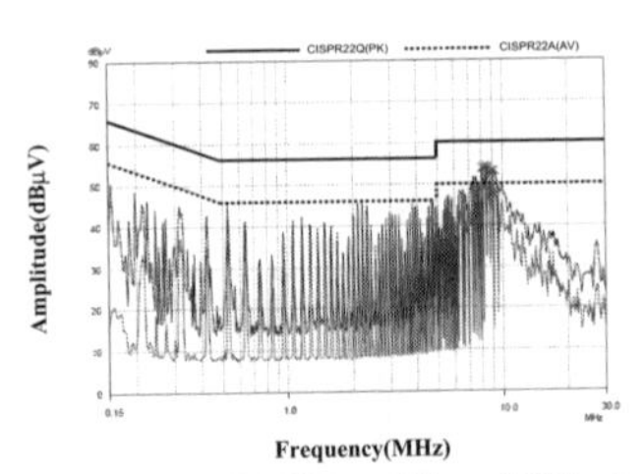

Figure 15. FSDH0165 Full Range EMI scan(100kHz, no Frequency Modulation) with charger set

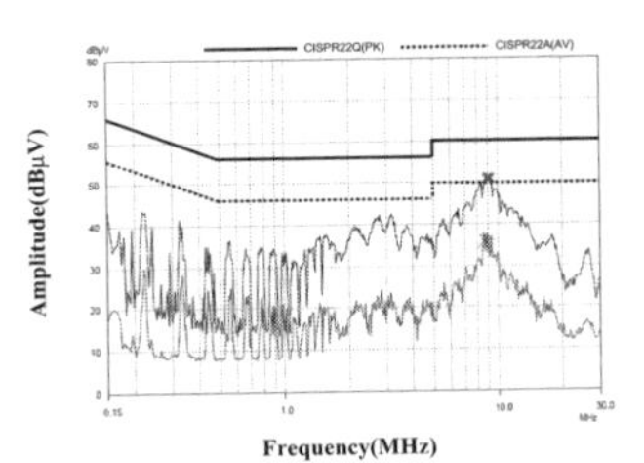

Figure 16. FSD210 Full Range EMI scan(134kHz, with Frequency Modulation) with charger set

Typical application circuit

Application	Output power	Input voltage	Output voltage (Max current)
Cellular Phone Charger	3.38W	Universal input (85-265Vac)	5.2V (650mA)

Features

- High efficiency (>67% at Universal Input)
- Low zero load power consumption (<100mW at 240Vac) with FSD210
- Low component count
- Enhanced system reliability through various protection functions
- Internal soft-start (3ms)
- Frequency Modulation for low EMI

Key Design Notes

- The constant voltage (CV) mode control is implemented with resistors, R8, R9, R10 and R11, shunt regulator, U2, feedback capacitor, C9 and opto-coupler, U3.
- The constant current (CC) mode control is designed with resistors, R8, R9, R15, R16, R17 and R19, NPN transistor, Q1 and NTC, TH1. When the voltage across current sensing resistors, R15,R16 and R17 is 0.7V, the NPN transistor turns on and the current through the opto coupler LED increases. This reduces the feedback voltage and duty ratio. Therefore, the output voltage decreases and the output current is regulated.
- The NTC(negative thermal coefficient) is used to compensate the temperature characteristics of the transistor Q1.

1. Schematic

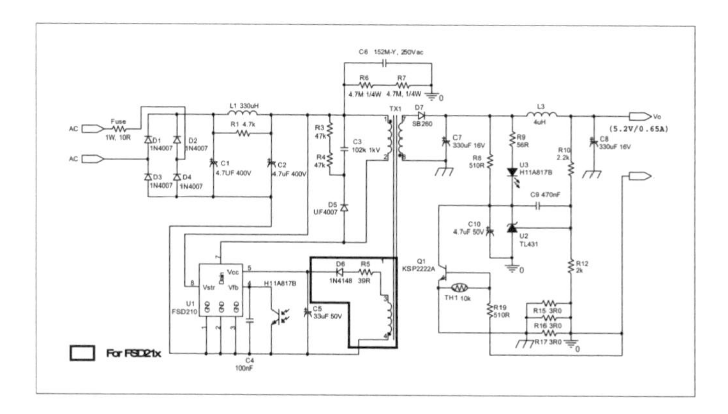

2. Demo Circuit Part List

Reference	Part #	Quantity	Description	Requirement/Comment
D1, D2, D3, D4	1N4007	4	1A/1000V Junction Rectifier	DO41 Type
D5	UF4008	1	1A/1000V Ultra Fast Diode	DO41 Type
D6	1N4148	1	10mA/100V Junction Diode	D0-213 Type
D7	SB260	1	2A/60V Schottky Diode	D0-41 Type
Q1	KSP2222A	1	Ic = 600mA, V_{CE} = 30V	TO-92 Type
U1	FSD210 (FSD200)	1	0.5A/700V	Iover = 0.3A, Fairchildsemi
U2	KA431AZ	1	V_{REF} = 2.495V (Typ.)	TO-92 Type, LM431
U3	H11A817A	1	CTR 80~160%	

3. Transformer Schematic Diagram

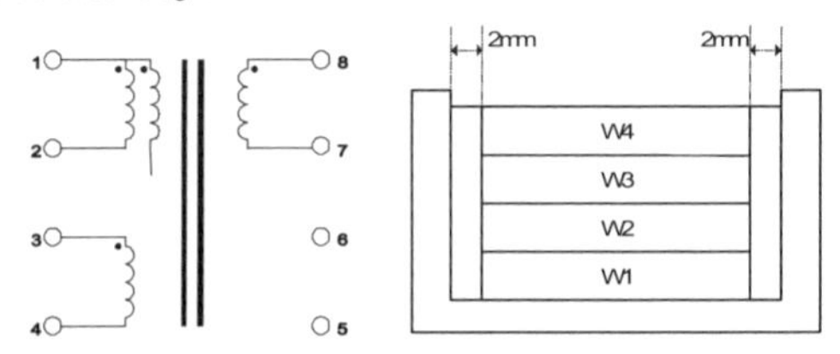

CORE: EE1616
BOBBIN: EE1616(H)

4. Winding Specification

No.	Pin (S→F)	Wire	Turns	Winding Method
W1	1→2	0.16 ϕ x 1	99 Ts	SOLENOID WINDING
INSULATION: POLYESTER TAPE t = 0.025mm/10mm, 2Ts				
W2	4→3	0.16 ϕ x 1	18 Ts	CENTER SOLENOID WINDING
INSULATION: POLYESTER TAPE t = 0.025mm/10mm, 2Ts				
W3	1→ open	0.16 ϕ x 1	50 Ts	SOLENOID WINDING
INSULATION: POLYESTER TAPE t = 0.025mm/10mm, 3Ts				
W4	8→7	0.40 ϕ x 1	9 Ts	SOLENOID WINDING
INSULATION: POLYESTER TAPE t = 0.025mm/10mm, 3Ts				

5. Electrical Characteristics

ITEM	TERMINAL	SPECIFICATION	REMARKS
INDUCTANCE	1–2	1.6 mH	1 kHz, 1 V
LEAKAGE L	1–2	50 uH	3, 4, 7, 8 short 100 kHz, 1 V

Typical application circuit

Application	Output power	Input voltage	Output voltage (Max current)
Non Isolation Buck	1.2W	Universal dc input (100 ~ 375Vac)	12V (100mA)

Features

- Non isolation buck converter
- Low component count
- Enhanced system reliability through various protection functions

Key Design Notes

- The output voltage(12V) is regulated with resistors, R1, R2 and R3, zener diode, D3, the transistor, Q1 and the capacitor, C2. While the FSD210 is off diodes, D1 and D2, are on. At this time the output voltage, 12V, can be sensed by the feedback components above. This output is also used with bias voltage for the FSD210.
- R, 680K, is to prevent the OLP(over load protection) at startup.
- R, 8.2K, is a dummy resistor to regulate output voltage in light load.

1. Schematic

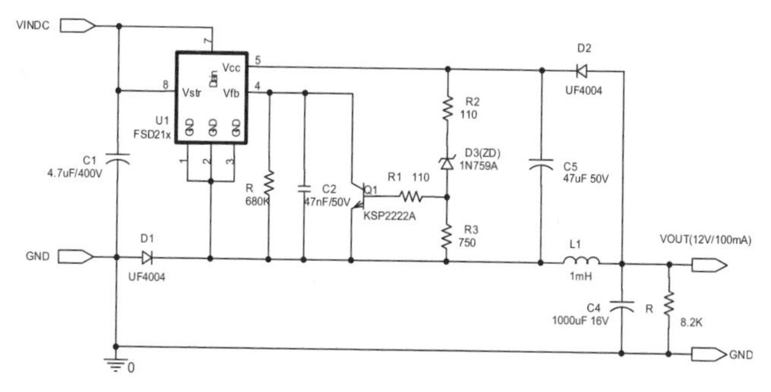

2. Demo Circuit Part List

Reference	Part #	Quantity	Description	Requirement/Comment
D1, D2	UF4007	2	1A/1000V Ultra Fast Diode	DO41 Type
Q1	KSP2222A	1	Ic=200mA, Vcc=40V	TO-92 Type
ZD1	1N759A	1	12VZD/0.5W	DO-35 Type
U1	FSD210	1	0.5A/700V	Iover=0.3A

Layout Considerations (for Flyback Convertor)

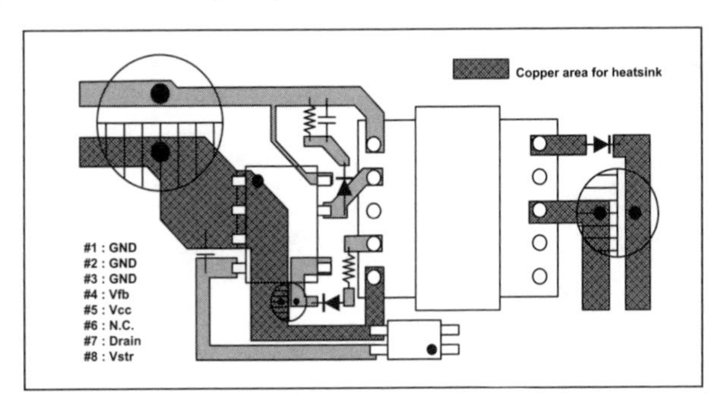

Figure 17. Layout Considerations for FSD2x0 using 7DIP

Package Dimensions

7-DIP

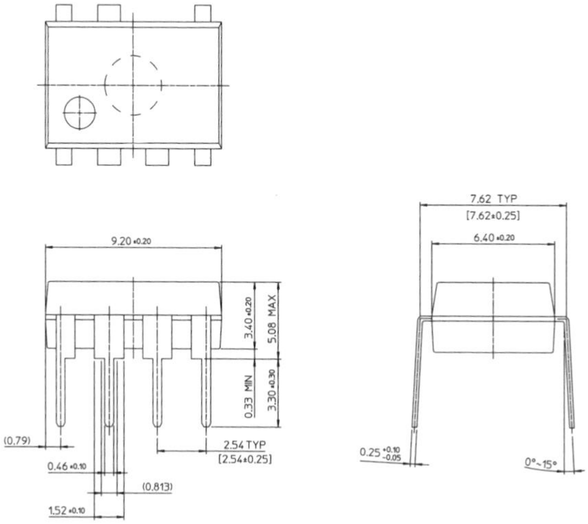

9.20 ±0.20
3.40 ±0.20
5.08 MAX
0.33 MIN
3.30 ±0.30
(0.79)
0.46 ±0.10
(0.813)
1.52 ±0.10
2.54 TYP
[2.54±0.25]
7.62 TYP
[7.62±0.25]
6.40 ±0.20
$0.25^{+0.10}_{-0.05}$
0°~15°

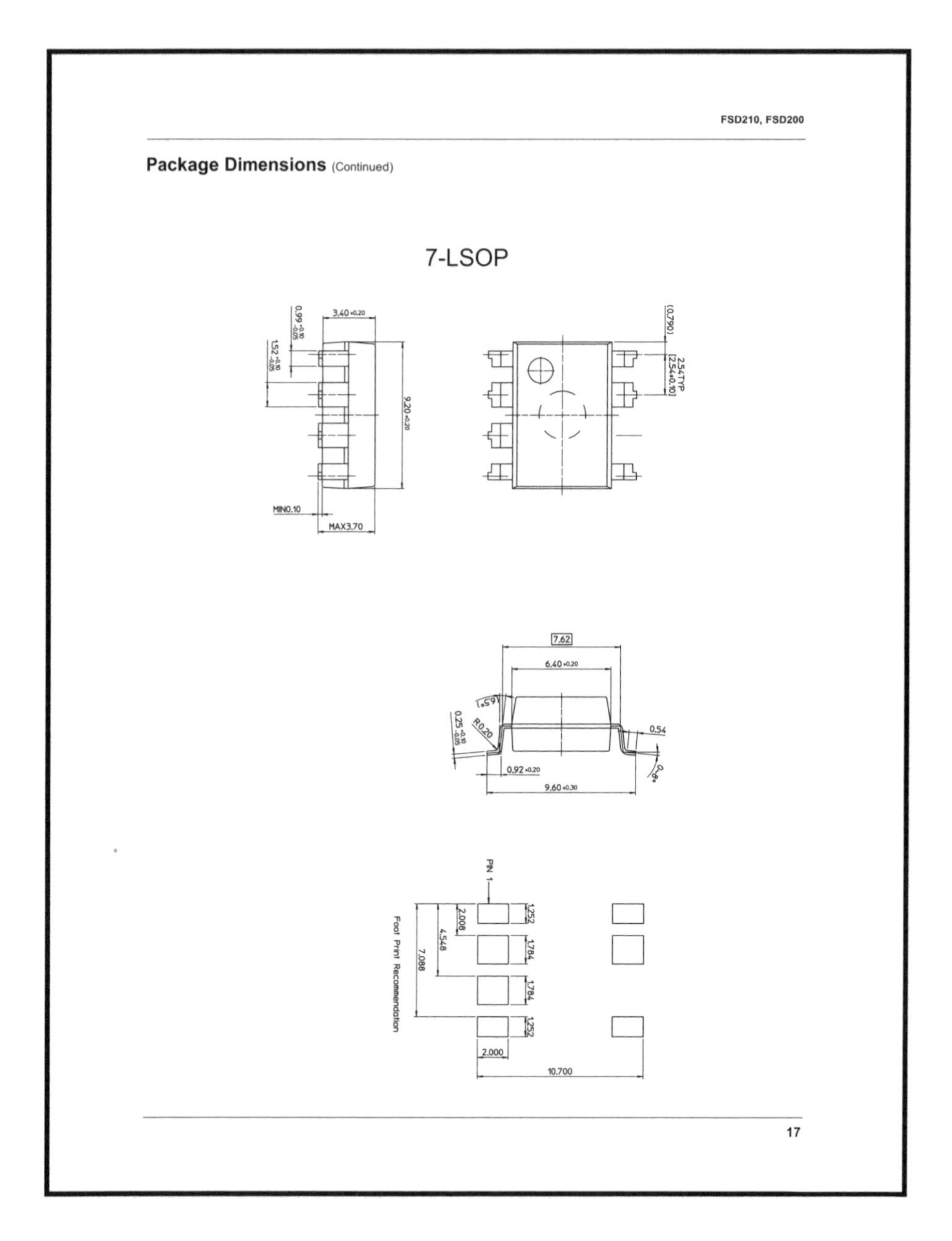

FSD210, FSD200

Package Dimensions (Continued)

7-LSOP

17

Ordering Information

Product Number	Package	Rating	Topr (°C)
FSD210	7DIP	700V, 0.5A	–25°C to +85°C
FSD200	7DIP	700V, 0.5A	–25°C to +85°C
FSD210M	7LSOP	700V, 0.5A	–25°C to +85°C
FSD200M	7LSOP	700V, 0.5A	–25°C to +85°C

Appendix D

Fairchild Specifications for FAN5307

www.fairchildsemi.com

FAN5307

High-Efficiency Step-Down DC-DC Converter

Features

- 95% Efficiency, Synchronous Operation
- Adjustable Output Voltage Option from 0.7V to V_{IN}
- 2.5V to 5.5V Input Voltage Range
- Customized Fixed Output Voltage Options
- Up to 300mA Output Current
- Fixed Frequency 1MHz PWM Operation
- High Efficiency Power Save Mode
- 100% Duty Cycle Low Dropout Operation
- Soft Start
- Dynamic Output Voltage Positioning
- 15μA Quiescent Current
- Excellent Load Transient Response
- 5-Lead SOT-23 Package
- 6-Lead MLP 3x3mm Package

Applications

- Pocket PCs, PDAs
- Cell Phones
- Battery-Powered Portable Devices
- Digital Cameras
- Low Power DSP Supplies

Description

The FAN5307, a high efficiency low noise synchronous PWM current mode and Pulse Skip (Power Save) mode DC-DC converter is designed for battery-powered applications. It provides up to 300mA of output current over a wide input range from 2.5V to 5.5V. The output voltage can be either internally fixed or externally adjustable over a wide range of 0.7V to 5.5V by an external voltage divider. Custom output voltages are also available.

At moderate and light loads pulse skipping modulation is used. Dynamic voltage positioning is applied, and the output voltage is shifted 0.8% above nominal value for increased headroom during load transients. At higher loads the system automatically switches to current mode PWM control, operating at 1 MHz. A current mode control loop with fast transient response ensures excellent line and load regulation. In Power Save mode, the quiescent current is reduced to 15μA in order to achieve high efficiency and to ensure long battery life. In shut-down mode, the supply current drops below 1μA. The device is available in 5-lead SOT-23 and 6-lead MLP 3x3mm packages.

Typical Application

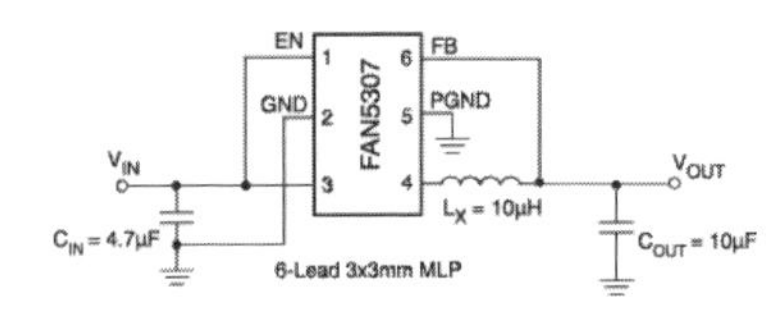

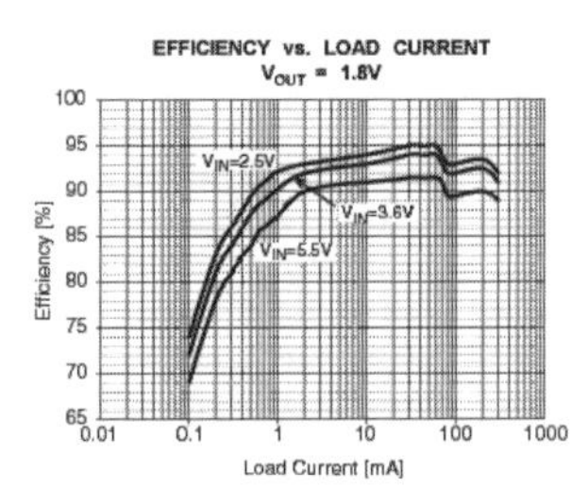

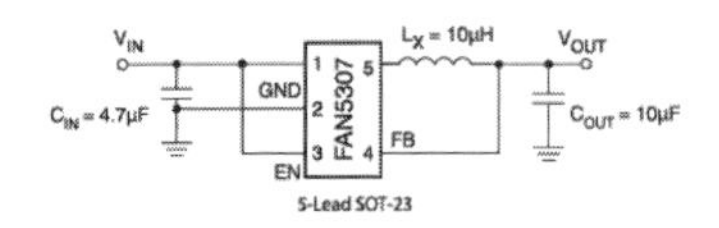

Pin Assignment

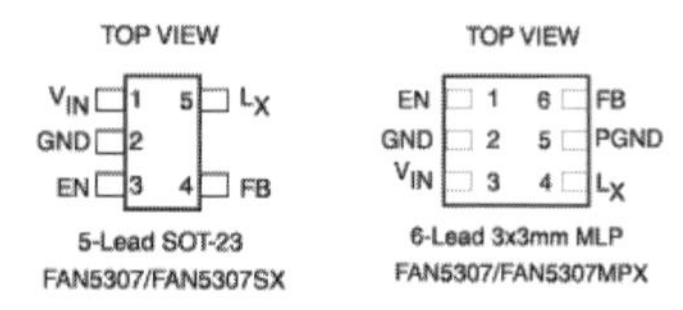

Pin Description 5SOT-23

Pin No.	Pin Name	Pin Description
1	V_{IN}	Supply voltage input.
2	GND	Ground.
3	EN	Enable Input. Logic high enables the chip and logic low disables the chip and reduces supply current to <1μA. Do not float this pin.
4	FB	Feedback Input. In case of fixed voltage options, connect this pin directly to the output. For an adjustable voltage option, connect this pin to the resistor divider.
5	L_X	Inductor pin. This pin is connected to the internal MOSFET switches.

Pin Description 6-Lead 3x3mm MLP

Pin No.	Pin Name	Pin Description
1	EN	Enable Input. Logic high enables the chip and logic low disables the chip and reduces supply current to <1μA. Do not float this pin.
2	GND	Reference ground.
3	V_{IN}	Supply voltage input.
4	L_X	Inductor pin. This pin is connected to the internal MOSFET switches.
5	PGND	Power ground. Internal N-channel MOSFET is connected to this pin.
6	FB	Feedback Input. In case of fixed voltage options, connect this pin directly to the output. For an adjustable voltage option, connect this pin to the resistor divider.

Absolute Maximum Ratings

Parameter		Min	Max	Unit
V_{IN}		-0.3	6.5	V
Voltage on any other pin		GND-0.3	V_{IN} + 0.3	V
Power Dissipation (Continuous, at T_A = 25°C) (Note 1)			357	mW
Lead Soldering Temperature (10 seconds)			260	°C
Storage Temperature		-65	150	°C
Electrostatic Discharge (ESD) Protection Level (Note 2)	HBM	4		kV
	CDM	1		

Recommended Operating Conditions

Parameter	Min	Typ	Max	Unit
Supply Voltage Range	2.5		5.5	V
Output Voltage Range, Adjustable Version	0.7		V_{IN}	V
Output Current			300	mA
Inductor (Note 3)		10		µH
Input Capacitor (Note 3)		4.7		µF
Output Capacitor (Note 3)		10		µF
Operating Ambient Temperature Range	-40		+85	°C
Operating Junction Temperature Range	-40		+125	°C

Notes:

1. Derate above 25°C at a rate of 35°C/W.
2. Using Mil Std. 883E, method 3015.7(Human Body Model) and EIA/JESD22C101-A (Charge Device Model).
3. Refer to the applications section for further details.

General Electrical Characteristics

V_{IN} = 2.5V to 5.5V, I_{OUT} = 200mA, EN = V_{IN}, C_{IN} = 4.7µF, C_{OUT} = 22µF, L_X = 10µH, T_A = -40°C to +85°C, unless otherwise noted. Typical values are at T_A = 25°C.

Parameter	Conditions	Min.	Typ.	Max.	Units
	Input Voltage	2.5		5.5	V
Quiescent Current	I_{OUT} = 0mA, Device is not switching		15	30	µA
Shutdown Supply Current	EN = GND		0.1	1	µA
Enable High Input Voltage		1.3			V
Enable Low Input Voltage				0.4	V
En Input Bias Current	EN = V_{IN} or GND		0.01	0.1	µA
PMOS On Resistance	V_{IN} = V_{GS} = 3.6V		530	690	mΩ
	V_{IN} = V_{GS} = 2.5V		670	850	
NMOS On Resistance	V_{IN} = V_{GS} = 3.6V		430	540	mΩ
	V_{IN} = V_{GS} = 2.5V		530	660	
P-channel Current Limit	2.5V < V_{IN} < 5.5V	400	520	700	mA
N-channel Leakage Current	V_{DS} = 5.5V		0.1	1	µA
P-channel Leakage Current	V_{DS} = 5.5V		0.1	1	µA
Switching Frequency		800	1000	1200	kHz
Line Regulation	V_{IN} = 2.5 to 5.5V, I_{OUT} = 10mA		0.16		%/V
Load Regulation 6-Lead 3x3mm MLP	100mA ≤ I_{OUT} ≤ 300mA		0.0014		%/mA
Load Regulation 5-Lead SOT-23	100mA ≤ I_{OUT} ≤ 300mA		0.0022		%/mA
Output Voltage Accuracy (5SOT-23)	V_{IN} = 2.5 to 4.5V, 0mA ≤ I_{OUT} ≤ 300mA	-3		3	%
	V_{IN} = 2.5 to 5.5V, 0mA ≤ I_{OUT} ≤ 300mA	-4		3	%
Leakage Current into SW Pin	V_{IN} > V_{OUT}, 0V ≤ V_{SW} ≤ V_{IN}		0.1	1	µA
Reverse Leakage Current into pin SW	V_{IN} = Open, EN = GND, V_{sw} = 5.5V		0.1	1	µA
Output Voltage Accuracy (6-Lead 3x3mm MLP)	V_{IN} = 2.5 to 5.5V, 0mA ≤ I_{OUT} ≤ 300mA	-3		3	%

Electrical Characteristics For Adjustable Version

V_{IN} = 2.5V to 5.5V, I_{OUT} = 200mA, EN = V_{IN}, C_{IN} = 4.7µF, C_{OUT} = 22µF, L_X = 10µH, T_A = -40°C to +85°C, unless otherwise noted. Typical values are at T_A = 25°C.

Parameter	Conditions	Min.	Typ.	Max.	Units
Feedback (FB) Voltage			0.5		V

Electrical Characteristics for Fixed V_{OUT} = 1.8V Version

V_{IN} = 2.5V to 5.5V, I_{OUT} = 200mA, EN = V_{IN}, C_{IN} = 4.7µF, C_{OUT} = 22µF, L_X = 10µH, T_A = -40°C to +85°C, unless otherwise noted. Typical values are at T_A = 25°C.

Parameter	Conditions	Min.	Typ.	Max.	Units
PFM to PWM Transition Voltage (Note 4)	V_{IN} = 3.7V, T_A = 25°C, 0.1mA ≤ I_{OUT} ≤ 300mA			72	mV
PFM to PWM Transition Voltage (Note 4)	V_{IN} = 4.2V, T_A = 25°C, 0.1mA ≤ I_{OUT} ≤ 300mA			72	mV
Output Voltage during Mode Transition (Note 5, 6)		1.7		1.93	V
Over Voltage Clamp Threshold	Incl. line, load, load transients, and temperature		1.878	1.93	V

Note:

4. Transition voltage is defined as the difference between the output voltage measured at 0.1m A (PFM mode) and 300mA (PWM mode), respectively.

5.

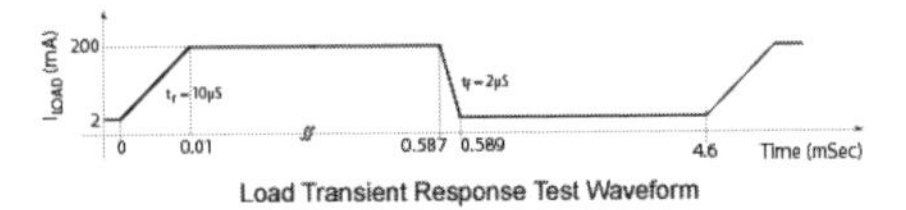

Load Transient Response Test Waveform

6. These limits also apply to any mode transition caused by any kind of load transition within specified output current range.

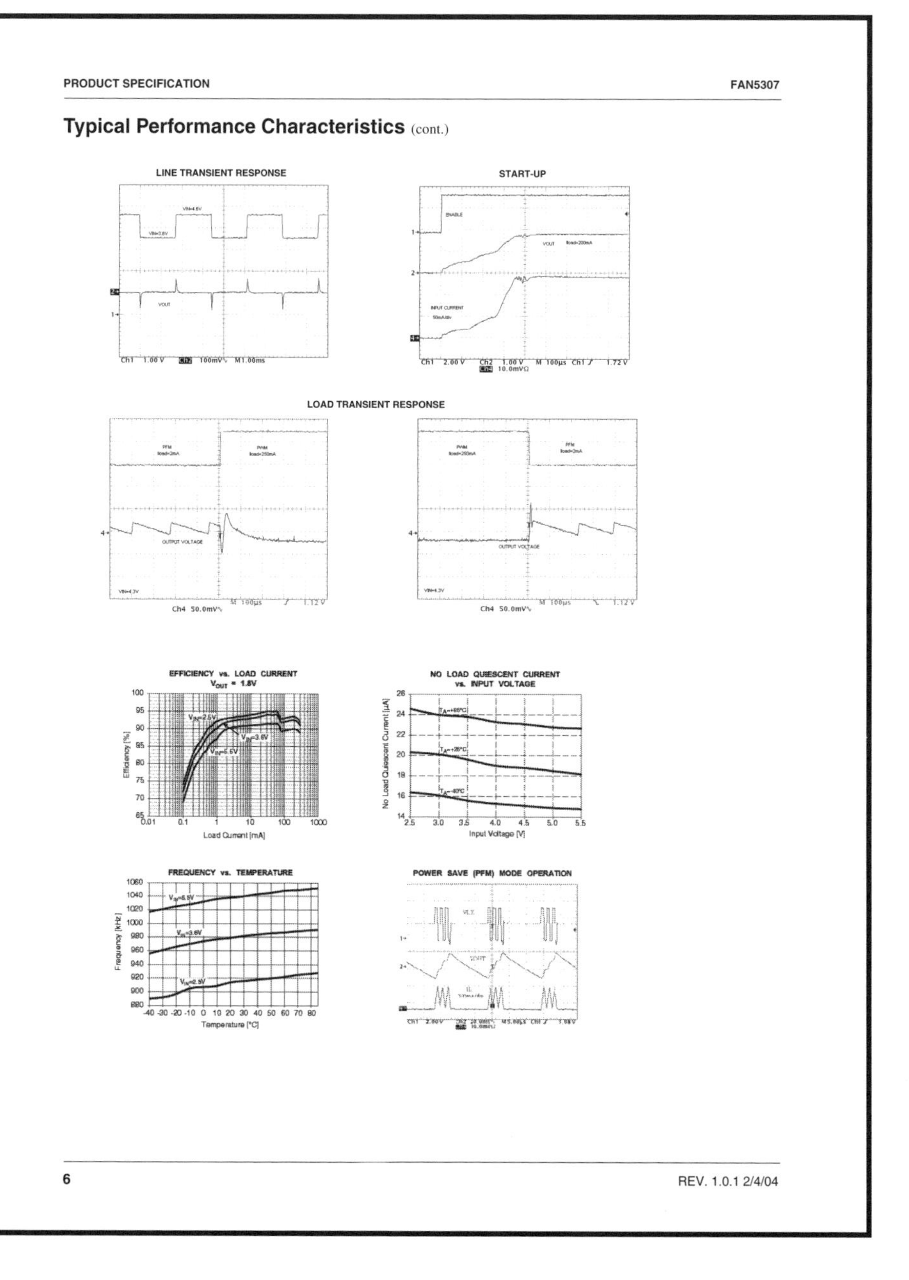

Typical Performance Characteristics (cont.)

Block Diagram

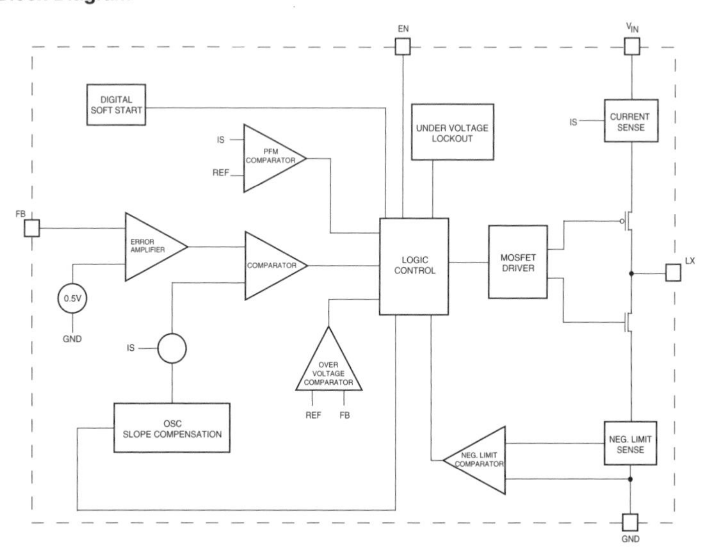

Detailed Operation Description

The FAN5307 is a step-down converter operating in a current-mode PFM/PWM architecture with a typical switching frequency of 1MHz. At moderate to heavy loads, the converter operates in pulse-width-modulation (PWM) mode. At light loads the converter enters a power-save mode (PFM pulse skipping) to keep the efficiency high.

PWM Mode

In PWM mode, the device operates at a fixed frequency of 1MHz. At the beginning of each clock cycle, the P-channel transistor is turned on. The inductor current ramps up and is monitored via an internal circuit. The P-channel switch is turned off when the sensed current causes the PWM comparator to trip when the output voltage is in regulation or when the inductor current reaches the current limit (set internally to typically 520mA). After a minimum dead time the N-channel transistor is turned on and the inductor current ramps down. As the clock cycle is completed, the N-channel switch is turned off and the next clock cycle starts.

PFM (Power Save) Mode

As the load current decreases and the peak inductor current no longer reaches the typical threshold of 80mA, the converter enters pulse-frequency-modulation (PFM) mode. In PFM mode the device operates with a variable frequency and constant peak current thus reducing the quiescent current to minimum. Consequently, the high efficiency is maintained at light loads. As soon as the output voltage falls below a threshold, set at 0.8% above the nominal value, the P-channel transistor is turned on and the inductor current ramps up. The P-channel switch turns off and the N-channel turns on as the peak inductor current is reached (typical 140mA).

The N-channel transistor is turned off before the inductor current becomes negative. At this time the P-channel is switched on again starting the next pulse. The converter continues these pulses until the high threshold (typical 1.6% above nominal value) is reached. A higher output voltage in

PFM mode gives additional headroom for the voltage drop during a load transient from light to full load. The voltage overshoot during this load transient is also minimized due to active regulation during turning on the N-channel rectifier switch. The device stays in sleep mode until the output voltage falls below the low threshold. The FAN5307 enters the PWM mode as soon as the output voltage can no longer be regulated in PFM with constant peak current.

100% Duty Cycle Operation

As the input voltage approaches the output voltage and the duty cycle exceeds the typical 95%, the converter turns the P-channel transistor continuously on. In this mode the output voltage is equal to the input voltage minus the voltage drop across the P-channel transistor:

$V_{OUT} = V_{IN} - I_{LOAD} \times (R_{dsON} + R_L)$, where

R_{dsON} = P-channel switch ON resistance
I_{LOAD} = Output current
R_L = Inductor DC resistance

Soft Start

The FAN5307 has an internal soft-start circuit that limits the inrush current during start-up. This prevents possible voltage drops of the input voltage and eliminates the output voltage overshoot. The soft-start is implemented as a digital circuit increasing the switch current in four steps to the P-channel current limit (520mA). Typical start-up time for a 10μF output capacitor and a load current of 200mA is 500μs.

Short-Circuit Protection

The switch peak current is limited cycle by cycle to a typical value of 520mA. In the event of a output voltage short circuit the device operates at minimum duty cycle, therefore the average input current is typically 100mA.

Application Information

Adjustable Output Voltage Version

The output voltage for the adjustable version is set by the external resistor divider, as shown below:

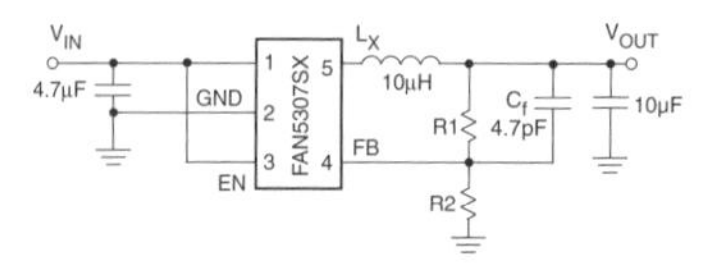

and is calculated as:

$$V_{OUT} = 0.5V \times \left[1 + \frac{R_1}{R_2}\right]$$

To reduce noise sensitivity, R1 + R2 should not exceed 800kΩ.

Inductor Selection

The inductor parameters directly related to device performances are saturation current and dc resistance. The FAN5307 operates with a typical inductor value of 10µH. The lower the dc resistance, the higher the efficiency. For saturation current, the inductor should be rated higher than the maximum load current plus half of the inductor ripple current that is calculated as follows:

$$\Delta I_L = V_{OUT} \times \frac{1 - (V_{OUT} / V_{IN})}{L \times f}$$

where:
f = Switching Frequency
L = Inductor Value
ΔI_L = Inductor Ripple Current

Inductor Value	Vendor	Part Number	Performance
10µH	Sumida	CDRH5D28-100	High Efficiency
		CDRH5D18-100	
		CDRH4D28-100	
	Murata	LQH66SN100M 01L	
6.8µH	Sumida	CDRH3D16-6R8	Smallest Solution
10µH		CDRH4D18-100	
		CR32-100	
		CR43-100	
	Murata	LQH4C100K04	

Table 1: Recommended Inductors

Input Capacitor Selection

For best performances, a low ESR input capacitor is required. A ceramic capacitor of at least 4.7µF, placed as close to the input pin of the device is recommended.

Output Capacitor Selection

The FAN5307's switching frequency of 1MHz allows the use of a low ESR ceramic capacitor with a value of 10µF to 22µF. This provides low output voltage ripple. In power save mode the output voltage ripple is independent of the output capacitor value and the ripple is determined by the internal comparator thresholds. The typical output voltage ripple at light load is 1% of the nominal output voltage.

Capacitor Value	Vendor	Part Number
4.7µF	Taiyo Yuden	JMK212BY475MG
10µF		JMK212BJ106MG
		JMK316BJ106KL
	TDK	C12012X5ROJ106K
		C3216X5ROJ106M
22µF	Murata	GRM32DR60J226K

Table 2: Recommended Capacitors

PCB Layout Recommendations

The inherently high peak currents and switching frequency of the power supplies require a careful PCB layout design. Therefore, use wide traces for the high current path and place the input capacitor, the inductor, and the output capacitor as close as possible to the integrated circuit terminals. For the adjustable version the resistor divider should be routed away from the inductor to avoid electromagnetic interference.

The 6-lead MLP version of the FAN5307 separates the high current ground from the reference ground, therefore it is more tolerant to the PCB layout design and shows better performance.

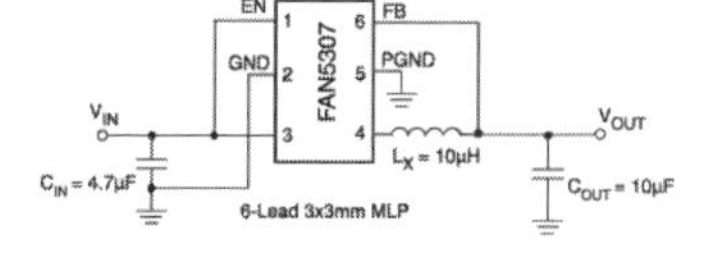

Mechanical Dimensions

6-Lead 3x3mm MLP Package

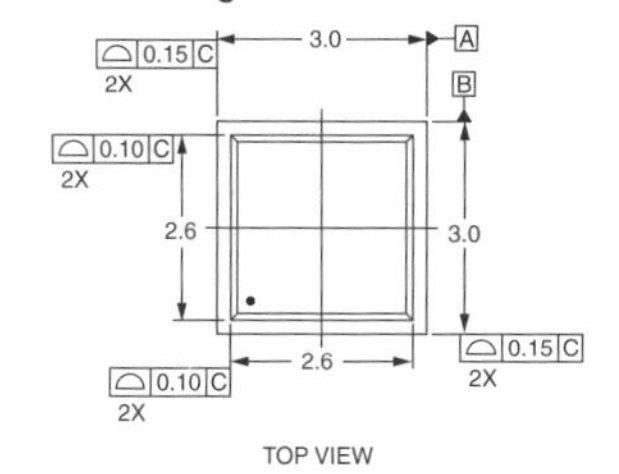

TOP VIEW

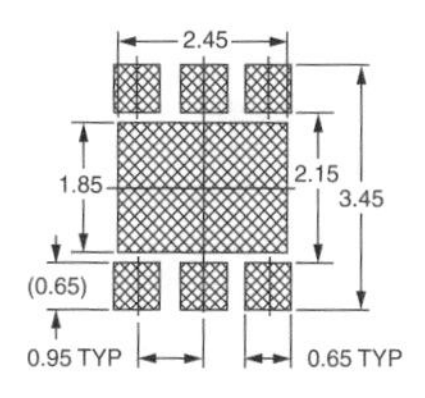

RECOMMENDED LAND PATTERN

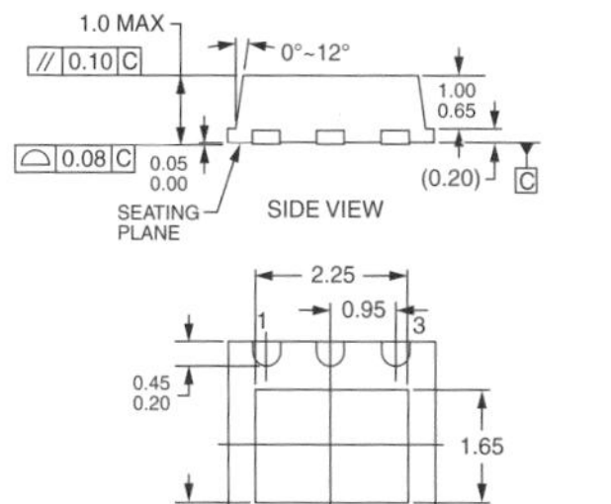

SIDE VIEW

BOTTOM VIEW

NOTES:

A. CONFORMS TO JEDEC REGISTRATION MO-229, VARIATION VEEA, DATED 11/2001
B. DIMENSIONS ARE IN MILLIMETERS.
C. DIMENSIONS AND TOLERANCES PER ASME Y14.5M, 1994

Mechanical Dimensions

5-Lead SOT-23 Package

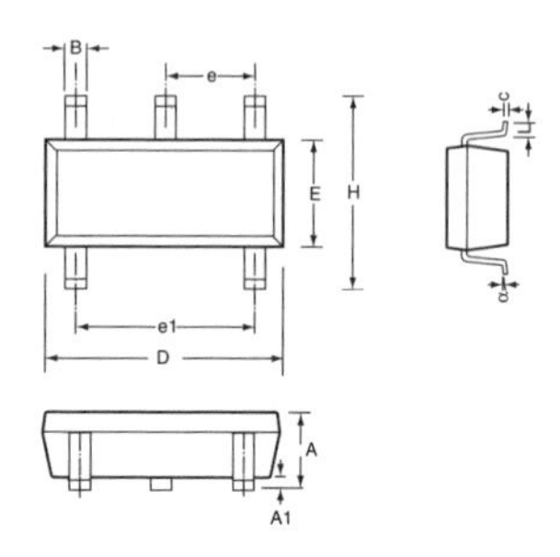

Symbol	Inches		Millimeters		Notes
	Min	Max	Min	Max	
A	.035	.057	.90	1.45	
A1	.000	.006	.00	.15	
B	.008	.020	.20	.50	
c	.003	.010	.08	.25	
D	.106	.122	2.70	3.10	
E	.059	.071	1.50	1.80	
e	.037 BSC		.95 BSC		
e1	.075 BSC		1.90 BSC		
H	.087	.126	2.20	3.20	
L	.004	.024	.10	.60	
α	0°	10°	0°	10°	

Notes:

1. Package outline exclusive of mold flash & metal burr.
2. Package outline exclusive of solder plating.
3. EIAJ Ref Number SC_74A

Ordering Information

Product Number	V_{OUT} (V)	Package Type	Order Code
FAN5307	1.8	5-Lead SOT-23 Tape and Reel	FAN5307S18X
	1.8	6-Lead 3x3mm MLP Tape and Reel	FAN5307MP18X
	Adjustable	5-Lead SOT-23 Tape and Reel	FAN5307SX
	Adjustable	6-Lead 3x3mm MLP Tape and Reel	FAN5307MPX

1/20/04 0.0m 000
Stock#DS30005307

Appendix E

Fairchild Specifications for ACE1502

FAIRCHILD
SEMICONDUCTOR®

October 2002

ACE1502 Product Family Arithmetic Controller Engine (ACEx™) for Low Power Applications

General Description

The ACE1502 (Arithmetic Controller Engine) family of microcontrollers is a dedicated programmable monolithic integrated circuit for applications requiring high performance, low power, and small size. It is a fully static part fabricated using CMOS technology.

The ACE1502 product family has an 8-bit microcontroller core, 64 bytes of RAM, 64 bytes of data EEPROM and 2K bytes of code EEPROM. Its on-chip peripherals include a multifunction 16-bit timer, a watchdog/idle timer, and programmable under-voltage detection circuitry. On-chip clock and reset functions reduce the number of required external components. The ACE1502 product family is available in 8- and 14-pin SOIC, TSSOP and DIP packages.

Features

- Arithmetic Controller Engine
- 2K bytes on-board code EEPROM
- 64 bytes data EEPROM
- 64 bytes RAM
- Watchdog
- Multi-input wake-up on all eight general purpose I/O pins
- 16-bit multifunction timer with difference capture
- Hardware Bit–Coder (HBC)
- On-chip oscillator
  - — No external components
  - — 1μs instruction cycle time +/-2% accuracy
- Instruction set geared for block encryption
- On-chip Power-on Reset
- Programmable read and write disable functions
- Memory mapped I/O
- 32-level Low Voltage Detection
- Brown-out Reset
- Software selectable I/O option
  - — Push-pull outputs with tri-state option
  - — Weak pull-up or high impedance inputs
- Fully static CMOS
  - — Low power HALT mode (100nA @ 2.7V)
  - — Power saving IDLE mode
- Single supply operation
  - — 1.8-3.6V
- 40 years data retention
- 1.8V data EEPROM min writing voltage
- 1,000,000 data changes
- 8- and 14-pin SOIC, TSSOP and DIP packages
- In-circuit programming

Block and Connection Diagram

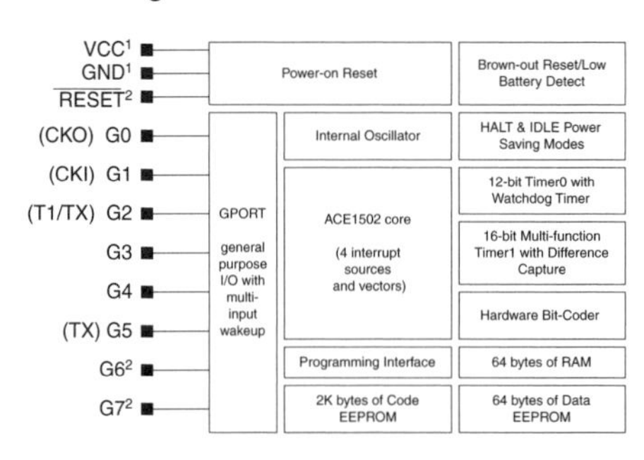

1. 100nf Decoupling capacitor recommended
2. Available only in the 14-pin package option

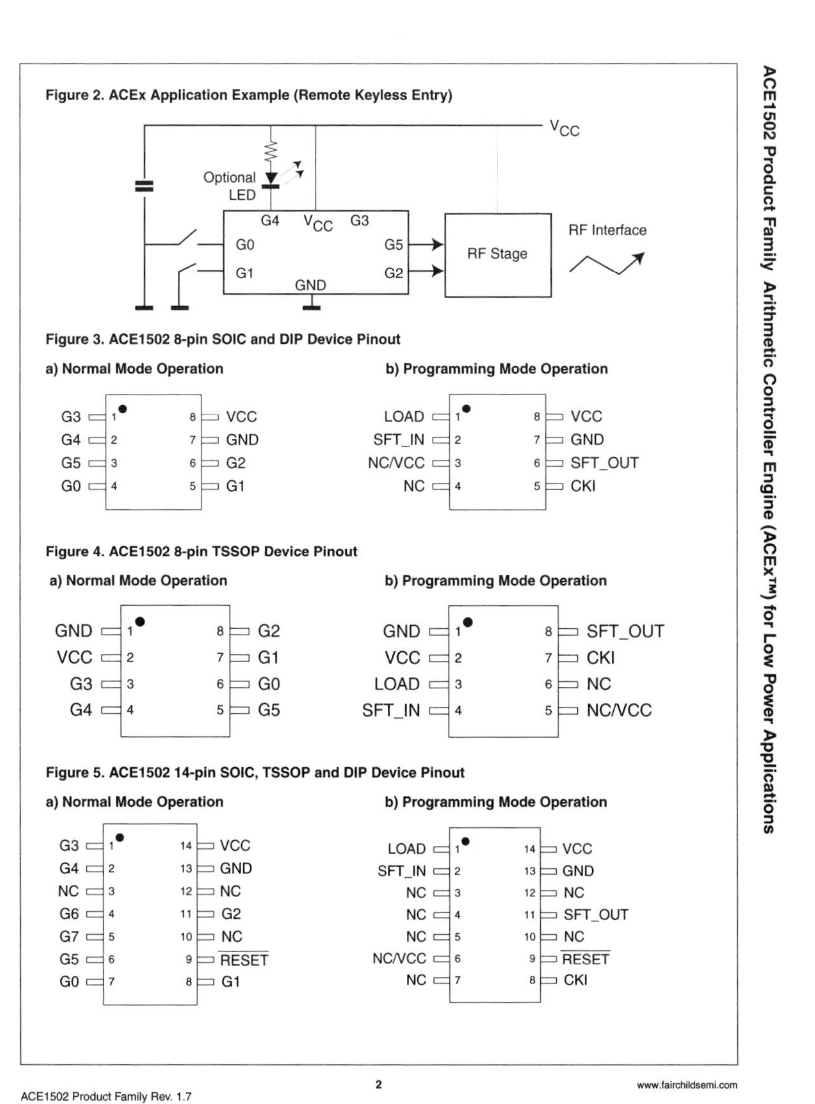

ACE1502 Product Family Arithmetic Controller Engine (ACEx™) for Low Power Applications

2. Electrical Characteristics

Absolute Maximum Ratings

Ambient Storage Temperature -65 °C to +150 °C
Input Voltage -0.3V to V_{CC} + 0.3V
Lead Temperature (10s max) +300°C
Electrostatic Discharge on all pins 2000V min

Operating Conditions

Relative Humidity (non-condensing) 95%
EEPROM write limits See DC Electrical Characteristics

Part Number	Operating Voltage	Ambient Operating Temperature
ACE1502E	1.8 to 3.6V	-40°C to +85°C
ACE1502V	1.8 to 3.6V	-40°C to +125°C

ACE1502 DC Electrical Characteristics, V_{CC} = 1.8 to 3.6V

All measurements are valid for ambient operating temperature unless otherwise stated.

Symbol	Parameter	Conditions	MIN	TYP	MAX	Units
Icc^{3}	Suppy Current - no data EEPROM write in progress	1.8V 2.2V 2.7V 3.6V		0.4 0.4 0.5 0.6	0.6 0.6 0.7 1.0	mA mA mA mA
Icc_H	HALT Mode current	2.7V @ 25°C 2.7V @ -40°C to +85°C		 100	400 5000	nA nA
		3.6V @ 25°C 3.6V @ -40°C to +85°C		 0.25	1000 10	nA μA
$Icc_L{}^{4}$	IDLE Mode current	1.8V 3.6V		210 250	 400	μA μA
Vcc_W	EEPROM write voltage	Code EEPROM in Programming Mode	3.0	3.3	3.6	V
		Data EEPROM in Operating Mode	1.8		3.6	V
S_{Vcc}	Power Supply Slope		1μs/V		10ms/V	
V_{IL}	Input Low with Schmitt Trigger buffer	Vcc = 2.2 - 3.6V			0.2Vcc	V
		Vcc < 2.2V			0.15Vcc	V
V_{IH}	Input High with Schmitt Trigger buffer	Vcc = 1.8 - 3.6V	0.8Vcc			V
I_{IP}	Input Pull-up Current	Vcc = 3.6V, V_{IN} = 0V	30	65	350	μA
I_{TL}	Tri-State Leakage	Vcc = 3.6V		2	200	nA
V_{OL}	Output Low Voltage: G0, G1, G2, G3, G4, G5, G6, G7	Vcc = 1.8 - 2.7V 2 mA sink			0.2Vcc	V
	Output Low Voltage: G0, G1, G2, G3, G4, G5, G6, G7	Vcc = 3.3 - 3.6V 7.0 mA sink			0.2Vcc	V
V_{OH}	Output High Voltage: G0, G1, G2, G3, G4, G5, G6, G7	Vcc = 2.2 - 2.7V 2 mA source	0.8Vcc			V
	Output High Voltage: G0, G1, G2, G3, G4, G5, G6, G7	Vcc = 3.3 - 3.6V 7 mA source	0.8Vcc			V

3. Icc active current is dependant on the program code.
4. Based on a continuous IDLE looping program.

ACE1502 AC Electrical Characteristics, Vcc = 1.8 to 3.6V

All measurements are valid for ambient operating temperature unless otherwise stated.

Parameter	Conditions	MIN	TYP	MAX	Units
Instruction cycle time from internal clock - setpoint	3.3V at +25°C	0.98	1.0	1.02	μs
Internal clock frequency variation	1.8V to 3.6V at constant temperature		1.2		%
	1.8V to 3.6V at full temperature range (Note 6)			6	%
Crystal oscillator frequency	(Note 5)			25	MHz
External clock frequency	(Note 5)			8	MHz
EEPROM write time			5.5	10	ms
Internal clock start up time	(Note 6)			2	ms
Oscillator start up time	(Note 6)			2400	cycles

5. The maximum permissible frequency is guaranteed by design but is not 100% tested

6. The parameter is characterized but is not 100% tested, contact Fairchild for additional characterization data.

ACE1502 Electrical Characteristics for programming

All data valid at ambient temperature between 3.0V and 3.6V. The following characteristics are guaranteed by design but are not 100% tested. See "EEPROM write time" in the AC Electrical Characteristics for definition of the programming ready time.

Parameter	Description	MIN	MAX	Units
t_{HI}	CLOCK high time	500	DC	ns
t_{LO}	CLOCK low time	500	DC	ns
t_{DIS}	SHIFT_IN setup time	100		ns
t_{DIH}	SHIFT_IN hold time	100		ns
t_{DOS}	SHIFT_OUT setup time	100		ns
t_{DOH}	SHIFT_OUT hold time	900		ns
T_{RESET}	Power On Reset time	3.2	4.5	ms
t_{LOAD1}, t_{LOAD2}, t_{LOAD3}, t_{LOAD4}	LOAD timing	5		μs

ACE1502 Low Battery Detect (LBD) Characteristics, Vcc = 1.8 to 3.6V

Parameter	Conditions	MIN	TYP	MAX	Units
LBD voltage threshold variation	-40°C to +85°C	-5		+5	%

ACE1502 Brown-out Reset (BOR) Characteristics, Vcc = 1.8 to 3.6V

Parameter	Conditions	MIN	TYP	MAX	Units
BOR voltage threshold variation	-40°C to +85°C	1.72	1.83	1.92	V

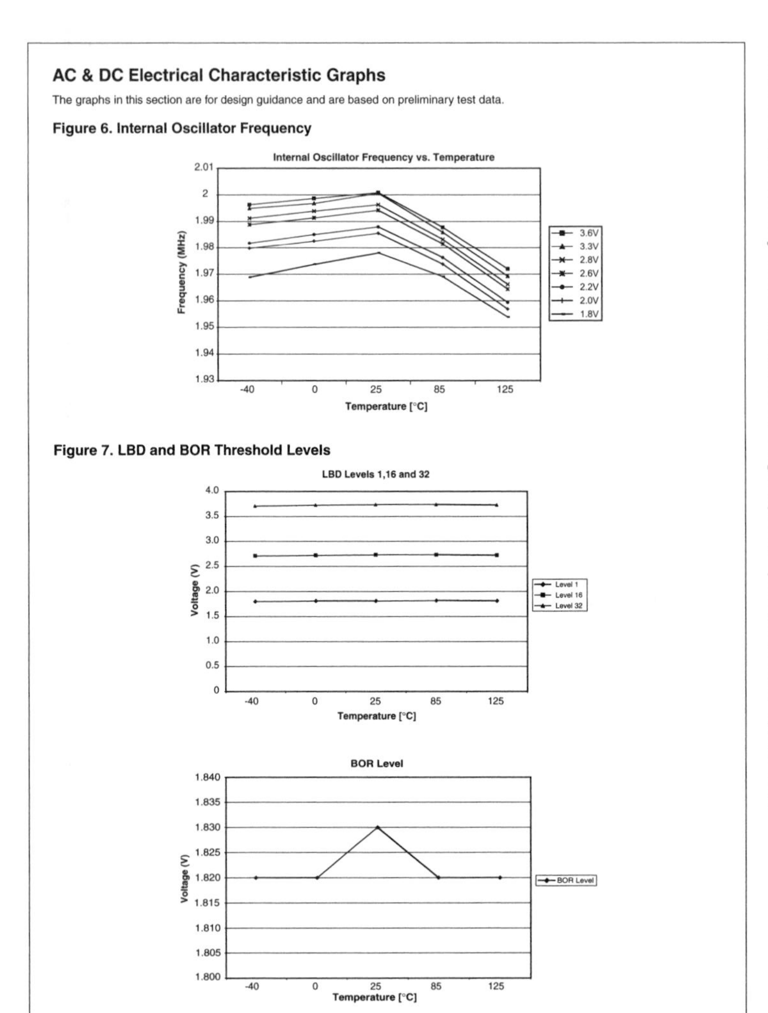

AC & DC Electrical Characteristic Graphs

The graphs in this section are for design guidance and are based on preliminary test data.

Figure 6. Internal Oscillator Frequency

Figure 7. LBD and BOR Threshold Levels

ACE1502 Product Family Arithmetic Controller Engine (ACEx™) for Low Power Applications

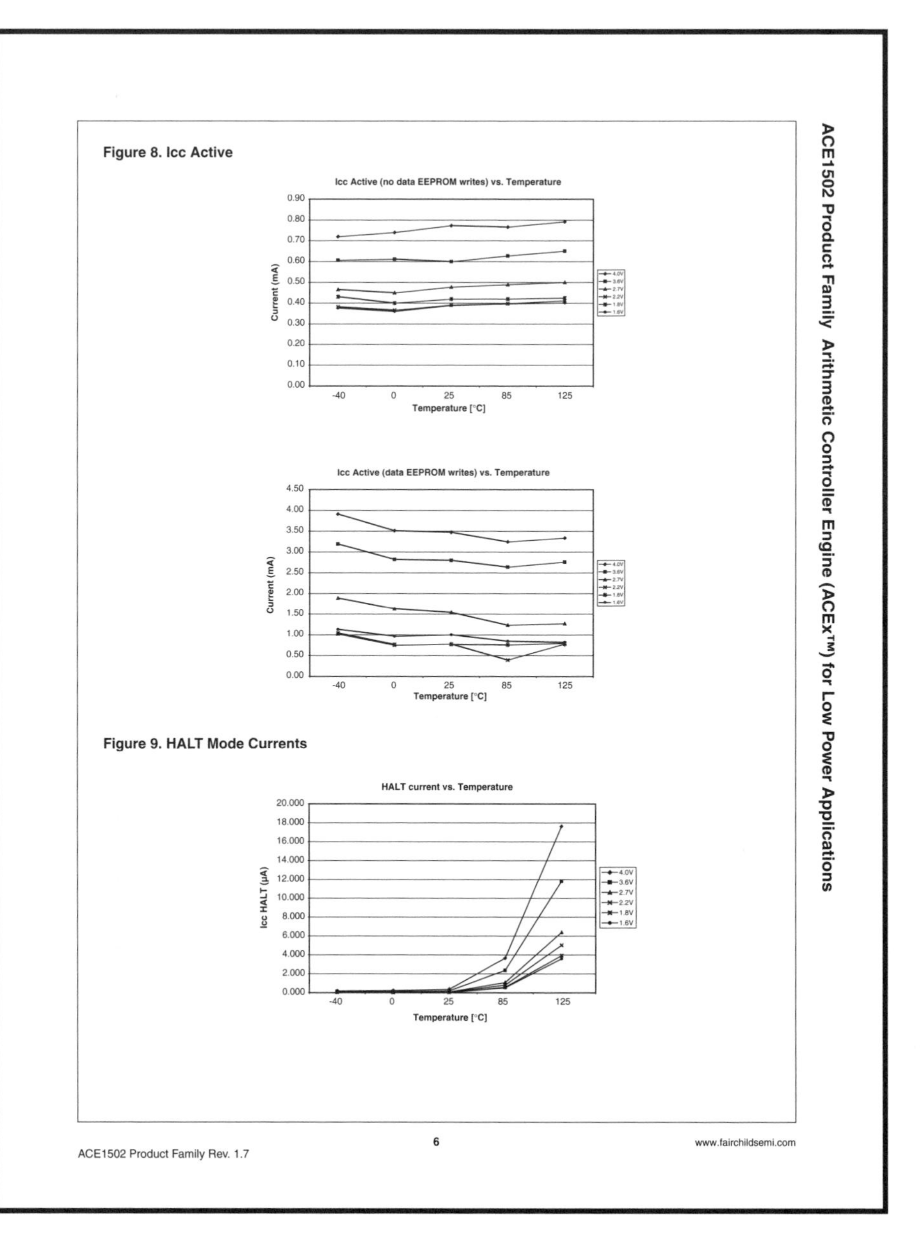

ACE1502 Product Family Arithmetic Controller Engine (ACEx™) for Low Power Applications

Figure 8. Icc Active

Figure 9. HALT Mode Currents

Figure 10. IDLE Mode Currents

Figure 11. V_{OL}/V_{OH} vs. Current

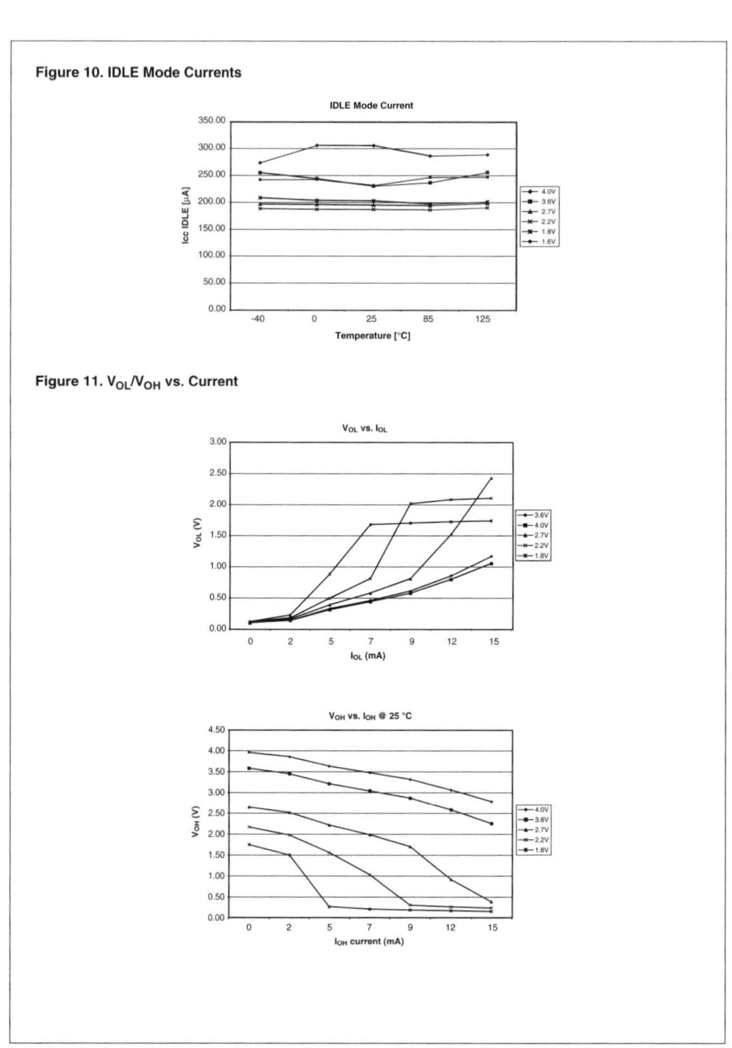

3. Arithmetic Controller Core

The ACEx microcontroller core is specifically designed for low cost applications involving bit manipulation, shifting and block encryption. It is based on a modified Harvard architecture meaning peripheral, I/O, and RAM locations are addressed separately from instruction data.

The core differs from the traditional Harvard architecture by aligning the data and instruction memory sequentially. This allows the X-pointer (12-bits) to point to any memory location in either segment of the memory map. This modification improves the overall code efficiency of the ACEx microcontroller and takes advantage of the flexibility found on Von Neumann style machines.

3.1 CPU Registers

The ACEx microcontroller has five general-purpose registers. These registers are the Accumulator (A), X-Pointer (X), Program Counter (PC), Stack Pointer (SP), and Status Register (SR). The X, SP, and SR registers are all memory-mapped.

Figure 12. Programming Model

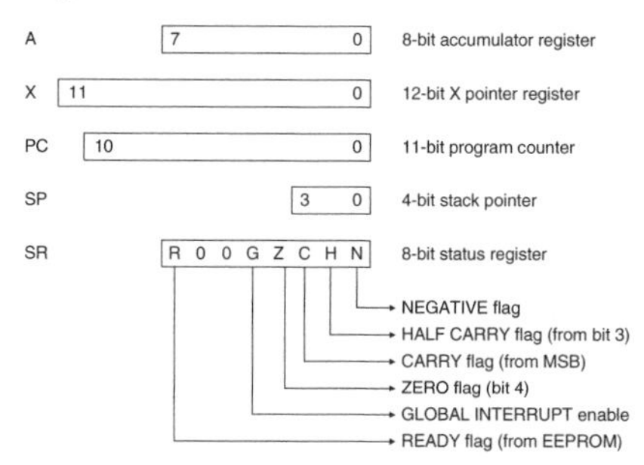

3.1.1 Accumulator (A)

The Accumulator is a general-purpose 8-bit register that is used to hold data and results of arithmetic calculations or data manipulations.

3.1.2 X-Pointer (X)

The X-Pointer register allows for a 12-bit indexing value to be added to an 8-bit offset creating an effective address used for reading and writing between the entire memory space. (Software can only read from code EEPROM.) This provides software with the flexibility of storing lookup tables in the code EEPROM memory space for the core's accessibility during normal operation.

The ACEx core allows software to access the entire 12-bit X-Pointer register using the special X-pointer instructions e.g. LD X, #000H. (See Table 8.) However, software may also access the register through any of the memory-mapped instructions using the XHI (X[11:8]) and XLO (X[7:0]) variables located at 0xBE and 0xBF, respectively. (See Table 10.)

The X register is divided into two sections. The 11 least significant bits (LSBs) of the register is the address of the program or data memory space. The most significant bit (MSB) of the register is write only and selects between the data (0x000 to 0x0FF) or program (0x800 to 0xFFF) memory space.

Example: If Bit 11 = 0, then the LD A, [00,X] instruction will take a value from address range 0x000 to 0x0FF and load it into A. If Bit 11 = 1, then the LD A, [00,X] instruction will take a value from address range 0x800 to 0xFFF and load it into A.

The X register can also serve as a counter or temporary storage register. However, this is true only for the 11-LSBs since the 12th bit is dedicated for memory space selection.

3.1.3 Program Counter (PC)

The 11-bit program counter register contains the address of the next instruction to be executed. After a reset, if in normal mode the program counter is initialized to 0x800.

3.1.4 Stack Pointer (SP)

The ACEx microcontroller has an automatic program stack with a 4-bit stack pointer. The stack can be initialized to any location between addresses 0x30-0x3F. Normally, the stack pointer is initialized by one of the first instructions in an application program. After a reset, the stack pointer is defaulted to 0xF pointing to address 0x3F.

The stack is configured as a data structure which decrements from high to low memory. Each time a new address is pushed onto the stack, the core decrements the stack pointer by two. Each time an address is pulled from the stack, the core increments the stack pointer is by two. At any given time, the stack pointer points to the next free location in the stack.

When a subroutine is called by a jump to subroutine (JSR) instruction, the address of the instruction is automatically pushed onto the stack least significant byte first. When the subroutine is finished, a return from subroutine (RET) instruction is executed. The RET instruction pulls the previously stacked return address from the stack and loads it into the program counter. Execution then continues at the recovered return address.

3.1.5 Status Register (SR)

The 8-bit Status register (SR) contains four condition code indicators (C, H, Z, and N), one interrupt masking bit (G), and an EEPROM write flag (R.) The condition codes are automatically updated by most instructions. (See Table 9.)

Carry/Borrow (C)

The carry flag is set if the arithmetic logic unit (ALU) performs a carry or borrow during an arithmetic operation and by its dedicated instructions. The rotate instruction operates with and through the carry bit to facilitate multiple-word shift operations. The LDC and INVC instructions facilitate direct bit manipulation using the carry flag.

Half Carry (H)

The half carry flag indicates whether an overflow has taken place on the boundary between the two nibbles in the accumulator. It is primarily used for Binary Coded Decimal (BCD) arithmetic calculation.

Zero (Z)

The zero flag is set if the result of an arithmetic, logic, or data manipulation operation is zero. Otherwise, it is cleared.

Negative (N)

The negative flag is set if the MSB of the result from an arithmetic, logic, or data manipulation operation is set to one. Otherwise, the flag is cleared. A result is said to be negative if its MSB is a one.

Interrupt Mask (G)

The interrupt request mask (G) is a global mask that disables all maskable interrupt sources. If the G Bit is cleared, interrupts can become pending, but the operation of the core continues uninterrupted. However, if the G Bit is set an interrupt is recognized. After any reset, the G bit is cleared by default and can only be set by a software instruction. When an interrupt is recognized, the G bit is cleared after the PC is stacked and the interrupt vector is fetched. Once the interrupt is serviced, a return from interrupt instruction is normally executed to restore the PC to the value that was present before the interrupt occurred. The G bit is the reset to one after a return from interrupt is executed. Although the G bit can be set within an interrupt service routine, "nesting" interrupts in this way should only be done when there is a clear understanding of latency and of the arbitration mechanism.

3.2 Interrupt handling

When an interrupt is recognized, the current instruction completes its execution. The return address (the current value in the program counter) is pushed onto the stack and execution continues at the address specified by the unique interrupt vector (see Table 10.). This process takes five instruction cycles. At the end of the interrupt service routine, a return from interrupt (RETI) instruction is executed. The RETI instruction causes the saved address to be pulled off the stack in reverse order. The G bit is set and instruction execution resumes at the return address.

The ACEx microcontroller is capable of supporting four interrupts. Three are maskable through the G bit of the SR and the fourth (software interrupt) is not inhibited by the G bit (Figure 13.) The software interrupt is generated by the execution of the INTR instruction. Once the INTR instruction is executed, the ACEx core will interrupt whether the G bit is set or not. The INTR interrupt is executed in the same manner as the other maskable interrupts where the program counter register is stacked and the G bit is cleared. This means, if the G bit was enabled prior to the software interrupt the RETI instruction must be used to return from interrupt in order to restore the G bit to its previous state. However, if the G bit was not enabled prior to the software interrupt the RET instruction must be used.

In case of multiple interrupts occurring at the same time, the ACEx microcontroller core has prioritized the interrupts. The interrupt priority sequence in shown in Table 7.

Table 7: Interrupt Priority Sequence

Priority (4 highest, 1 lowest)	Interrupt
4	MIW (EDGEI)
3	Timer0 (TMRI0)
2	Timer1 (TMRI1)
1	Software (INTR)

Figure 13. Basic Interrupt Structure

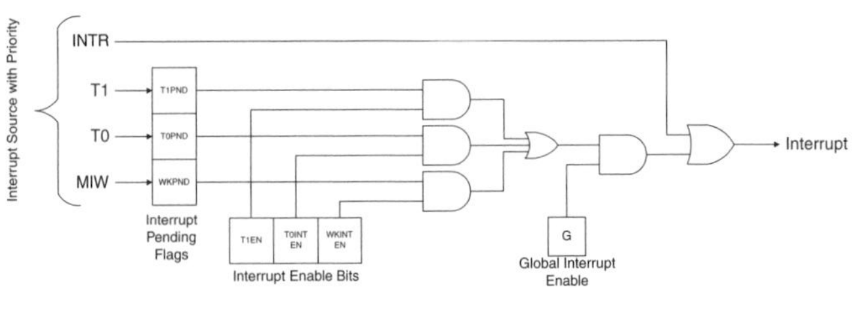

3.3 Addressing Modes

The ACEx microcontroller has seven addressing modes indexed, indirect, direct, immediate, absolute jump, and relative jump.

Indexed

The instruction allows an 8-bit unsigned offset value to be added to the 11-LSBs of the X-pointer yielding a new effective address. This mode can be used to address either data or program memory space.

Indirect

The instruction allows the X-pointer to address any location within the data memory space.

Direct

The instruction contains an 8-bit address field that directly points to the data memory space as an operand.

Immediate

The instruction contains an 8-bit immediate field as an operand.

Inherent

This instruction has no operands associated with it.

Absolute

The instruction contains an 11-bit address that directly points to a location in the program memory space. There are two operands associated with this addressing mode. Each operand contains a byte of an address. This mode is used only for the long jump (JMP) and JSR instructions.

Relative

This mode is used for the short jump (JP) instructions where the operand is a value relative to the current PC address. With this instruction, software is limited to the number of bytes it can jump, -31 or +32.

Table 8. Instruction Addressing Modes

Instruction	Immediate			Direct		Indexed	Indirect	Inherent		Relative	Absolute
ADC	A, #			A, M		A, [#, X]	A, [X]				
ADD	A, #			A, M		A, [#, X]	A, [X]				
AND	A, #			A, M		A, [#, X]	A, [X]				
OR	A, #			A, M		A, [#, X]	A, [X]				
SUBC	A, #			A, M		A, [#, X]	A, [X]				
XOR	A, #			A, M		A, [#, X]	A, [X]				
CLR				M				A	X		
INC				M				A	X		
DEC				M				A	X		
IFEQ	A, #	X, #	M,#	A, M		A, [#, X]	A, [X]				
IFGT	A, #	X, #		A, M		A, [#, X]	A, [X]				
IFNE	A, #	X, #	M,#	A, M		A, [#, X]	A, [X]				
IFLT		X, #									
SC								no-op			
RC								no-op			
IFC								no-op			
IFNC								no-op			
INVC								no-op			
LDC				#, M							
STC				#, M							
RLC				M				A			
RRC				M				A			
LD	A, #	M, #	X, #	A, M	M, M	A, [#, X]	A, [X]				
ST				A, M		A, [#, X]	A, [X]				
NOP								no-op			
IFBIT	#, A			#, M			[#, X]				
IFNBIT	#, A			#, M			[#, X]				
SBIT				#, M			[#, X]				
RBIT				#, M			[#, X]				
JP										Rel	
JSR						[#, X]					M
JMP						[#, X]					M
RET								no-op			
RETI								no-op			
INTR								no-op			

Table 9. Instruction Cycles and Bytes

Mnemonic	Operand	Bytes	Cycles	Flags affected
ADC	A, [X]	1	1	C,H,Z,N
ADC	A, [#,X]	2	3	C,H,Z,N
ADC	A, M	2	2	C,H,Z,N
ADC	A, #	2	2	C,H,Z,N
ADD	A, [X]	1	1	Z,N
ADD	A, [#,X]	2	3	Z,N
ADD	A, M	2	2	Z,N
ADD	A, #	2	2	Z,N
AND	A, [X]	1	1	Z,N
AND	A, [#,X]	2	3	Z,N
AND	A, M	2	2	Z,N
AND	A, #	2	2	Z,N
CLR	X	1	1	Z
CLR	A	1	1	C,H,Z,N
CLR	M	2	1	C,H,Z,N
DEC	X	1	1	Z
DEC	A	1	1	Z,N
DEC	M	2	2	Z,N
IFBIT	#, A	1	1	None
IFBIT	#, M	2	2	None
IFBIT	#, [X]	1	1	None
IFC		1	1	None
IFEQ	A, [#, X]	2	3	None
IFEQ	A, [X]	1	1	None
IFEQ	A, #	2	2	None
IFEQ	A, M	2	2	None
IFEQ	M, #	3	3	None
IFEQ	X, #	3	3	None
IFGT	A, [#, X]	2	3	None
IFGT	A, [X]	1	1	None
IFGT	A, #	2	2	None
IFGT	A, M	2	2	None
IFGT	X, #	3	3	None
IFLT	X, #	3	3	None
IFNBIT	#, A	1	1	None
IFNBIT	#, M	2	2	None
IFNBIT	#, [X]	1	1	None
IFNC		1	1	None
IFNE	A, [#, X]	2	3	None
IFNE	A, [X]	1	1	None
IFNE	A, #	2	2	None
IFNE	A, M	2	2	None
IFNE	X, #	3	3	None
IFNE	M, #	3	3	None
INC	A	1	1	Z,N
INC	M	2	2	Z,N

Mnemonic	Operand	Bytes	Cycles	Flags affected
INC	X	1	1	Z
INTR		1	5	None
INVC		1	1	C
JMP	M	3	4	None
JMP	[#, X]	2	3	None
JP		1	1	None
JSR	M	3	5	None
JSR	[#, X]	2	5	None
LD	A, #	2	2	None
LD	A, [#,X]	2	3	None
LD	A, [X]	1	1	None
LD	A, M	2	2	None
LD	M, #	3	3	None
LD	M, M	3	3	None
LD	X, #	3	3	None
LDC	#, M	2	2	C
NOP		1	1	None
OR	A, [X]	1	1	Z, N
OR	A, [#,X]	2	3	Z,N
OR	A, M	2	2	Z,N
OR	A, #	2	2	Z,N
RBIT	#, [X]	1	2	Z,N
RBIT	#, M	2	2	Z,N
RC		1	1	C,H
RET		1	5	None
RETI		1	5	None
RLC	A	1	1	C,Z,N
RLC	M	2	2	C,Z,N
RRC	A	1	1	C,Z,N
RRC	M	2	2	C,Z,N
SBIT	#, [X]	1	2	Z,N
SBIT	#, M	2	2	Z,N
SC		1	1	C,H
ST	A, [#,X]	2	3	None
ST	A, [X]	1	1	None
ST	A, M	2	2	None
STC	#, M	2	2	Z,N
SUBC	A, [X]	1	1	C,H,Z,N
SUBC	A, [#,X]	2	3	C,H,Z,N
SUBC	A, M	2	2	C,H,Z,N
SUBC	A, #	2	2	C,H,Z,N
XOR	A, [X]	1	1	Z,N
XOR	A, [#,X]	2	3	Z,N
XOR	A, M	2	2	Z,N
XOR	A, #	2	2	Z,N

3.4 Memory Map

All I/O ports, peripheral registers, and core registers (except the accumulator and the program counter) are mapped into the memory space.

Table 10. Memory Mapped Registers

Address	Memory Space	Block	Contents
0x00 - 0x3F	Data	SRAM	Data RAM
0x40 - 0x7F	Data	EEPROM	Data EEPROM
0x80-0x9F	Data	Reserved	
0xA0	Data	HBC	HBCNTRL register
0xA1	Data	HBC	PSCALE register
0xA2	Data	HBC	HPATTERN register
0xA3	Data	HBC	LPATTERN register
0xA4	Data	HBC	BPSEL register
0xA7	Data	Timer1	T1RBLO register
0xA8	Data	Timer1	T1RBHI register
0xA9	Data	HBC	DAT0 register
0xAA	Data	Timer1	T1RALO register
0xAB	Data	Timer1	T1RAHI register
0xAC	Data	Timer1	TMR1LO register
0xAD	Data	Timer1	TMR1HI register
0xAE	Data	Timer1	T1CNTRL register
0xAF	Data	MIW	WKEDG register
0xB0	Data	MIW	WKPND register
0xB1	Data	MIW	WKEN register
0xB2	Data	I/O	PORTGD register
0xB3	Data	I/O	PORTGC register
0xB4	Data	I/O	PORTGP register
0xB5	Data	Timer0	WDSVR register
0xB6	Data	Timer0	T0CNTRL register
0xB7	Data	Clock	HALT mode register
0xB8-0xBA	Data	Reserved	
0xBB	Data	Init. Register	Initialization Register 1
0xBC	Data	Init. Register	Initialization Register 2
0xBD	Data	LBD	LBD register
0xBE	Data	Core	XHI register
0xBF	Data	Core	XLO register
0xC0	Data	Clock	Power Mode Clear (PMC) Register
0xCE	Data	Core	SP register
0xCF	Data	Core	Status register (SR)
0xD0 - 0xFF	Data	Reserved	
0x800 - 0xFF5	Program	EEPROM	Code EEPROM
0xFF6 - 0xFF7	Program	Core	Timer0 Interrupt vector
0xFF8 - 0xFF9	Program	Core	Timer1 Interrupt vector
0xFFA - 0xFFB	Program	Core	MIW Interrupt vector
0xFFC - 0xFFD	Program	Core	Soft Interrupt vector
0xFFE - 0xFFF	Program	Reserved	

3.5 Memory

The ACEx microcontroller has 64 bytes of SRAM and 64 bytes of EEPROM available for data storage. The device also has 2K bytes of EEPROM for program storage. Software can read and write to SRAM and data EEPROM but can only read from the code EEPROM. While in normal mode, the code EEPROM is protected from any writes. The code EEPROM can only be rewritten when the device is in program mode and if the write disable (WDIS) bit of the initialization register is not set to 1.

While in normal mode, the user can write to the data EEPROM array by 1) polling the ready (R) flag of the SR, then 2) executing the appropriate instruction. If the R flag is 1, the data EEPROM block is ready to perform the next write. If the R flag is 0, the data EEPROM is busy. The data EEPROM array will reset the R flag after the completion of a write cycle. Attempts to read, write, or enter HALT/IDLE mode while the data EEPROM is busy (R = 0) can affect the current data being written.

3.6 Initialization Registers

The ACEx microcontroller has two 8-bit wide initialization registers. These registers are read from the memory space on power-up to initialize certain on-chip peripherals. Figure 14 provides a detailed description of Initialization Register 1. The Initialization Register 2 is used to trim the internal oscillator to its appropriate frequency. This register is pre-programmed in the factory to yield an internal instruction clock of 1MHz.

The Initialization Registers 1 and 2 can be read from and written to during programming mode. However, re-trimming the internal oscillator (writing to the Initialization Register 2) once it has left the factory is *discouraged.*

Figure 14. Initialization Register 1

Bit 7	Bit 6	Bit 5	Bit 4	Bit 3	Bit 2	Bit 1	Bit 0
CMODE[0]	CMODE[1]	WDEN	BOREN	LDBEN	UBD	WDIS	RDIS

(0) RDIS	If set, disables attempts to read the contents from the memory while in programming mode. Once this bit is set, it is no longer possible to unset this option even though the write disable option is not enabled.
(1) WDIS	If set, disables attempts to write new contents to the memory while in programming mode
(2) UBD	If set, the device will not allow any writes to occur in the upper block of data EEPROM (0x60-0x7F)
(3) LBDEN	If set, the Low Battery Detection circuit is enabled
(4) BOREN	If set, allows a BOR to occur if Vcc falls below the voltage reference level
(5) WDEN	If set, enables the on-chip processor watchdog circuit
(6) CMODE[1]	Clock mode select bit 1 (See Table 16)
(7) CMODE[0]	Clock mode select bit 0 (See Table 16)

4. Timer 1

Timer 1 is a versatile 16-bit timer that can operate in one of four modes:

- **Pulse Width Modulation** (PWM) mode, which generates pulses of a specified width and duty cycle
- **External Event Counter** mode, which counts occurrences of an external event
- **Standard Input Capture** mode, which measures the elapsed time between occurrences of external events
- **Difference Input Capture** mode, which automatically measures the difference between edges.

Timer 1 contains a 16-bit timer/counter register (TMR1), a 16-bit auto-reload/capture register (T1RA), a secondary 16-bit auto-reload register (T1RB), and an 8-bit control register (T1CNTRL). All register are memory-mapped for simple access through the core with both the 16-bit registers organized as a pair of 8-bit register bytes {TMR1HI, TMR1LO}, {T1RAHI, T1RALO}, and {T1RBHI, T1RBLO}. Depending on the operating mode, the timer contains an external input or output (T1) that is multiplexed with the I/O pin G2. By default, the TMR1 is reset to 0xFFFF, T1RA/T1RB is reset to 0x0000, and T1CNTRL is reset to 0x00.

The timer can be started or stopped through the T1CNTRL register bit T1C0. When running, the timer counts down (decrements) every clock cycle. Depending on the operating mode, the timer's clock is either the instruction clock or a transition on the T1 input. In addition, occurrences of timer underflow (transitions from 0x0000 to 0xFFFF/T1RA/T1RB value) can either generate an interrupt and/or toggle the T1 output pin.

Timer 1's interrupt (TMRI1) can be enabled by interrupt enable (T1EN) bit in the T1CNTRL register. When the timer interrupt is enabled, depending on the operating mode, the source of the interrupt is a timer underflow and/or a timer capture.

4.1 Timer control bits

Reading and writing to the T1CNTRL register controls the timer's operation. By writing to the control bits, the user can enable or disable the timer interrupts, set the mode of operation, and start or stop the timer. The T1CNTRL register bits are described in Table 11 and Table 12.

Table 11. Timer 1 Control Register (T1CNTRL)

T1CNTRL Register	Bit Name	Function
Bit 7	T1C3	Timer TIMER1 control bit 3 (see Table 12)
Bit 6	T1C2	Timer TIMER1 control bit 2 (see Table 12)
Bit 5	T1C1	Timer TIMER1 control bit 1 (see Table 12)
Bit 4	T1C0	Timer TIMER1 run: 1= Start timer, 0 = Stop timer; or Timer TIMER1 underflow interrrupt pending flag in input capture mode
Bit 3	T1PND	Timer1 interrupt pending flag: 1 = Timer1 interrupt Pending, 0 = Timer1 interrupt not pending
Bit 2	T1EN	Timer1 interrupt enable bit: 1 = Timer1 interrupt enabled, 0 = Timer1 interrupt disabled
Bit 1	M1S1	Capture type: 0 = Pulse capture, 1 = Cycle capture (see Table 12)
Bit 0	T1RBEN	PWM Mode: 0 = Timer1 reload on T1RA, 1 = TIMER1 reload on T1RA and T1RB (always starting with T1RA)

Table 12. Timer 1 Operating Modes

T1 C3	T1 C2	T1 C1	M4 S1	T1 RB	Timer Mode Source	Interrupt	Timer Counts-on
0	0	0	X	X	MODE 2	TIMER1 Underflow	T1 Pos. Edge
0	0	1	X	X	MODE 2	TIMER1 Underflow	T1 Neg. Edge
1	0	1	X	0	MODE 1 T1 Toggle	Autoreload T1RA	Instruction Clock
1	0	0	X	0	MODE 1 No T1 Toggle	Autoreload T1RA	Instruction Clock
1	0	1	X	1	MODE 1 T1 Toggle	Autoreload T1RA/T1RB	Instruction Clock
1	0	0	X	1	MODE 1 No T1 Toggle	Autoreload T1RA/T1RB	Instruction Clock
0	1	0	X	X	MODE 3 Captures: T1 Pos Edge	Pos. T1 Edge	Instruction Clock
0	1	1	X	X	MODE 3 Captures: T1 Neg Edge	Neg. T1 Edge	Instruction Clock
1	1	0	0	X	MODE 4	Pos. to Neg.	Instruction Clock
1	1	0	1	X	MODE 4	Pos. to Pos.	Instruction Clock
1	1	1	0	X	MODE 4	Neg. to Pos.	Instruction Clock
1	1	1	1	X	MODE 4	Neg. to Neg.	Instruction Clock

4.2 Mode 1: Pulse Width Modulation (PWM) Mode

In the PWM mode, the timer counts down at the instruction clock rate. When an underflow occurs, the timer register is reloaded from T1RA/T1RB and the count down proceeds from the loaded value. At every underflow, a pending flag (T1PND) located in the T1CNTRL register is set. Software must then clear the T1PND flag and load the T1RA/T1RB register with an alternate PWM value (if desired.) In addition, the timer can be configured to toggle the T1 output bit upon underflow. Configuring the timer to toggle T1 results in the generation of a signal outputted from port G2 with the width and duty cycle controlled by the values stored in the T1RA/T1RB. A block diagram of the timer's PWM mode of operation is shown in Figure 15.

The PWM timer can be configured to use the T1RA register only for auto-reloading the timer registers or can be configured to use both T1RA and T1RB alternately. If the T1RBEN bit of the T1CNTRL register is 0, the PWM timer will reload using only T1RA ignoring any value store in the T1RB register. However, if the T1RBEN bit is 1 the PWM timer will be reloaded using both the T1RA and T1RB registers. A hardware select logic is implemented to select between T1RA and T1RB alternately, always starting with T1RA, every timer underflows to auto-reload the timer registers. This feature is useful when a signal with variable duty cycle needs to be generated without software intervention.

The timer has one interrupt (TMRI1) that is maskable through the T1EN bit of the T1CNTRL register. However, the core is only interrupted if the T1EN bit and the G (Global Interrupt enable) bit of the SR is set. If interrupts are enabled, the timer will generate an interrupt each time T1PND flags is set (whenever the timer underflows provided that the pending flag was cleared.) The interrupt service routine is responsible for proper handling of the T1PND flag and the T1EN bit.

The interrupt will be synchronous with every rising and falling edge of the T1 output signal. Generating interrupts only on rising or falling edges of T1 is achievable through appropriate handling of the T1EN bit or T1PND flag through software.

The following steps show how to properly configure Timer 1 to operate in the PWM mode. For this example, the T1 output signal is toggled with every timer underflow and the "high" and "low" times for the T1 output can be set to different values. The T1 output signal can start out either high or low depending on the configuration of G2; the instructions below are for starting with the T1 output high. Follow the instructions in parentheses to start the T1 output low.

1. Configure T1 as an output by setting bit 2 of PORTGC.
   - SBIT 2, PORTGC ; Configure G2 as an output
2. Initialize T1 to 1 (or 0) by setting (or clearing) bit 2 of PORTGD.
   - SBIT 2, PORTGD ; Set G2 high
3. Load the initial PWM high (low) time into the timer register.
   - LD TMR1LO, #6FH ; High (Low) for 1.391ms (1MHz clock)
   - LD TMR1HI, #05H
4. Load the PWM low (high) time into the T1RA register.
   - LD T1RALO, #2FH ; Low (High) for .303ms (1MHz clock)
   - LD T1RAHI, #01H
5. Write the appropriate control value to the T1CNTRL register to select PWM mode with T1 toggle, to clear the enable bit and pending flag, and to start the timer. (See Table 11 and Table 12.)
   - LD T1CNTRL, #0B0H ; Setting the T1C0 bit starts the timer
6. After every underflow, load T1RA with alternate values. If the user wishes to generate an interrupt on a T1 output transition, reset the pending flags and then enable the interrupt using T1EN. The G bit must also be set. The interrupt service routine must reset the pending flag and perform whatever processing is desired.
   - RBIT T1PND, T1CNTRL ; T1PND equals 3
   - LD T1RALO, #6FH ; High (Low) for 1.391ms (1MHz clock)
   - LD T1RAHI, #05H

Figure 15. Pulse Width Modulation Mode

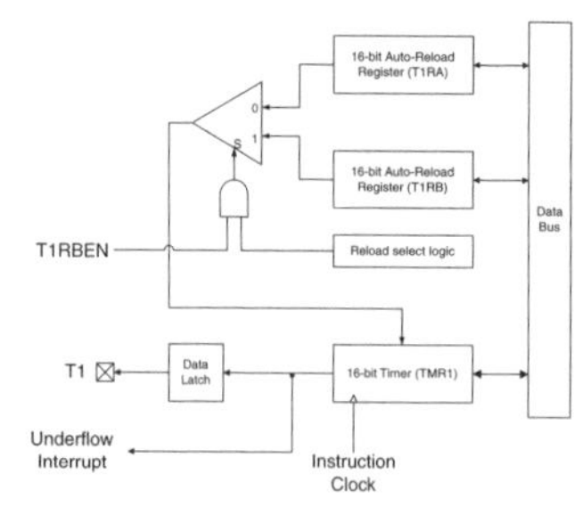

4.3 Mode 2: External Event Counter Mode

The External Event Counter mode operates similarly to the PWM mode; however, the timer is not clocked by the instruction clock but by transitions of the T1 input signal. The edge is selectable through the T1C1 bit of the T1CNTRL register. A block diagram of the timer's External Event Counter mode of operation is shown in Figure 16.

The T1 input should be connected to an external device that generates a positive/negative-going pulse for each event. By clocking the timer through T1, the number of positive/negative transitions can be counted therefore allowing software to capture the number of events that occur. The input signal on T1 must have a pulse width equal to or greater than one instruction clock cycle.

The counter can be configured to sense either positive-going or negative-going transitions on the T1 pin. The maximum frequency at which transitions can be sensed is one-half the frequency of the instruction clock.

As with the PWM mode, when the counter underflows the counter is reloaded from the T1RA register and the count down proceeds from the loaded value. At every underflow, a pending flag (T1PND) located in the T1CNTRL register is set. Software must then clear the T1PND flag and can then load the T1RA register with an alternate value.

The counter has one interrupt (TMRI1) that is maskable through the T1EN bit of the T1CNTRL register. However, the core is only interrupted if the T1EN bit and the G (Global Interrupt enable) bit of the SR is set. If interrupts are enabled, the counter will generate an interrupt each time the T1PND flag is set (whenever timer underflows provided that the pending flag was cleared.) The interrupt service routine is responsible for proper handling of the T1PND flag and the T1EN bit.

The following steps show how to properly configure Timer 1 to operate in the External Event Counter mode. For this example, the counter is clocked every falling edge of the T1 input signal. Follow the instructions in parentheses to clock the counter every rising edge.

1. Configure T1 as an input by clearing bit 2 of PORTGC.
   - RBIT 2, PORTGC ; Configure G2 as an input
2. Initialize T1 to input with pull-up by setting bit 2 of PORTGD.
   - SBIT 2, PORTGD ; Set G2 high
3. Enable the global interrupt enable bit.
   - SBIT 4, STATUS
4. Load the initial count into the TMR1 and T1RA registers. When the number of external events is detected, the counter will reach zero; however, it will not underflow until the next event is detected. To count N pulses, load the value N-1 into the registers. If it is only necessary to count the number of occurrences and no action needs to be taken at a particular count, load the value 0xFFFF into the registers.
   - LD TMR1LO, #0FFH
   - LD TMR1HI, #0FFH
   - LD T1RALO, #0FFH
   - LD T1RAHI, #0FFH

5. Write the appropriate control value to the T1CNTRL register to select External Event Counter mode, to clock every falling edge, to set the enable bit, to clear the pending flag, and to start the counter. (See Table 11 and Table 12)
   - LD T1CNTRL, #34H (#00h) ;Setting the T1C0 bit starts the timer
6. When the counter underflows, the interrupt service routine must clear the T1PND flag and take whatever action is required once the number of events occurs. If the software wishes to merely count the number of events and the anticipated number may exceed 65,536, the interrupt service routine should record the number of underflows by incrementing a counter in memory. Software can then calculate the correct event count.
   - RBIT T1PND, T1CNTRL ; T1PND equals 3

Figure 16. External Event Counter Mode

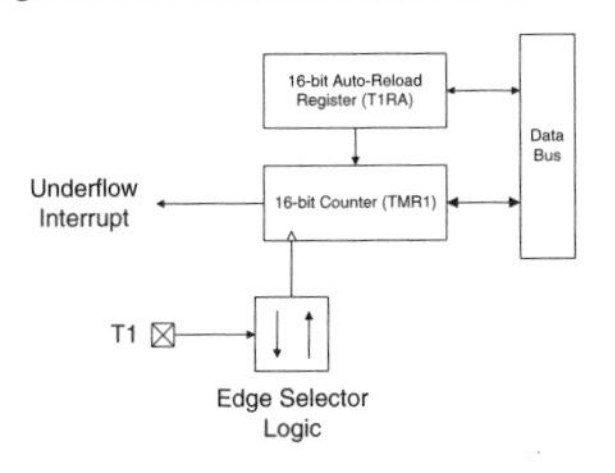

4.4 Mode 3: Input Capture Mode

In the Input Capture mode, the timer is used to measure elapsed time between edges of an input signal. Once the timer is configured for this mode, the timer starts counting down immediately at the instruction clock rate. The Timer 1 will then transfer the current value of the TMR1 register into the T1RA register as soon as the selected edge of T1 is sensed. The input signal on T1 must have a pulse width equal to or greater than one instruction clock cycle. At every T1RA capture, software can then store the values into RAM to calculate the elapsed time between edges on T1. At any given time (with proper consideration of the state of T1) the timer can be configured to capture on positive-going or negative-going edges. A block diagram of the timer's Input Capture mode of operation is shown in Figure 17.

The timer has one interrupt (TMRI1) that is maskable through the T1EN bit of the T1CNTRL register. However, the core is only interrupted if the T1EN bit and the G (Global Interrupt enable) bit of the SR is set. The Input Capture mode contains two interrupt pending flags 1) the TMR1 register capture in T1RA (T1PND) and 2) timer underflow (T1C0). If interrupts are enabled, the timer will generate an interrupt each time a pending flag is set (provided that the pending flag was previously cleared.) The interrupt service routine is responsible for proper handling of the T1PND flag, T1C0 flag, and the T1EN bit.

For this operating mode, the T1C0 control bit serves as the timer underflow interrupt pending flag. The Timer 1 interrupt service routine must read both the T1PND and T1C0 flags to determine the cause of the interrupt. A set T1C0 flag means that a timer underflow occurred whereas a set T1PND flag means that a capture occurred in T1RA. It is possible that both flags will be found set, meaning that both events occurred at the same time. The interrupt service routine should take this possibility into consideration.

Because the T1C0 bit is used as the underflow interrupt pending flag, it is not available for use as a start/stop bit as in the other modes.

The TMR1 register counts down continuously at the instruction clock rate starting from the time that the input capture mode is selected. (See Table 11 and Table 12) To stop the timer from running, you must change the mode to an alternate mode (PWM or External Event Counter) while resetting the T1C0 bit.

The input pins can be independently configured to sense positive-going or negative-going transitions. The edge sensitivity of pin T1 is controlled by bit T1C1 as indicated in Table 12.

The edge sensitivity of a pin can be changed without leaving the input capture mode even while the timer is running. This feature allows you to measure the width of a pulse received on an input pin.

For example, the T1 pin can be programmed to be sensitive to a positive-going edge. When the positive edge is sensed, the TMR1 register contents is transferred to the T1RA register and a Timer 1 interrupt is generated. The Timer 1 interrupt service routine records the contents of the T1RA register, changes the edge sensitivity from positive to negative-going edge, and clears the T1PND flag. When the negative-going edge is sensed another Timer 1 interrupt is generated. The interrupt service routine reads the T1RA register again. The difference between the previous reading and the current reading reflects the elapsed time between the positive edge and negative edge of the T1 input signal i.e. the width of the positive-going pulse.

Remember that the Timer1 interrupt service routine must test the T1C0 and T1PND flags to determine the cause of the interrupt. If the T1C0 flag caused the interrupt, the interrupt service routine should record the occurrence of an underflow by incrementing a counter in memory or by some other means. The software that calculates the elapsed time between captures should take into account the number of underflow that occurred when making its calculation.

The following steps show how to properly configure Timer 1 to operate in the Input Capture mode.

1. Configure T1 as an input by clearing bit 2 of PORTGC.
   - RBIT 2, PORTGC ; Configure G2 as an input
2. Initialize T1 to input with pull-up by setting bit 2 of PORTGD.
   - SBIT 2, PORTGD ; Set G2 high
3. Enable the global interrupt enable bit.
   - SBIT 4, STATUS
4. With the timer stopped, load the initial time into the TMR1 register (typically the value is 0xFFFF.)
   - LD TMR1LO, #0FFH
   - LD TMR1HI, #0FFH
5. Write the appropriate control value to the T1CNTRL register to select Input Capture mode, to sense the appropriate edge, to set the enable bit, and to clear the pending flags. (See

Table 11 and Table 12)
- LD T1CNTRL, #64H ; T1C1 is the edge select bit

6. As soon as the input capture mode is enabled, the timer starts counting. When the selected edge is sensed on T1, the T1RA register is loaded and a Timer 1 interrupt is triggered.

Figure 17. Input Capture Mode

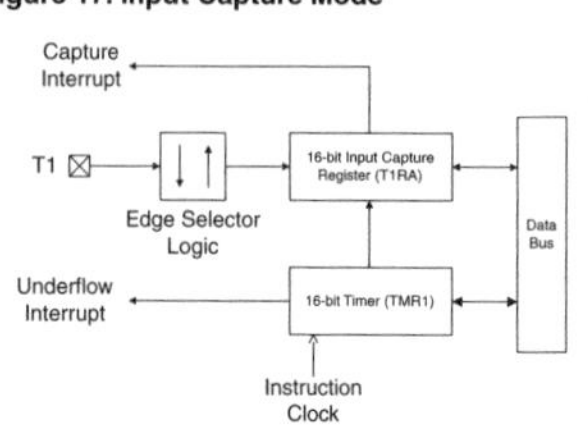

4.5 Mode 4: Difference Input Capture Mode

The Difference Input Capture mode works similarly to the standard Input Capture mode. However, for the Difference Input Capture the timer automatically captures the elapsed time between the selected edges without the core needing to perform the calculation.

For example, the standard Input Capture mode requires that the timer be configured to capture a particular edge (rising or falling) at which time the timer's value is copied into the capture register. If the elapsed time is required, software must move the captured data into RAM and reconfigure the Input Capture mode to capture on the next edge (rising or falling). Software must then subtract the difference between the two edges to yield useful information.

The Difference Capture mode eliminates the need for software intervention and allows for capturing very short pulse or cycle widths. It can be configured to capture the elapsed time between:

1. rising edge to falling edge
2. rising edge to rising edge
3. falling edge to rising edge
4. falling edge to falling edge

Once configured, the Difference Capture timer waits for the first selected edge. When the edge transition has occurred, the 16-bit timer starts counting up based every instruction clock cycle. It will continue to count until the second selected edge transition occurs at which time the timer stops and stores the elapse time into the T1RA register.

Software can now read the difference between transitions directly without using any processor resources. However, like the standard Input Capture mode both the capture (T1PND) and the underflow (T1C0) flags must be monitored and handled appropriately. This feature allows the ACEx microcontroller to capture very small pulses where standard microcontrollers might have missed cycles due to the limited bandwidth.

Figure 18. Difference Capture Mode

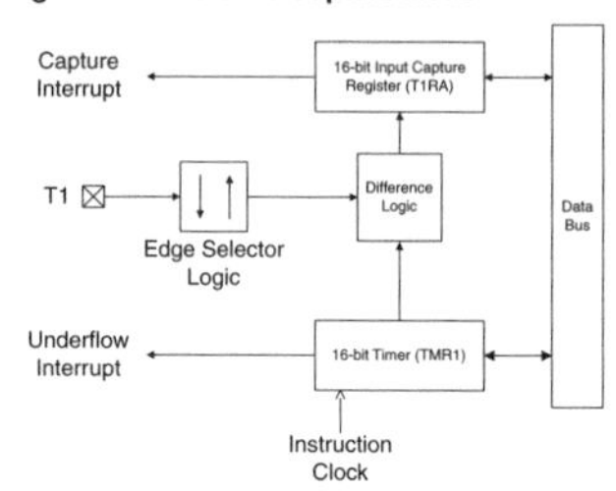

5. Timer 0

Timer 0 is a 12-bit free running idle timer. Upon power-up or any reset, the timer is reset to 0x000 and then counts up continuously based on the instruction clock of 1MHz (1 µs). Software cannot read from or write to this timer. However, software can monitor the timer's pending (T0PND) bit that is set every 8192 cycles (initially 4096 cycles after a reset). The T0PND flag is set every other time the timer overflows (transitions from 0xFFF to 0x000) through a divide-by-2 circuit. After an overflow, the timer will reset and restart its counting sequence.

Software can either poll the T0PND bit or vector to an interrupt subroutine. In order to interrupt on a T0PND, software must be sure to enable the Timer 0 interrupt enable (T0INTEN) bit in the Timer 0 control (T0CNTRL) register and also make sure the G bit is set in SR. Once the timer interrupt is serviced, software should reset the T0PND bit before exiting the routine. Timer 0 supports the following functions:

1. Exiting from IDLE mode (See Section 16 for details.)
2. Start up delay from HALT mode
3. Watchdog pre-scalar (See Section 6 for details.)

The T0INTEN bit is a read/write bit. If set to 0, interrupt requests from the Timer 0 are ignored. If set to 1, interrupt requests are accepted. Upon reset, the T0INTEN bit is reset to 0.

The T0PND bit is a read/write bit. If set to 1, it indicates that a Timer 0 interrupt is pending. This bit is set by a Timer 0 overflow and is reset by software or system reset.

The WKINTEN bit is used in the Multi-input Wakeup/Interrupt block. See Section 8 for details.

Figure 19. Timer 0 Control Register Definition (T0CNTRL)

Bit 7	Bit 6	Bit 5	Bit 4	Bit 3	Bit 2	Bit 1	Bit 0
WKINTEN	x	x	x	x	x	T0PND	T0INTEN

6. Watchdog

The Watchdog timer is used to reset the device and safely recover in the rare event of a processor "runaway condition." The 12-bit Timer 0 is used as a pre-scalar for Watchdog timer. The Watchdog timer must be serviced before every 61,440 cycles but no sooner than 4096 cycles since the last Watchdog reset. The Watchdog is serviced through software by writing the value 0x1B to the Watchdog Service (WDSVR) register (see Figure 20). The part resets automatically if the Watchdog is serviced too frequent, or not frequent enough.

The Watchdog timer must be enabled through the Watchdog enable bit (WDEN) in the initialization register. The WDEN bit can only be set while the device is in programming mode. Once set, the Watchdog will always be powered-up enabled. Software cannot disable the Watchdog. The Watchdog timer can only be disabled in programming mode by resetting the WDEN bit as long as the memory write protect (WDIS) feature is not enabled.

WARNING

Ensure that the Watchdog timer has been serviced before entering IDLE mode because it remains operational during this time.

Figure 20. Watchdog Service Register (WDSVR)

Bit 7	Bit 6	Bit 5	Bit 4	Bit 3	Bit 2	Bit 1	Bit 0
0	0	0	1	1	0	1	1

7. Hardware Bit-Coder

The Hardware Bit-Coder is a dedicated hardware bit-encoding peripheral block, Hardware Bit-Coder (HBC), for IR/RF data transmission (see Figure 21.) The HBC is completely software programmable and can be configured to emulate various bit-encoding formats. The software developer has the freedom to encode each bit of data into a desired pattern and output the encoded data at the desired frequency through either the G2 or G5 output (TX) ports.

The HBC contains six 8-bit memory-mapped configuration registers PSCALE, HPATTERN, LPATTERN, BPSEL, HBCNTRL, and DATO. The registers are used to select the transmission frequency, store the data bit-encoding patterns, configure the data bit-pattern/frame lengths, and control the data transmission flow.

To select the IR/RF transmission frequency, an 8-bit divide constant must be written into the IR/RF Pre-scalar (PSCALE) register. The IR/RF transmission frequency generator divides the 1MHz instruction clock down by 4 and the PSCALE register is used to select the desired IR/RF frequency shift. Together, the transmission frequency range can be configured between 976Hz (PSCALE = 0xFF) and 125kHz (PSCALE = 0x01). Upon a reset, the PSCALE register is initialized to zero disabling the IR/RF transmission frequency generator. However, once the PSCALE register is programmed, the desired IR/RF frequency is maintained as long as the device is powered.

Once the transmission frequency is selected, the data bit-encoding patterns must be stored in the appropriate registers. The HBC contains two 8-bit bit-encoding pattern registers, High-pattern (HPATTERN) and Low-pattern (LPATTERN). The encoding pattern stored in the HPATTERN register is transmitted when the data bit value to be encoded is a 1. Similarly, the pattern stored in the LPATTERN register is transmitted when the data bit value to be encoded is a 0. The HBC transmits each encoded pattern MSB first.

The number of bits transmitted from the HPATTERN and LPATTERN registers is software programmable through the Bit Period Configuration (BPSEL) register (see Figure 22). During the transmission of HPATTERN, the number of bits transmitted is configured by BPH[2:0] (BPSEL[2:0]) while BPL[2:0] (BPSEL[5:3]) configures the number of transmitted bits for the LPATTERN. The HBC allows from 2 (0x1) to 8 (0x7) encoding pattern bits to be transmitted from each register. Upon a reset, BPSEL is initially 0 disabling the HBC from transmitting pattern bits from either register.

The Data (DATO) register is used to store up to 8 bits of data to be encoded and transmitted by the HBC. This data is shifted, bit by bit, MSB to LSB into a 1-bit decision register. If the active bit shifted into the decision register is 1, the pattern in the HPATTERN register is shifted out of the output port. Similarly, if the active bit is 0 the pattern in the LPATTERN register is shifted out.

The HBC control (HBCNTRL) register is used to configure and control the data transmission. HBCNTRL is divided in 5 different controlling signal FRAME[2:0], IOSEL, TXBUSY, START / STOP, and OCFLAG (see Figure 23.)

FRAME[2:0] selects the number of bits of DATO to encode and transmit. The HBC allows from 2 (0x1) to 8 (0x7) DATO bits to be encoded and transmitted. Upon a reset, FRAME is initialized to zero disabling the DATO's decision register transmitting no data.

The IOSEL signal selects the transmission to output (TX) through either port G2 or G5. If IOSEL is 1, G5 is selected as the output port otherwise G2 is selected.

The TXBUSY signal is read only and is used to inform software that a transmission is in progress. TXBUSY goes high when the encoded data begins to shift out of the output port and will remains high during each consecutive DATO frame bit transmission (see Figure 25). The HBC will clear the TXBUSY signal when the last DATO encoded bit of the frame is transmitted and the STOP signal is 0.

The START / STOP signal controls the encoding and transmission process for each data frame. When software sets the START / STOP bit the DATO frame transmission process begins. The START signal will remain high until the beginning of the last encoded DATO frame bit transmission. The HBC then clears the START / STOP bit allowing software to elect to either continue with a new DATO frame transmission or stop the transmission all together (see Figure 25). If TXBUSY is 0 when the START signal is enabled, a synchronization period occurs before any data is transmitted lasting the amount of time to transmit a 0 encoded bit (see Figure 24).

The OCFLAG signal is read only and goes high when the last encoded bit of the DAT0 frame is transmitting. The OCFLAG signal is used to inform software that the DAT0 frame transmission operation is completing (see Figure 25). If multiple DAT0 frames are to be transmitted consecutively, software should poll the OCFLAG signal for a 1. Once OCFLAG is 1, DAT0 must be reload and the START / STOP bit must be restored to 1 in order to begin the new frame transmission without interruptions (the synchronization period). Since OCFLAG remains high during the entire last encoded DAT0 frame bit transmission, software should wait for the HBC to clear the OCFLAG signal before polling for the new OCFLAG high pulse. If new data is not reloaded into DAT0 and the START signal (STOP is active) is not set before the OCFLAG is 0, the transmission process will end (TXBUSY is cleared) and a new process will begin starting with the synchronization period.

Figure 24 and Figure 25 shows how the HBC performs its data encoding. In the example, two frames are encoded and transmitted consecutively with the following bit encoding format specification:

1. Transmission frequency = 62.5KHz
2. Data to be encoded = 0x52, 0x92 (all 8-bits)
3. Each bit should be encoded as a 3-bit binary value, '1' = 110b and '0' = 100b
4. Transmission output port : G2

To perform the data transmission, software must first initialize the PSCALE, BPSEL, HPATTERN, LPATTERN, and DAT0 registers with the appropriate values.

```
LD  PSCALE, #03H       ; (1MHz ?? 4) ?? 4 = 62.5KHz
LD  BPSEL, #012H       ; BPH = 2, BPL = 2 (3 bits each)
LD  HPATTERN, #0C0H    ; HPATTERN = 0xC0
LD  LPATTERN, #090H    ; LPATTERN = 0x90
LD  DAT0, #052H        ; DAT0 = 0x52
```

Once the basic registers are initialized, the HBC can be started. (At the same time, software must set the number of data bits per data frame and select the desired output port.)

```
LD  HBCNTRL, #27H      ; START / STOP = 1,
                         FRAME = 7, IOSEL = 0
```

After the HBC has started, software must then poll the OCFLAG for a high pulse and restore the DAT0 register and the START signal to continue with the next data transmission.

```
LOOP_HI:
  IFBIT  OCFLAG, HBCNTRL     ; Wait for OCFLAG = 1
  JP     NXT_FRAME
  JP     LOOP_HI

NXT_FRAME:
  LD     DAT0, #092H         ; DAT0 = 0x92
  SBIT   START, HBCNTRL      ; START / STOP = 1
```

If software is to proceed with another data transmission, the OCFLAG must be zero before polling for the next OCFLAG high pulse. However, since the specification in the example requires no other data transmission software can proceed as desired.

```
LOOP_LO:
  IFBIT  OCFLAG, HBCNTRL     ; Wait for OCFLAG = 0
  JP     LOOP_LO
  Etc.                       ; Program proceeds
                               as desired
```

Figure 21. Hardware Bit-coder (HBC) Block Diagram

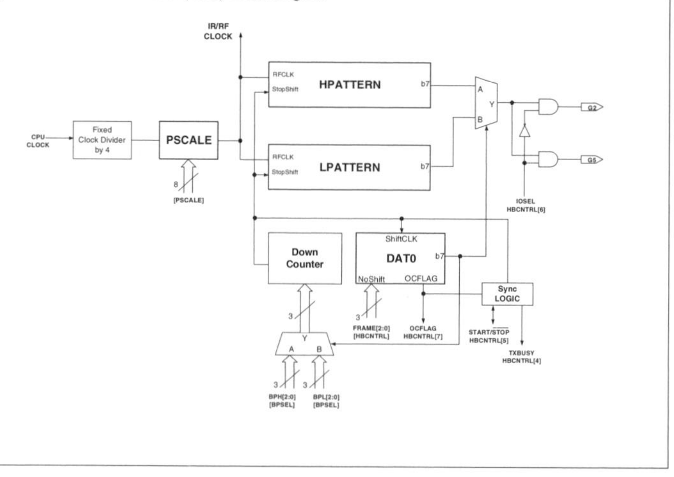

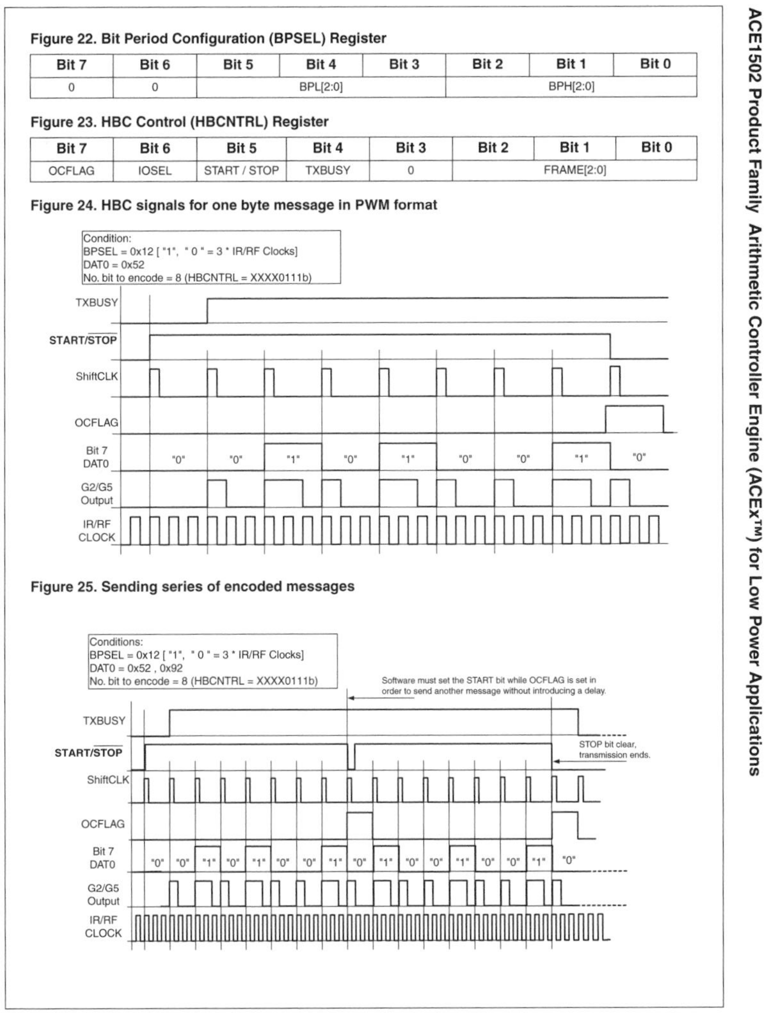

ACE1502 Product Family Arithmetic Controller Engine (ACEx™) for Low Power Applications

Figure 22. Bit Period Configuration (BPSEL) Register

Bit 7	Bit 6	Bit 5	Bit 4	Bit 3	Bit 2	Bit 1	Bit 0
0	0	BPL[2:0]			BPH[2:0]		

Figure 23. HBC Control (HBCNTRL) Register

Bit 7	Bit 6	Bit 5	Bit 4	Bit 3	Bit 2	Bit 1	Bit 0
OCFLAG	IOSEL	START / STOP	TXBUSY	0	FRAME[2:0]		

Figure 24. HBC signals for one byte message in PWM format

Figure 25. Sending series of encoded messages

20 www.fairchildsemi.com

ACE1502 Product Family Rev. 1.7

8. Multi-Input Wakeup/Interrupt Block

The Multi-Input Wakeup (MIW)/Interrupt contains three memory-mapped registers associated with this circuit: WKEDG (Wakeup Edge), WKEN (Wakeup Enable), and WKPND (Wakeup Pending). Each register has 8-bits with each bit corresponding to an input pins as shown in Figure 27. All three registers are initialized to zero upon reset.

The WKEDG register establishes the edge sensitivity for each of the wake-up input pin: either positive going-edge (0) or negative-going edge (1).

The WKEN register enables (1) or disables (0) each of the port pins for the Wakeup/Interrupt function. The wakeup I/Os used for the Wakeup/Interrupt function must also be configured as an input pin in its associated port configuration register. However, an interrupt of the core will not occur unless interrupts are enabled for the block via bit 7 of the T0CNTRL register (see Figure 19) and the G (global interrupt enable) bit of the SR is set.

The WKPND register contains the pending flags corresponding to each of the port pins (1 for wakeup/interrupt pending, 0 for wakeup/interrupt not pending). If an I/O is not selected to become a wakeup input, the pending flag will not be generated.

To use the Multi-Input Wakeup/Interrupt circuit, perform the steps listed below making sure the MIW edge is selected before enabling the I/O to be used as a wakeup input thus preventing false pending flag generation. This same procedure should be used following any type of reset because the wakeup inputs are left floating after resets resulting in unknown data on the port inputs.

1. Clear the WKEN register.
   - CLR WKEN
2. Clear the WKPND register to cancel any pending bits.
   - CLR WKPND
3. If necessary, write to the port configuration register to select the desired port pins to be configured as inputs.
   - RBIT 4, PORTGC ; G4
4. If necessary, write to the port data register to select the desired port pins input state.
   - SBIT 4, PORTGD ; Pull-up
5. Write the WKEDG register to select the desired type of edge sensitivity for each of the pins used.
   - LD WKEDG, #0FFH ; All negative-going edges
6. Set the WKEN bits associated with the pins to be used, thus enabling those pins for the Wakeup/Interrupt function.
   - LD WKEN, #10H ; Enabling G4

Once the Multi-Input Wakeup/Interrupt function has been configured, a transition sensed on any of the I/O pins will set the corresponding bit in the WKPND register. The WKPND bits, where the corresponding enable (WKEN) bits are set, will bring the device out of the HALT mode and can also trigger an interrupt if interrupts are enabled. The interrupt service routine can read the WKPND register to determine which pin sensed the interrupt.

The interrupt service routine or other software should clear the pending bit. The device will not enter HALT mode as long as a WKPND pending bit is pending and enabled. The user has the responsibility of clearing the pending flags before attempting to enter the HALT mode.

Upon reset, the WKEDG register is configured to select positive-going edge sensitivity for all wakeup inputs. If the user wishes to change the edge sensitivity of a port pin, use the following procedure to avoid false triggering of a Wakeup/Interrupt condition.

1. Clear the WKEN bit associated with the pin to disable that pin.
2. Clear the WKPND bit associated with the pin.
3. Write the WKEDG register to select the new type of edge sensitivity for the pin.
4. Set the WKEN bit associated with the pin to re-enable it.

PORTG provides the user with three fully selectable, edge sensitive interrupts that are all vectored into the same service subroutine. The interrupt from PORTG shares logic with the wakeup circuitry. The WKEN register allows interrupts from PORTG to be individually enabled or disabled. The WKEDG register specifies the trigger condition to be either a positive or a negative edge. The WKPND register latches in the pending trigger conditions.

Since PORTG is also used for exiting the device from the HALT mode, the user can elect to exit the HALT mode either with or without the interrupt enabled. If the user elects to disable the interrupt, then the device restarts execution from the point at which it was stopped (first instruction cycle of the instruction following HALT mode entrance instruction). In the other case, the device finishes the instruction that was being executed when the part was stopped and then branches to the interrupt service routine. The device then reverts to normal operation.

Figure 26. Multi-input Wakeup (MIW) Register bit assignments

WKEDG, WKEN, WKPND							
Bit 7	**Bit 6**	**Bit 5**	**Bit 4**	**Bit 3**	**Bit 2**	**Bit 1**	**Bit 0**
[9] G7	[9] G6	G5	G4	G3	G2	G1	G0

9. Available only on the 14-pin package option

Figure 27. Multi-input Wakeup (MIW) Block Diagram

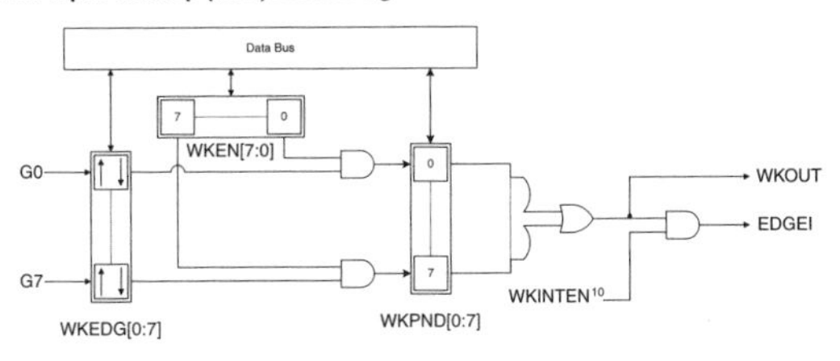

10. WKINTEN: Bit 7 of T0CNTRL

9. I/O Port

The eight I/O pins (six on 8-pin package option) are bi-directional (see Figure 28). The bi-directional I/O pins can be individually configured by software to operate as high-impedance inputs, as inputs with weak pull-up, or as push-pull outputs. The operating state is determined by the contents of the corresponding bits in the data and configuration registers. Each bi-directional I/O pin can be used for general purpose I/O, or in some cases, for a specific alternate function determined by the on-chip hardware.

Figure 28. PORTGD Logic Diagram

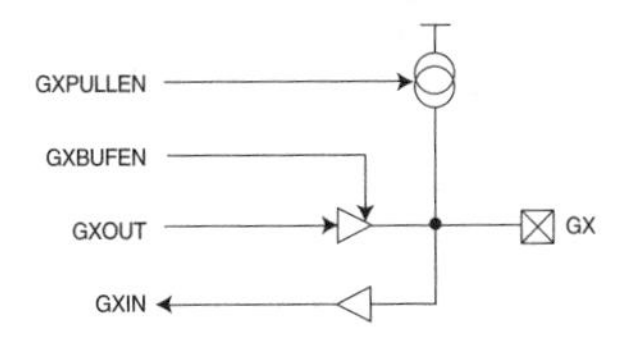

9.1 I/O registers

The I/O pins (G0-G7) have three memory-mapped port registers associated with the I/O circuitry: a port configuration register (PORTGC), a port data register (PORTGD), and a port input register (PORTGP). PORTGC is used to configure the pins as inputs or outputs. A pin may be configured as an input by writing a 0 or as an output by writing a 1 to its corresponding PORTGC bit. If a pin is configured as an output, its PORTGD bit represents the state of the pin (1 = logic high, 0 = logic low). If the pin is configured as an input, its PORTGD bit selects whether the pin is a weak pull-up or a high-impedance input. Table 13 provides details of the port configuration options. The port configuration and data registers can both be read from or written to. Reading PORTGP returns the value of the port pins regardless of how the pins are configured. Since this device supports MIW, PORTG inputs have Schmitt triggers.

Figure 29. I/O Register bit assignments

PORTGC, PORTGD, PORTGD							
Bit 7	**Bit 6**	**Bit 5**	**Bit 4**	**Bit 3**	**Bit 2**	**Bit 1**	**Bit 0**
[11]G7	[11]G6	G5	G4	[12]G3	G2	G1	G0

11. Available only on the 14-pin package option
12. G3 after reset is an input with weak pull-up

Table 13. I/O configuration options

Configuration Bit	Data Bit	Port Pin Configuration
0	0	High-impedence input (TRI-STATE input)
0	1	Input with pull-up (weak one input)
1	0	Push-pull zero output
1	1	Push-pull one output

10. In-circuit Programming Specification

The ACEx microcontroller supports in-circuit programming of the internal data EEPROM, code EEPROM, and the initialization registers.

In order to enter into program mode a 10-bit opcode (0x34B) must be shifted into the ACE1502 while the device is executing the internal power on reset (T_{RESET}). The shifting protocol follows the same timing rules as the programming protocol defined in Figure 30.

The opcode is shifted into the ACE1502 serially, MSB first, with the data being valid by the rising edge of the clock. Once the pattern is shifted into the device, the current 10-bit pattern is matched to protocol entrance opcode of 0x34B. If the 10-bit pattern is a match, the device will enable the internal program mode flag so that the device will enter into program mode once reset has completed (see Figure 30.)

The opcode must be shifted in after Vcc settles to the nominal level and should end before the power on reset sequence (T_{reset}) completes; otherwise, the device will start normal execution of the program code. If the external reset is applied by bringing the reset pin low, once the reset pin is release the opcode may now be shifted in and again should end before the reset sequence completes.

10.3 Programming Protocol

After placing the device in program, the programming protocol and commands may be issued.

An externally controlled four-wire interface consisting of a LOAD control pin (G3), a serial data SHIFT-IN input pin (G4), a serial data SHIFT-OUT output pin (G2), and a CLOCK pin (G1) is used to access the on-chip memory locations. Communication between the ACEx microcontroller and the external programmer is made through a 32-bit command and response word described in Table 14. Be sure to either float or tie G5 to Vcc for proper programming functionality.

The serial data timing for the four-wire interface is shown in Figure 31 and the programming protocol is shown in Figure 30.

10.3.1 Write Sequence

The external programmer brings the ACEx microcontroller into programming then needs to set the LOAD pin to Vcc before shifting in the 32-bit serial command word using the SHIFT_IN and CLOCK signals. By definition, bit 31 of the command word is shifted in first. At the same time, the ACEx microcontroller shifts out the 32-bit serial response to the last command on the SHIFT_OUT pin. It is recommended that the external programmer samples this signal t_{ACCESS} (500 ns) after the rising edge of the CLOCK signal. The serial response word, sent immediately after entering programming mode, contains indeterminate data.

After 32 bits have been shifted into the device, the external programmer must set the LOAD signal to 0V, and then apply two clock pulses as shown in Figure 30 to complete program cycle.

The SHIFT_OUT pin acts as the handshaking signal between the device and programming hardware once the LOAD signal is brought low. The device sets SHIFT_OUT low by the time the programmer has sent the second rising edge during the LOAD = 0V phase (if the timing specifications in Figure 30 are obeyed).

The device will set the R bit of the Status register when the write operation has completed. The external programmer must wait for the SHIFT_OUT pin to go high before bringing the LOAD signal to Vcc to initiate a normal command cycle.

10.3.2 Read Sequence

When reading the device after a write, the external programmer must set the LOAD signal to Vcc before it sends the new command word. Next, the 32-bit serial command word (for during a READ) should be shifted into the device using the SHIFT_IN and the CLOCK signals while the data from the previous command is serially shifted out on the SHIFT_OUT pin. After the Read command has been shifted into the device, the external programmer must, once again, set the LOAD signal to 0V and apply two clock pulses as shown in Figure 30 to complete READ cycle. Data from the selected memory location, will be latched into the lower 8 bits of the command word shortly after the second rising edge of the CLOCK signal.

Writing a series of bytes to the device is achieved by sending a series of Write command words while observing the devices handshaking requirements.

Reading a series of bytes from the device is achieved by sending a series of Read command words with the desired addresses in sequence and reading the following response words to verify the correct address and data contents.

The addresses of the data EEPROM and code EEPROM locations are the same as those used in normal operation.

Powering down the device will cause the part to exit programming mode.

Table 14 32-Bit Command and Response Word

Bit Number	Input Command Word	Output Response Word
bits 31-30	Must be set to 0	X
bit 29	Set to 1 to read/write data EEPROM, or the initialization registers, otherwise 0	X
bit 28	Set to 1 to read/write code EEPROM, otherwise 0	X
bits 27-25	Must be set to 0	X
bit 24	Set to 1 to read, 0 to write	X
bits 23-19	Must be set to 0	X
bits 18 -8	Address of the byte to be read or written	Same as Input command word
bits 7-0	Data to be programmed or zero if data is to be read	Programmed data or data read at specified address

Figure 30. Programming Protocol[13]

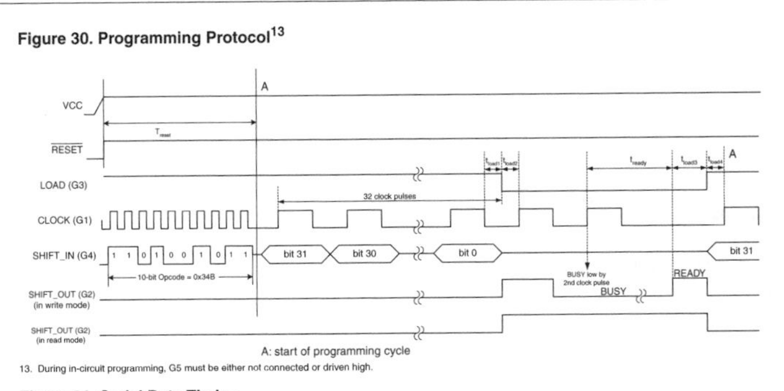

13. During in-circuit programming, G5 must be either not connected or driven high.

Figure 31. Serial Data Timing

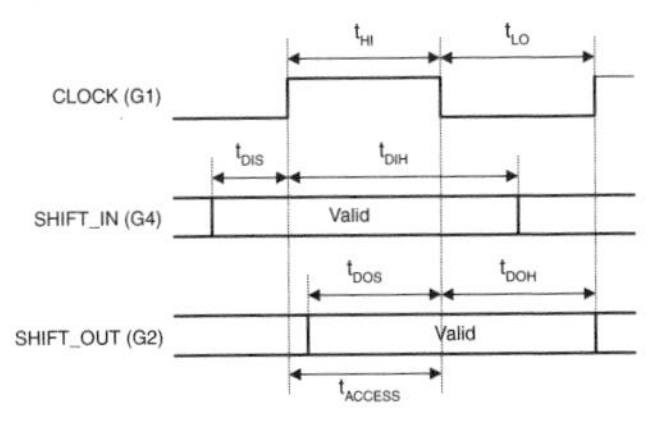

11. Brown-out/Low Battery Detect Circuit

The Brown-out Reset (BOR) and Low Battery Detect (LBD) circuits on the ACEx microcontroller have been designed to offer two types of voltage reference comparators. The sections below will describe the functionality of both circuits.

Figure 32. BOR/LBD Block Diagram

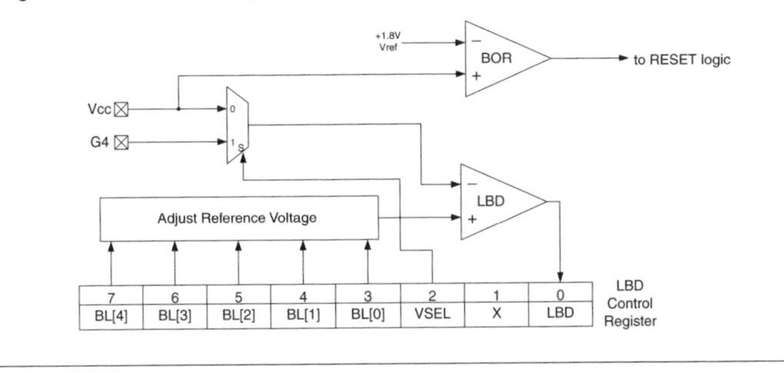

11.1 Brown-out Reset

The Brown-out Reset (BOR) function is used to hold the device in reset when Vcc drops below a fixed threshold (1.83V.) While in reset, the device is held in its initial condition until Vcc rises above the threshold value. Shortly after Vcc rises above the threshold value, an internal reset sequence is started. After the reset sequence, the core fetches the first instruction and starts normal operation.

The BOR should be used in situations when Vcc rises and falls slowly and in situations when Vcc does not fall to zero before rising back to operating range. The Brown-out Reset can be thought of as a supplement function to the Power-on Reset if Vcc does not fall below ~1.5V. The Power-on Reset circuit works best when Vcc starts from zero and rises sharply. In applications where Vcc is not constant, the BOR will give added device stability.

The BOR circuit must be enabled through the BOR enable bit (BOREN) in the initialization register. The BOREN bit can only be set while the device is in programming mode. Once set, the BOR will always be powered-up enabled. Software cannot disable the BOR. The BOR can only be disabled in programming mode by resetting the BOREN bit as long as the global write protect (WDIS) feature is not enabled.

Figure 33. BOR and POR Circuit Relationship Diagram

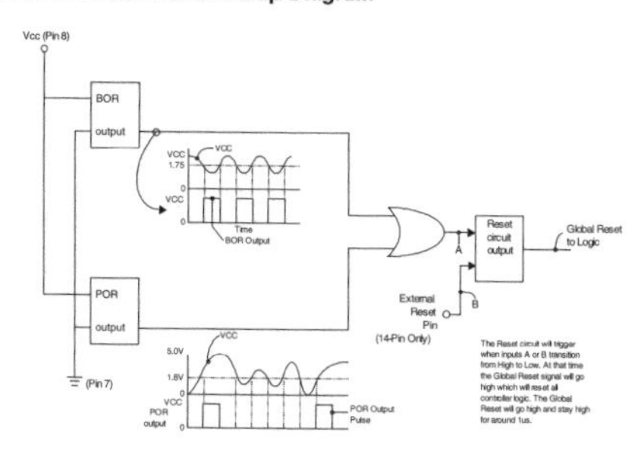

11.2 Low Battery Detect

The Low Battery Detect (LBD) circuit allows software to monitor the Vcc level at the lower voltage ranges. LBD has a 32-level software programmable voltage reference threshold that can be changed on the fly. Once Vcc falls below the selected threshold, the LBD flag in the LBD control register is set. The LBD flag will hold its value until Vcc rises above the threshold. (See Table 15)

The LBD bit is read only. If LBD is 0, it indicates that the Vcc level is higher than the selected threshold. If LBD is 1, it indicates that the Vcc level is below the selected threshold. The threshold level can be adjusted up to eight levels using the three trim bits (BL[4:0]) of the LBD control register. The LBD flag does not cause any hardware actions or an interruption of the processor. It is for software monitoring only.

The VSEL bit of the LBD control register can be used to select an external voltage source rather than Vcc. If VSEL is 1, the voltage source for the LBD comparator will be an input voltage provided through G4. If VSEL is 0, the voltage source will be Vcc.

The LBD circuit must be enabled through the LBD enable bit (LBDEN) in the initialization register. The LBDEN bit can only be set while the device is in programming mode. Once set, the LBD will always be powered-up enabled. Software cannot disable the LBD. The LBD can only be disabled in programming mode by resetting the LBDEN bit as long as the global write protect (WDIS) feature is not enabled.

The LBD circuit is disabled during HALT/IDLE mode. After exiting HALT/IDLE, software must wait at lease 10 μs before reading the LBD bit to ensure that the internal circuit has stabilized.

Table 15. LBD Control Register Definition

Bit 7	Bit 6	Bit 5	Bit 4	Bit 3	Bit 2	Bit 1	Bit 0
BL[4:0]					VSEL	X	LBD

Level	BL[4]	BL[3]	BL[2]	BL[1]	BL[0]	Voltage Reference Range (Typical)
1	0	0	0	0	0	1.81V
2	0	0	0	0	1	1.87V
3	0	0	0	1	0	1.93V
4	0	0	0	1	1	1.99V
5	0	0	1	0	0	2.05V
6	0	0	1	0	1	2.11V
7	0	0	1	1	0	2.17V
8	0	0	1	1	1	2.23V
9	0	1	0	0	0	2.29V
10	0	1	0	0	1	2.36V
11	0	1	0	1	0	2.42V
12	0	1	0	1	1	2.48V
13	0	1	1	0	0	2.54V
14	0	1	1	0	1	2.60V
15	0	1	1	1	0	2.66V
16	0	1	1	1	1	2.72V
17	1	0	0	0	0	2.77V
18	1	0	0	0	1	2.84V
19	1	0	0	1	0	2.91V
20	1	0	0	1	1	2.97V
21	1	0	1	0	0	3.03V
22	1	0	1	0	1	3.09V
23	1	0	1	1	0	3.16V
24	1	0	1	1	1	3.22V
25	1	1	0	0	0	3.28V
26	1	1	0	0	1	3.34V
27	1	1	0	1	0	3.41V
28	1	1	0	1	1	3.47V
29	1	1	1	0	0	3.54V
30	1	1	1	0	1	3.60V
31	1	1	1	1	0	3.67V
32	1	1	1	1	1	3.73V

12. RESET block

When a RESET sequence is initiated, all I/O registers will be reset setting all I/Os to high-impedence inputs. The system clock is restarted after the required clock start-up delay. A reset is generated by any one of the following four conditions:

- Power-on Reset (as described in Section 13)
- Brown-out Reset (as described in Section 11.1)
- Watchdog Reset (as described in Section 6)
- External Reset [18] (as described in Section 13)

18. Available only on the 14-pin package option

13. Power-On Reset

The Power-On Reset (POR) circuit is guaranteed to work if the rate of rise of Vcc is no slower than 10ms/1volt. The POR circuit was designed to respond to fast low to high transitions between 0V and Vcc. The circuit will not work if Vcc does not drop to 0V before the next power-up sequence. In applications where 1) the Vcc rise is slower than 10ms/1 volt or 2) Vcc does not drop to 0V before the next power-up sequence the external reset option should be used.

The external reset provides a way to properly reset the ACEx microcontroller if POR cannot be used in the application. The external reset pin contains an internal pull-up resistor. Therefore, to reset the device the reset pin should be held low for at least 2ms so that the internal clock has enough time to stabilize.

14. CLOCK

The ACEx microcontroller has an on-board oscillator trimmed to a frequency of 2MHz who is divided down by two yielding a 1MHz frequency. (See AC Electrical Characteristics) Upon power-up, the on-chip oscillator runs continuously unless entering HALT mode or using an external clock source.

If required, an external oscillator circuit may be used depending on the states of the CMODE bits of the initialization register. (See Table 16) When the device is driven using an external clock, the clock input to the device (G1/CKI) can range between DC to 4MHz. For external crystal configuration, the output clock (CKO) is on the G0 pin. (See Figure 34.) If the device is configured for an external square clock, it will not be divided.

Table 16. CMODEx Bit Definition

CMODE [1]	CMODE [0]	Clock Type
0	0	Internal 1 MHz clock
0	1	External square clock
1	0	External crystal/resonator
1	1	Reserved

Figure 34. Crystal

15. HALT Mode

The HALT mode is a power saving feature that almost completely shuts down the device for current conservation. The device is placed into HALT mode by setting the HALT enable bit (EHALT) of the HALT register through software using only the "LD M, #" instruction. EHALT is a write only bit and is automatically cleared upon exiting HALT. When entering HALT, the internal oscillator and all the on-chip systems including the LBD and the BOR circuits are shut down.

The device can exit HALT mode only by the MIW circuit. Therefore, prior to entering HALT mode, software must configure the MIW circuit accordingly. (See Section 8) After a wakeup from HALT, a 1ms start-up delay is initiated to allow the internal oscillator to stabilize before normal execution resumes. Immediately after exiting HALT, software must clear the Power Mode Clear (PMC) register by only using the "LD M, #" instruction. (See Figure 36)

Figure 35. HALT Register Definition

Bit 7	Bit 6	Bit 5	Bit 4	Bit 3	Bit 2	Bit 1	Bit 0
Undefined	undefined	undefined	undefined	undefined	undefined	EIDLE	EHALT

Figure 36. Recommended HALT Flow

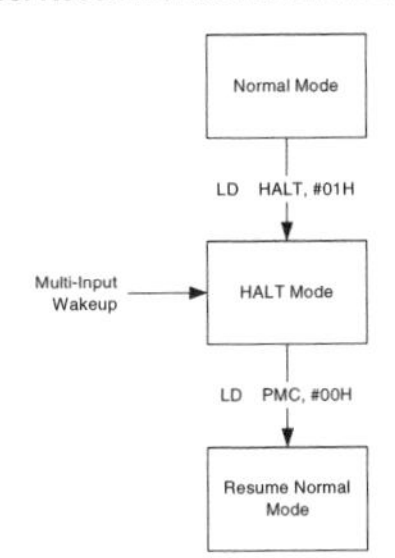

16. IDLE Mode

In addition to the HALT mode power saving feature, the device also supports an IDLE mode operation. The device is placed into IDLE mode by setting the IDLE enable bit (EIDLE) of the HALT register through software using only the "LD M, #" instruction. EIDLE is a write only bit and is automatically cleared upon exiting IDLE. The IDLE mode operation is similar to HALT except the internal oscillator, the Watchdog, and the Timer 0 remain active while the other on-chip systems including the LBD and the BOR circuits are shut down.

The device automatically wakes from IDLE mode by the Timer 0 overflow every 8192 cycles (see Section 5). Before entering IDLE mode, software must clear the WKEN register to disable the MIW block. Once a wake from IDLE mode is triggered, the core will begin normal operation by the next clock cycle. Immediately after exiting IDLE mode, software must clear the Power Mode Clear (PMC) register by using only the "LD M, #" instruction. (See Figure 37.)

Figure 37. Recommended IDLE Flow

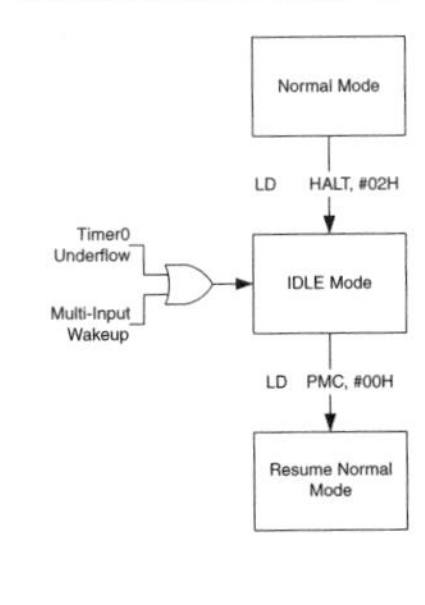

ACE1502 Product Family Arithmetic Controller Engine (ACEx™) for Low Power Applications

Ordering Information

Part Number	Core Type				Max. # I/Os	Program Memory Size		Operating Voltage Range	Package								Tape & Reel
	0	1	2	5	8	1K	2K	1.8 – 3.6V	-40 to +85°C	-40 to +125°C	8-pin SOIC	14-pin SOIC	8-pin DIP	14-pin DIP	8-pin TSSOP	14-pin TSSOP	
ACE1502EM8				X	X		X	X	X		X						
ACE1502EM8X				X	X		X	X	X		X						X
ACE1502EM				X	X		X	X	X			X					
ACE1502EMX				X	X		X	X	X			X					X
ACE1502EMT8				X	X		X	X	X						X		
ACE1502EMT8X				X	X		X	X	X						X		X
ACE1502EMT				X	X		X	X	X							X	
ACE1502EMTX				X	X		X	X	X							X	X
ACE1502EN				X	X		X	X	X				X				
ACE1502EN14				X	X		X	X	X					X			
ACE1502VM8				X	X		X	X		X	X						
ACE1502VM8X				X	X		X	X		X	X						X
ACE1502VM				X	X		X	X		X		X					
ACE1502VMX				X	X		X	X		X		X					X
ACE1502VMT8				X	X		X	X		X					X		
ACE1502VMT8X				X	X		X	X		X					X		X
ACE1502VMT				X	X		X	X		X						X	
ACE1502VMTX				X	X		X	X		X						X	X
ACE1502VN				X	X		X	X		X			X				
ACE1502VN14				X	X		X	X		X				X			

Physical Dimensions inches (millimeters) unless otherwise noted)

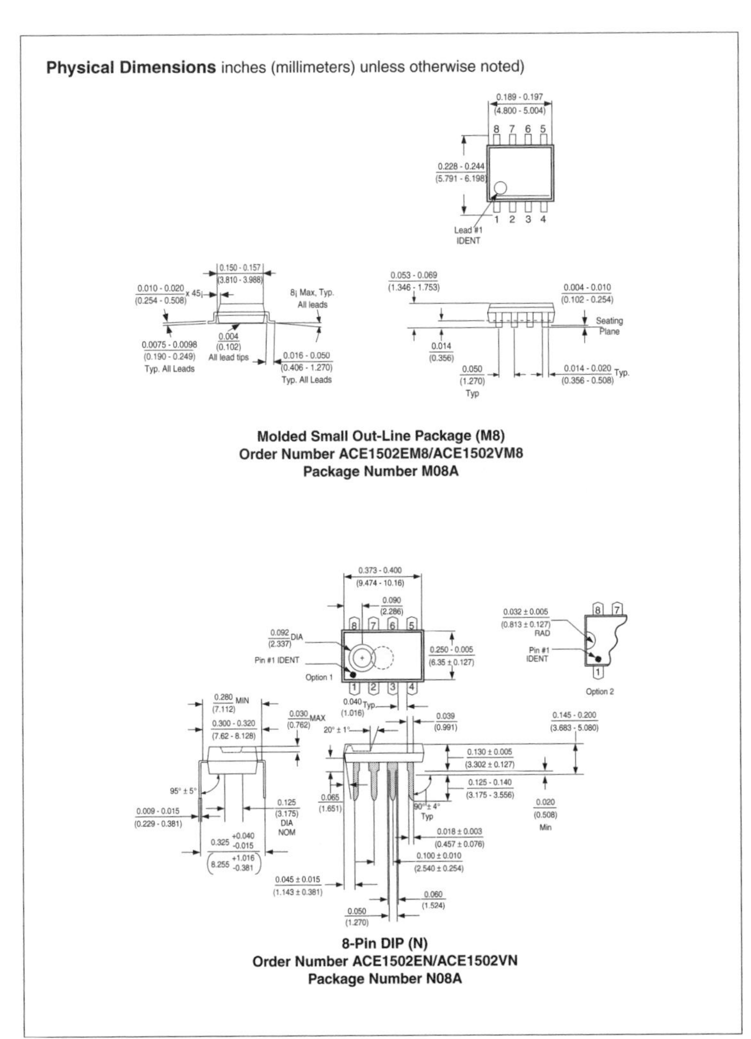

Molded Small Out-Line Package (M8)
Order Number ACE1502EM8/ACE1502VM8
Package Number M08A

8-Pin DIP (N)
Order Number ACE1502EN/ACE1502VN
Package Number N08A

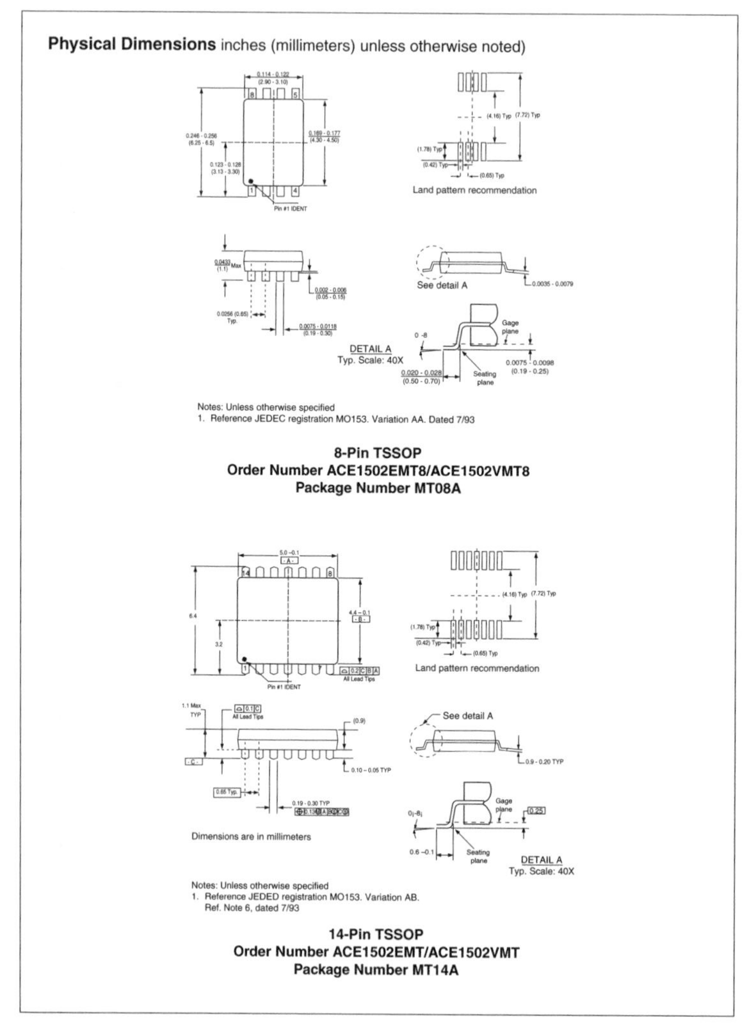

Physical Dimensions inches (millimeters) unless otherwise noted)

Land pattern recommendation

See detail A

DETAIL A
Typ. Scale: 40X

Notes: Unless otherwise specified
1. Reference JEDEC registration MO153. Variation AA. Dated 7/93

8-Pin TSSOP
Order Number ACE1502EMT8/ACE1502VMT8
Package Number MT08A

Land pattern recommendation

See detail A

Dimensions are in millimeters

DETAIL A
Typ. Scale: 40X

Notes: Unless otherwise specified
1. Reference JEDED registration MO153. Variation AB.
Ref. Note 6, dated 7/93

14-Pin TSSOP
Order Number ACE1502EMT/ACE1502VMT
Package Number MT14A

ACE1502 Product Family Rev. 1.7

31

www.fairchildsemi.com

ACE1502 Product Family Arithmetic Controller Engine (ACEx™) for Low Power Applications

Physical Dimensions inches (millimeters) unless otherwise noted

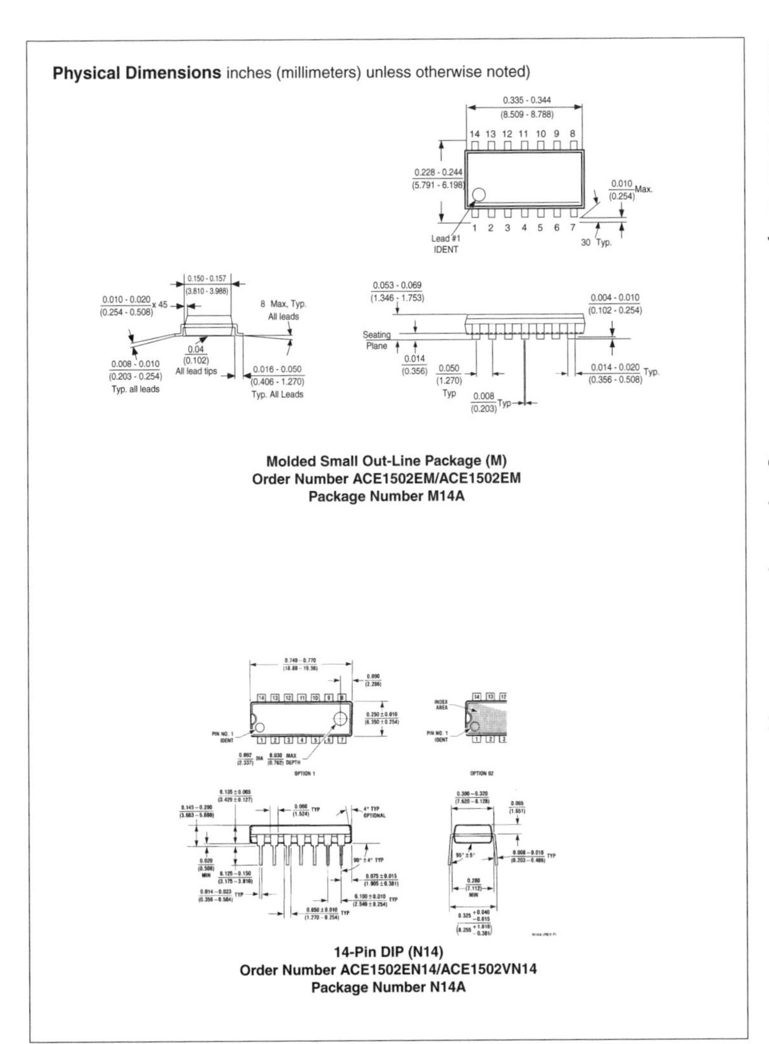

Molded Small Out-Line Package (M)
Order Number ACE1502EM/ACE1502EM
Package Number M14A

14-Pin DIP (N14)
Order Number ACE1502EN14/ACE1502VN14
Package Number N14A

ACEx Development Tools

General Information:

Fairchild Semiconductor offers different possibilities to evaluate and emulate software written for ACEx.

Simulator: Is a Windows program able to load, assemble, and debug ACEx programs. It is possible to place as many breakpoints as needed, trace the program execution in symbolic format, and program a device with the proper options. The ACEx Simulator is available free-of-charge and can be downloaded from Fairchild's web site at www.fairchildsemi.com/products/memory/ace

ACEx Emulator Kit: Fairchild also offers a low cost real-time in-circuit emulator kit that includes:

- Emulator board
- Emulator software
- Assembler and Manuals
- Power supply
- DIP14 target cable
- PC cable

The ACEx emulator allows for debugging the program code in a symbolic format. It is possible to place one breakpoint and watch various data locations. It also has built-in programming capability.

Prototype Board Kits: Fairchild offers two solutions for the simplification of the breadboard operation so that ACEx Applications can be quickly tested.

1) ACEDEMO can be used for general purpose applications

2) ACETXRX is for transmitting / receiving (RF, IR, RS232, RS485) applications.

ACEDEMO has 8 switches, 8 LEDs, RS232 voltage translator, buzzer, and a lamp with a small breadboard area.

Factory Programming:

Fairchild offers factory pre-programming and serialization (for justified quantities) for a small additional cost. Please refer to your local distributor for details regarding factory programming.

Ordering P/Ns

Emulator Kit and Programming adapters:

Please refer to your local distributor for details regarding development tools.

Life Support Policy

Fairchild's products are not authorized for use as critical components in life support devices or systems without the express written approval of the President of Fairchild Semiconductor Corporation. As used herein:

1. Life support devices or systems are devices or systems which, (a) are intended for surgical implant into the body, or (b) support or sustain life, and whose failure to perform, when properly used in accordance with instructions for use provided in the labeling, can be reasonably expected to result in a significant injury to the user.

2. A critical component is any component of a life support device or system whose failure to perform can be reasonably expected to cause the failure of the life support device or system, or to affect its safety or effectiveness.

Fairchild Semiconductor
Americas
Customer Response Center
Tel. 1-888-522-5372

Fairchild Semiconductor
Europe
Fax: +44 (0) 1793-856858
Deutsch Tel: +49 (0) 8141-6102-0
English Tel: +44 (0) 1793-856856
Français Tel: +33 (0) 1-6930-3696
Italiano Tel: +39 (0) 2-249111-1

Fairchild Semiconductor
Hong Kong
8/F, Room 808, Empire Centre
68 Mody Road, Tsimshatsui East
Kowloon. Hong Kong
Tel; +852-2722-8338
Fax: +852-2722-8383

Fairchild Semiconductor
Japan Ltd.
4F, Natsume Bldg.
2-18-6, Yushima, Bunkyo-ku
Tokyo, 113-0034 Japan
Tel: 81-3-3818-8840
Fax: 81-3-3818-8841

Appendix F

Fairchild Specifications for FAN5236

www.fairchildsemi.com

FAN5236

Dual Mobile-Friendly DDR / Dual-output PWM Controller

Features

- Highly flexible dual synchronous switching PWM controller includes modes for:
  - DDR mode with in-phase operation for reduced channel interference
  - 90° phase shifted two-stage DDR Mode for reduced input ripple
  - Dual Independent regulators 180° phase shifted
- Complete DDR Memory power solution
  - V_{TT} Tracks VDDQ/2
  - VDDQ/2 Buffered Reference Output
- Lossless current sensing on low-side MOSFET or precision over-current using sense resistor
- V_{CC} Under-voltage Lockout
- Converters can operate from +5V or 3.3V or Battery power input (5 to 24V)
- Excellent dynamic response with Voltage Feed-Forward and Average Current Mode control
- Power-Good Signal
- Also supports DDR-II and HSTL
- Light load Hysteretic mode maximizes efficiency
- QSOP28, TSSOP28

Applications

- DDR V_{DDQ} and V_{TT} voltage generation
- Mobile PC dual regulator
- Server DDR power
- Hand-Held PC power

General Description

The FAN5236 PWM controller provides high efficiency and regulation for two output voltages adjustable in the range from 0.9V to 5.5V that are required to power I/O, chip-sets, and memory banks in high-performance notebook computers, PDAs and Internet appliances. Synchronous rectification and hysteretic operation at light loads contribute to a high efficiency over a wide range of loads. The hysteretic mode of operation can be disabled separately on each PWM converter if PWM mode is desired for all load levels. Efficiency is even further enhanced by using MOSFET's $R_{DS(ON)}$ as a current sense component.

Feed-forward ramp modulation, average current mode control scheme, and internal feedback compensation provide fast response to load transients. Out-of-phase operation with 180 degree phase shift reduces input current ripple. The controller can be transformed into a complete DDR memory power supply solution by activating a designated pin. In DDR mode of operation one of the channels tracks the output voltage of another channel and provides output current sink and source capability — features essential for proper powering of DDR chips. The buffered reference voltage required by this type of memory is also provided. The FAN5236 monitors these outputs and generates separate PGx (power good) signals when the soft-start is completed and the output is within ±10% of its set point. A built-in over-voltage protection prevents the output voltage from going above 120% of the set point. Normal operation is automatically restored when the over-voltage conditions go away. Under-voltage protection latches the chip off when either output drops below 75% of its set value after the soft-start sequence for this output is completed. An adjustable over-current function monitors the output current by sensing the voltage drop across the lower MOSFET. If precision current-sensing is required, an external current-sense resistor may optionally be used.

Generic Block Diagrams

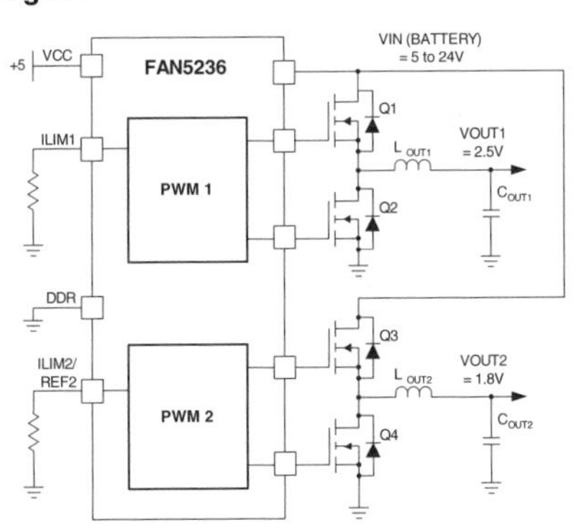

Figure 1. Dual output regulator

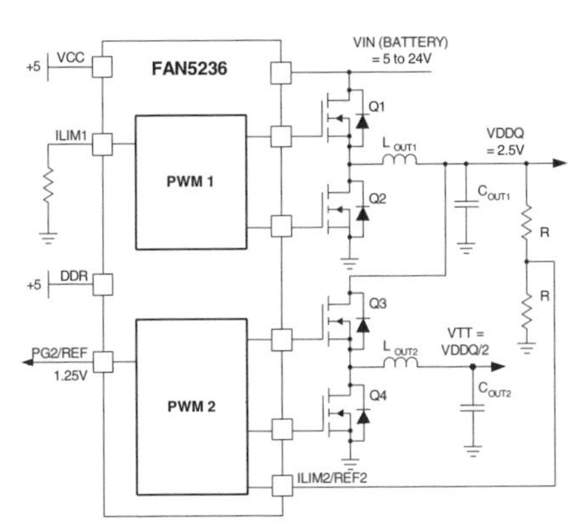

Figure 2. Complete DDR Memory Power Supply

Pin Configurations

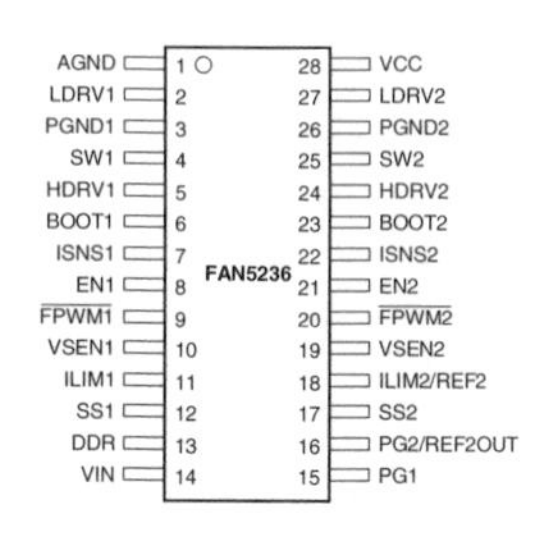

QSOP-28 or TSSOP-28
$\theta_{JA} = 90°C/W$

Pin Definitions

Pin Number	Pin Name	Pin Function Description
1	AGND	**Analog Ground.** This is the signal ground reference for the IC. All voltage levels are measured with respect to this pin.
2 27	LDRV1 LDRV2	**Low-Side Drive.** The low-side (lower) MOSFET driver output. Connect to gate of low-side MOSFET.
3 26	PGND1 PGND2	**Power Ground.** The return for the low-side MOSFET driver. Connect to source of low-side MOSFET.
4 25	SW1 SW2	**Switching node.** Return for the high-side MOSFET driver and a current sense input. Connect to source of high-side MOSFET and low-side MOSFET drain.
5 24	HDRV1	**High-Side Drive.** High-side (upper) MOSFET driver output. Connect to gate of high-side MOSFET.
6 23	BOOT1 BOOT2	**BOOT.** Positive supply for the upper MOSFET driver. Connect as shown in Figure 3.
7 22	ISNS1 ISNS2	**Current Sense input.** Monitors the voltage drop across the lower MOSFET or external sense resistor for current feedback.
8 21	EN1 EN2	**Enable**. Enables operation when pulled to logic high. Toggling EN will also reset the regulator after a latched fault condition. These are CMOS inputs whose state is indeterminate if left open.
9 20	$\overline{\text{FPWM1}}$ $\overline{\text{FPWM2}}$	**Forced PWM mode.** When logic low, inhibits the regulator from entering hysteretic mode. Otherwise tie to VOUT. The regulator uses VOUT on this pin to ensure a smooth transition from Hysteretic mode to PWM mode. When VOUT is expected to exceed VCC, tie to VCC.
10 19	VSEN1 VSEN2	**Output Voltage Sense.** The feedback from the outputs. Used for regulation as well as PG, under-voltage and over-voltage protection and monitoring.
11	ILIM1	**Current Limit 1.** A resistor from this pin to GND sets the current limit.
12 17	SS1 SS2	**Soft Start.** A capacitor from this pin to GND programs the slew rate of the converter during initialization. During initialization, this pin is charged with a 5μA current source.
13	DDR	**DDR Mode Control.** High = DDR mode. Low = 2 separate regulators operating 180° out of phase.

Pin Definitions (continued)

Pin Number	Pin Name	Pin Function Description
14	VIN	**Input Voltage.** Normally connected to battery, providing voltage feed-forward to set the amplitude of the internal oscillator ramp. When using the IC for 2-step conversion from 5V input, connect through 100K to ground, which will set the appropriate ramp gain and synchronize the channels 90° out of phase.
15	PG1	**Power Good Flag.** An open-drain output that will pull LOW when VSEN is outside of a ±10% range of the 0.9V reference.
16	PG2 / REF2OUT	**Power Good 2.** When not in DDR Mode: Open-drain output that pulls LOW when the VOUT is out of regulation or in a fault condition **Reference Out 2.** When in DDR Mode, provides a buffered output of REF2. Typically used as the VDDQ/2 reference.
18	ILIM2 / REF2	**Current Limit 2.** When not in DDR Mode, A resistor from this pin to GND sets the current limit. **Reference** for reg #2 when in DDR Mode. Typically set to VOUT1 / 2.
28	VCC	**VCC.** This pin powers the chip as well as the LDRV buffers. The IC starts to operate when voltage on this pin exceeds 4.6V (UVLO rising) and shuts down when it drops below 4.3V (UVLO falling).

Absolute Maximum Ratings

Absolute maximum ratings are the values beyond which the device may be damaged or have its useful life impaired. Functional operation under these conditions is not implied.

Parameter	Min.	Typ.	Max.	Units
VCC Supply Voltage:			6.5	V
VIN			27	V
BOOT, SW, ISNS, HDRV			33	V
BOOTx to SWx			6.5	V
All Other Pins	–0.3		VCC+0.3	V
Junction Temperature (T_J)	–40		150	°C
Storage Temperature	–65		150	°C
Lead Soldering Temperature, 10 seconds			300	°C

Recommended Operating Conditions

Parameter	Conditions	Min.	Typ.	Max.	Units
Supply Voltage VCC		4.75	5	5.25	V
Supply Voltage VIN				24	V
Ambient Temperature (T_A)	Note 1	–10		85	°C

Note 1: Industrial temperature range (–40 to + 85°C) may be special ordered from Fairchild. Please contact your authorized Fairchild representative for more information.

Electrical Specifications

Recommended operating conditions, unless otherwise noted.

Parameter	Conditions	Min.	Typ.	Max.	Units
Power Supplies					
VCC Current	LDRV, HDRV Open, VSEN forced above regulation point		2.2	3.0	mA
	Shut-down (EN=0)			30	μA
VIN Current – Sinking	VIN = 24V	10		30	μA
VIN Current – Sourcing	VIN = 0V		–15	–30	μA
VIN Current – Shut-down				1	μA
UVLO Threshold	Rising VCC	4.3	4.55	4.75	V
	Falling	4.1	4.25	4.45	V
UVLO Hysteresis			300		mV
Oscillator					
Frequency		255	300	345	KHz
Ramp Amplitude, pk–pk	VIN = 16V		2		V
Ramp Amplitude, pk–pk	VIN = 5V		1.25		V
Ramp Offset			0.5		V
Ramp / VIN Gain	VIN Š 3V		125		mV/V
Ramp / VIN Gain	1V < VIN < 3V		250		mV/V
Reference and Soft Start					
Internal Reference Voltage		0.891	0.9	0.909	V
Soft Start current (I_{SS})	at start-up		5		μA
Soft Start Complete Threshold			1.5		V
PWM Converters					
Load Regulation	I_{OUTX} from 0 to 5A, VIN from 5 to 24V	-2		+2	%
VSEN Bias Current		50	80	120	nA
VOUT pin input impedance		45	55	65	KΩ
Under-voltage Shutdown	as % of set point. 2μS noise filter	70	75	80	%
Over-voltage threshold	as % of set point. 2μS noise filter	115	120	125	%
I_{SNS} Over-Current threshold	R_{ILIM}= 68.5KΩ see Figure 11.	112	140	168	μA
Output Drivers					
HDRV Output Resistance	Sourcing		12	15	¾
	Sinking		2.4	4	¾
LDRV Output Resistance	Sourcing		12	15	¾
	Sinking		1.2	2	¾
PG (Power Good Output) and Control pins					
Lower Threshold	as % of set point, 2μS noise filter	–86		–94	%
Upper Threshold	as % of set point, 2μS noise filter	108		116	%
PG Output Low	IPG = 4mA			0.5	V
Leakage Current	V_{PULLUP} = 5V			1	μA
PG2/REF2OUT Voltage	DDR = 1, 0 mA < $I_{REF2OUT}$ ð10mA	99		1.01	% VREF2

Electrical Specifications Recommended operating conditions, unless otherwise noted. (continued)

Parameter	Conditions	Min.	Typ.	Max.	Units
DDR, EN Inputs					
Input High		2			V
Input Low				0.8	V
FPWM Inputs					
FPWM Low				0.1	V
FPWM High	FPWM connected to output	0.9			V

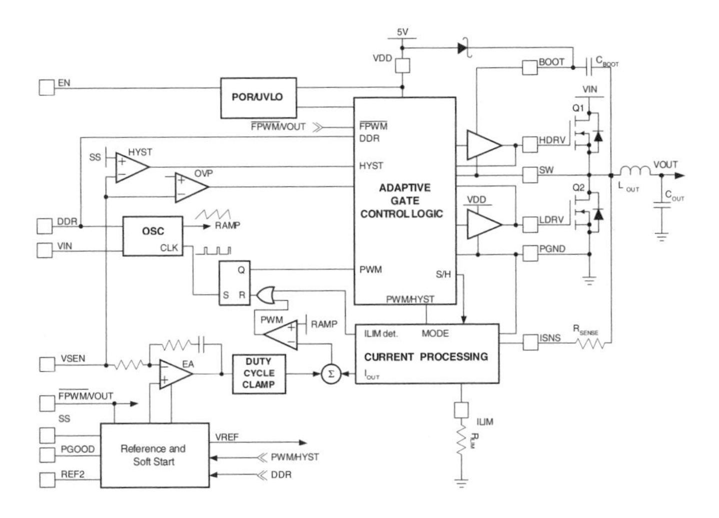

Figure 3. IC Block Diagram

Typical Applications

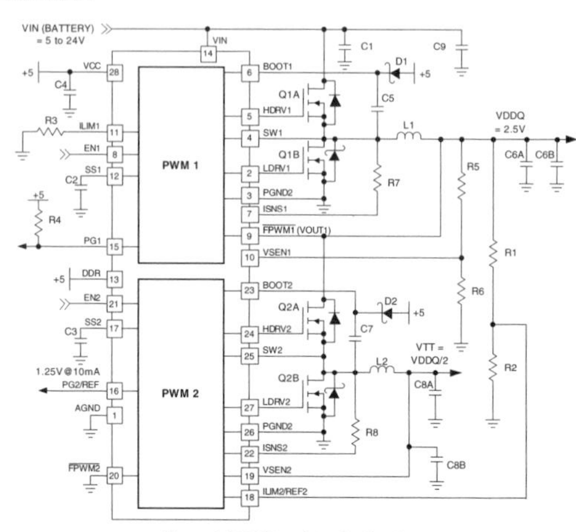

Figure 4. DDR Regulator Application

Table 1. DDR Regulator BOM

Description	Qty	Ref.	Vendor	Part Number
Capacitor 68µf, Tantalum, 25V, ESR 150mΩ	1	C1	AVX	TPSV686*025#0150
Capacitor 10nf, Ceramic	2	C2, C3	Any	
Capacitor 68µf, Tantalum, 6V, ESR 1.8Ω	1	C4	AVX	TAJB686*006
Capacitor 150nF, Ceramic	2	C5, C7	Any	
Capacitor 180µf, Specialty Polymer 4V, ESR 15mΩ	2	C6A, C6B	Panasonic	EEFUE0G181R
Capacitor 1000µf, Specialty Polymer 4V, ESR 10mΩ	1	C8	Kemet	T510E108(1)004AS4115
Capacitor 0.1µF, Ceramic	2	C9	Any	
18.2KΩ, 1% Resistor	3	R1, R2	Any	
1.82KΩ, 1% Resistor	1	R6	Any	
56.2KΩ, 1% Resistor	2	R3	Any	
10KΩ, 5% Resistor	2	R4	Any	
3.24KΩ, 1% Resistor	1	R5	Any	
1.5KΩ, 1% Resistor	2	R7, R8	Any	
Schottky Diode 30V	2	D1, D2	Fairchild	BAT54
Inductor 6.4µH, 6A, 8.64mΩ	1	L1,	Panasonic	ETQ-P6F6R4HFA
Inductor 0.8µH, 6A, 2.24mΩ	1	L2	Panasonic	ETQ-P6F0R8LFA
Dual MOSFET with Schottky	1	Q1, Q2	Fairchild	FDS6986S (note 1)
DDR Controller	1	U1	Fairchild	FAN5236

Note 1: Suitable for typical notebook computer application of 4A continuous, 6A peak for VDDQ. If continuous operation above 6A is required use single SO-8 packages for Q1A (FDS6612A) and Q1B (FDS6690S) respectively. Using FDS6690S, change R7 to 1200¾. Refer to Power MOSFET Selection, page 15 for more information.

Typical Applications (continued)

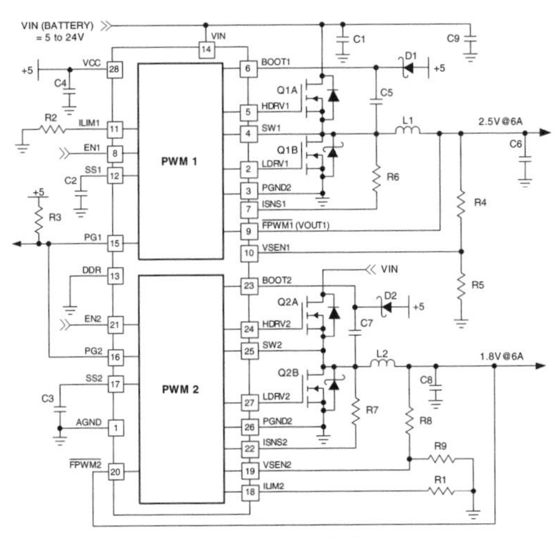

Figure 5. Dual Regulator Application

Table 2. Dual Regulator BOM

Item	Description	Qty	Ref.	Vendor	Part Number
1	Capacitor 68µf, Tantalum, 25V, ESR 95mΩ	1	C1	AVX	TPSV686*025#095
2	Capacitor 10nf, Ceramic	2	C2, C3	Any	
3	Capacitor 68µf, Tantalum, 6V, ESR 1.8Ω	1	C4	AVX	TAJB686*006
4	Capacitor 150nF, Ceramic	2	C5, C7	Any	
5	Capacitor 330µf, Poscap, 4V, ESR 40mΩ	2	C6, C8	Sanyo	4TPB330ML
5	Capacitor 0.1µF, Ceramic	2	C9	Any	
11	56.2KΩ, 1% Resistor	2	R1, R2	Any	
12	10KΩ, 5% Resistor	2	R3	Any	
13	3.24KΩ, 1% Resistor	1	R4	Any	
14	1.82KΩ, 1% Resistor	3	R5, R8, R9	Any	
15	1.5KΩ, 1% Resistor	2	R6, R7	Any	
27	Schottky Diode 30V	2	D1, D2	Fairchild	BAT54
28	Inductor 6.4µH, 6A, 8.64mΩ	1	L1, L2	Panasonic	ETQ-P6F6R4HFA
29	Dual MOSFET with Schottky	1	Q1	Fairchild	FDS6986S (note 1)
30	DDR Controller	1	U1	Fairchild	FAN5236

Note 1: If currents above 4A continuous required, use single SO-8 packages for Q1A/Q2A (FDS6612A) and Q1B/Q2B (FDS6690S) respectively. Using FDS6690S, change R6/R7 as required. Refer to Power MOSFET Selection, page 15 for more information.

Circuit Description

Overview

The FAN5236 is a multi-mode, dual channel PWM controller intended for graphic chipset, SDRAM, DDR DRAM or other low voltage power applications in modern notebook, desktop, and sub-notebook PCs. The IC integrates a control circuitry for two synchronous buck converters. The output voltage of each controller can be set in the range of 0.9V to 5.5V by an external resistor divider.

The two synchronous buck converters can operate from either an unregulated DC source (such as a notebook battery) with voltage ranging from 5.0V to 24V, or from a regulated system rail of 3.3V to 5V. In either mode of operation the IC is biased from a +5V source. The PWM modulators use an average current mode control with input voltage feed-forward for simplified feedback loop compensation and improved line regulation. Both PWM controllers have integrated feedback loop compensation that dramatically reduces the number of external components.

Depending on the load level, the converters can operate either in fixed frequency PWM mode or in a hysteretic mode. Switch-over from PWM to hysteretic mode improves the converters' efficiency at light loads and prolongs battery run time. In hysteretic mode, comparators are synchronized to the main clock that allows seamless transition between the operational modes and reduced channel-to-channel interaction. The hysteretic mode of operation can be inhibited independently for each channel if variable frequency operation is not desired.

The FAN5236 can be configured to operate as a complete DDR solution. When the DDR pin is set high, the second channel can provide the capability to track the output voltage of the first channel. The PWM2 converter is prevented from going into hysteretic mode if the DDR pin is set high. In DDR mode, a buffered reference voltage (buffered voltage of the REF2 pin), required by DDR memory chips, is provided by the PG2 pin.

Converter Modes and Synchronization

Table 3. Converter modes and Synchronization

Mode	VIN	VIN Pin	DDR Pin	PWM 2 w.r.t. PWM1
DDR1	Battery	VIN	HIGH	IN PHASE
DDR2	+5V	R to GND	HIGH	+ 90°
DUAL	ANY	VIN	LOW	+ 180°

When used as a dual converter (as in Figure 5), out-of-phase operation with 180 degree phase shift reduces input current ripple.

For the "2-step" conversion (where the VTT is converted from VDDQ as in Figure 4) used in DDR mode, the duty cycle of the second converter is nominally 50% and the optimal phasing depends on VIN. The objective is to keep noise generated from the switching transition in one converter from influencing the "decision" to switch in the other converter.

When VIN is from the battery, it's typically higher than 7.5V. As shown in Figure 6, 180° operation is undesirable since the turn-on of the VDDQ converter occurs very near the decision point of the VTT converter.

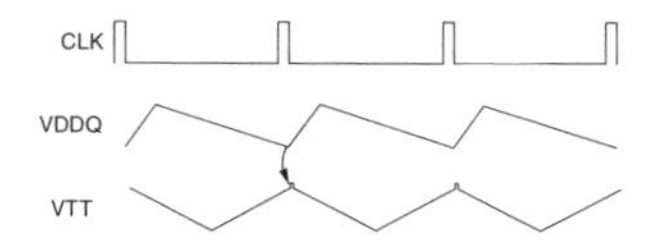

Figure 6. Noise-susceptible 180° phasing for DDR1

In-phase operation is optimal to reduce inter-converter interference when VIN is higher than 5V, (when VIN is from a battery), as can be seen in Figure 7. Since the duty cycle of PWM1 (generating VDDQ) is short, it's switching point occurs far away from the decision point for the VTT regulator, whose duty cycle is nominally 50%.

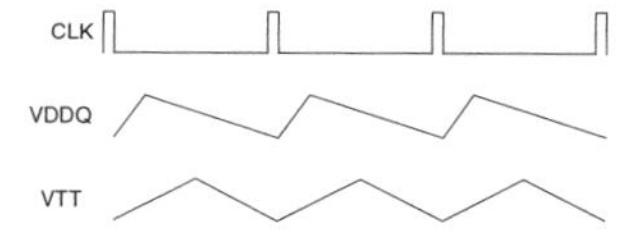

Figure 7. Optimal In-Phase operation for DDR1

When VIN ≈ 5V, 180° phase shifted operation can be rejected for the same reasons demonstrated Figure 6. In-phase operation with VIN ≈ 5V is even worse, since the switch point of either converter occurs near the switch point of the other converter as seen in Figure 8. In this case, as VIN is a little higher than 5V it will tend to cause early termination of the VTT pulse width. Conversely, VTT's switch point can cause early termination of the VDDQ pulse width when VIN is slightly lower than 5V.

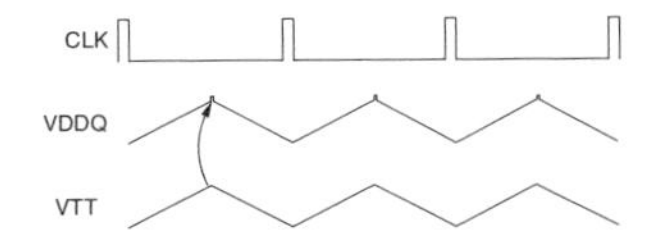

Figure 8. Noise-susceptible In-Phase operation for DDR2

These problems are nicely solved by delaying the 2nd converter's clock by 90° as shown in Figure 9. In this way, all switching transitions in one converter take place far away from the decision points of the other converter.

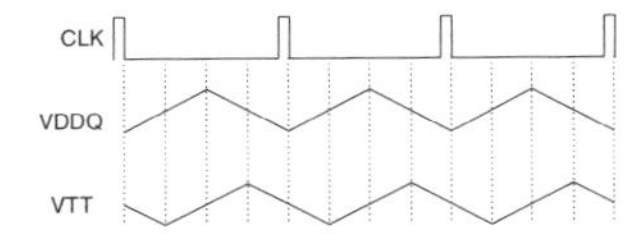

Figure 9. Optimal 90° phasing for DDR2

Initialization and Soft Start

Assuming EN is high, FAN5236 is initialized when VCC exceeds the rising UVLO threshold. Should VCC drop below the UVLO threshold, an internal Power-On Reset function disables the chip.

The voltage at the positive input of the error amplifier is limited by the voltage at the SS pin which is charged with a 5μA current source. Once C_{SS} has charged to VREF (0.9V) the output voltage will be in regulation. The time it takes SS to reach 0.9V is:

$$T_{0.9} = \frac{0.9 \times C_{SS}}{5} \qquad (1)$$

where $T_{0.9}$ is in seconds if C_{SS} is in μF.

When SS reaches 1.5V, the Power Good outputs are enabled and hysteretic mode is allowed. The converter is forced into PWM mode during soft start.

Operation Mode Control

The mode-control circuit changes the converter's mode of operation from PWM to Hysteretic and visa versa, based on the voltage polarity of the SW node when the lower MOSFET is conducting and just before the upper MOSFET turns on. For continuous inductor current, the SW node is negative when the lower MOSFET is conducting and the converters operate in fixed-frequency PWM mode as shown in Figure 10. This mode of operation achieves high efficiency at nominal load. When the load current decreases to the point where the inductor current flows through the lower MOSFET in the 'reverse' direction, the SW node becomes positive, and the mode is changed to hysteretic, which achieves higher efficiency at low currents by decreasing the effective switching frequency.

To prevent accidental mode change or "mode chatter" the transition from PWM to Hysteretic mode occurs when the SW node is positive for eight consecutive clock cycles (see Figure 10). The polarity of the SW node is sampled at the end of the lower MOSFET's conduction time. At the transition between PWM and hysteretic mode both the upper and lower MOSFETs are turned off. The phase node will 'ring' based on the output inductor and the parasitic capacitance on the phase node and settle out at the value of the output voltage.

The boundary value of inductor current, where current becomes discontinuous, can be estimated by the following expression.

$$I_{LOAD(DIS)} = \frac{(V_{IN} - V_{OUT})V_{OUT}}{2F_{SW}L_{OUT}V_{IN}} \qquad (2)$$

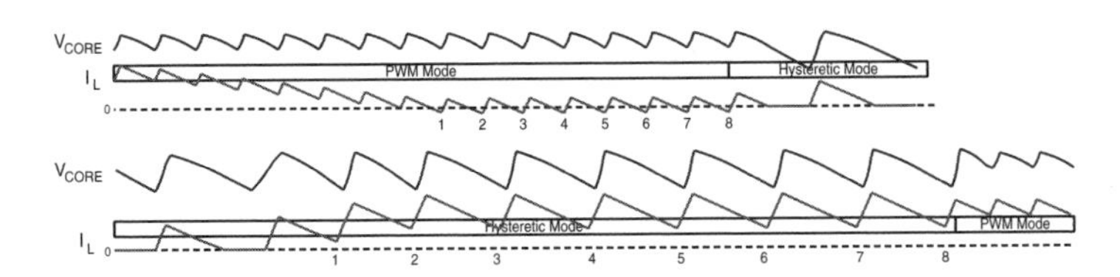

Figure 10. Transitioning between PWM and Hysteretic Mode

Hysteretic Mode

Conversely, the transition from Hysteretic mode to PWM mode occurs when the SW node is negative for 8 consecutive cycles.

A sudden increase in the output current will also cause a change from hysteretic to PWM mode. This load increase causes an instantaneous decrease in the output voltage due to the voltage drop on the output capacitor ESR. If the load causes the output voltage (as presented at VSNS) to drop below the hysteretic regulation level (20mV below VREF), the mode is changed to PWM on the next clock cycle.

In hysteretic mode, the PWM comparator and the error amplifier that provide control in PWM mode are inhibited and the hysteretic comparator is activated. In hysteretic mode the low side MOSFET is operated as a synchronous rectifier, where the voltage across ($V_{DS(ON)}$) it is monitored, and it is switched off when $V_{DS(ON)}$ goes positive (current flowing back from the load) allowing the diode to block reverse conduction.

The hysteretic comparator initiates a PFM signal to turn on HDRV at the rising edge of the next oscillator clock, when the output voltage (at VSNS) falls below the lower threshold (10mV below VREF) and terminates the PFM signal when VSNS rises over the higher threshold (5mV above VREF).

The switching frequency is primarily a function of:

1. Spread between the two hysteretic thresholds
2. I_{LOAD}
3. Output Inductor and Capacitor ESR

A transition back to PWM (Continuous Conduction Mode or CCM) mode occurs when the inductor current rises sufficiently to stay positive for 8 consecutive cycles. This occurs when:

$$I_{LOAD(CCM)} = \frac{\Delta V_{HYSTERESIS}}{2\ ESR} \quad (3)$$

where $\Delta V_{HYSTERESIS}$ = 15mV and ESR is the equivalent series resistance of C_{OUT}.

Because of the different control mechanisms, the value of the load current where transition into CCM operation takes place is typically higher compared to the load level at which transition into hysteretic mode occurs. Hysteretic mode can be disabled by setting the $\overline{FPWM}$ pin low.

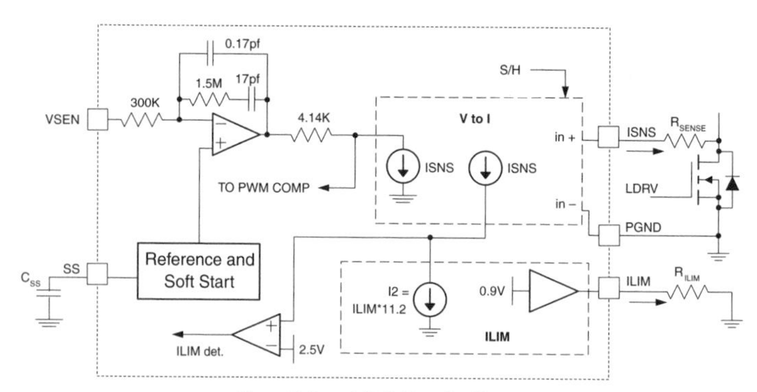

Figure 11. Current Limit / Summing Circuits

Current Processing Section

The following discussion refers to Figure 11.

The current through R_{SENSE} resistor (ISNS) is sampled shortly after Q2 is turned on. That current is held, and summed with the output of the error amplifier. This effectively creates a current mode control loop. The resistor connected to ISNSx pin (R_{SENSE}) sets the gain in the current feedback loop. For stable operation, the voltage induced by the current feedback at the PWM comparator input should be set to 30% of the ramp amplitude at maximum load currrent and line voltage. The following expression estimates the recommended value of R_{SENSE} as a function of the maximum load current ($I_{LOAD(MAX)}$) and the value of the MOSFET's $R_{DS(ON)}$:

$$R_{SENSE} = \frac{I_{LOAD(MAX)} \bullet R_{DS(ON)} \bullet 4.1K}{0.30 \bullet 0.125 \bullet V_{IN(MAX)}} - 100 \qquad (4a)$$

R_{SENSE} must, however, be kept higher than:

$$R_{SENSE(MIN)} = \frac{I_{LOAD(MAX)} \bullet R_{DS(ON)}}{150\mu A} - 100 \qquad (4b)$$

Setting the Current Limit

A ratio of ISNS is also compared to the current established when a 0.9 V internal reference drives the ILIM pin:

$$R_{ILIM} = \frac{11.2}{I_{LIMIT}} \times \frac{(100 + R_{SENSE})}{R_{DS(ON)}} \qquad (5)$$

Since the tolerance on the current limit is largely dependent on the ratio of the external resistors it is fairly accurate if the voltage drop on the Switching Node side of R_{SENSE} is an accurate representation of the load current. When using the MOSFET as the sensing element, the variation of $R_{DS(ON)}$ causes proportional variation in the ISNS. This value not only varies from device to device, but also has a typical junction temperature coefficient of about 0.4% / °C (consult the MOSFET datasheet for actual values), so the actual current limit set point will decrease propotional to increasing MOSFET die temperature. A factor of 1.6 in the current limit setpoint should compensate for all MOSFET $R_{DS(ON)}$ variations, assuming the MOSFET's heat sinking will keep its operating die temperature below 125°C.

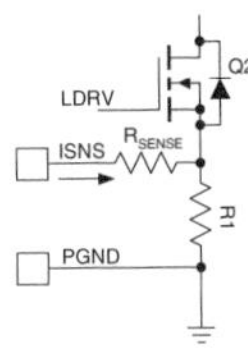

Figure 12. Improving current sensing accuracy

More accurate sensing can be achieved by using a resistor (R1) instead of the $R_{DS(ON)}$ of the FET as shown in Figure 12. This approach causes higher losses, but yields greater accuracy in both V_{DROOP} and I_{LIMIT}. R1 is a low value (e.g. 10mΩ) resistor.

Current limit (I_{LIMIT}) should be set sufficiently high as to allow inductor current to rise in response to an output load transient. Typically, a factor of 1.2 is sufficient. In addition, since I_{LIMIT} is a peak current cut-off value, we will need to multiply $I_{LOAD(MAX)}$ by the inductor ripple current (we'll use 25%). For example, in Figure 5 the target for I_{LIMIT} would be:

$$I_{LIMIT} > 1.2 \times 1.25 \times 1.6 \times 6A \text{ Ý}14A \qquad (6)$$

Duty Cycle Clamp

During severe load increase, the error amplifier output can go to its upper limit pushing a duty cycle to almost 100% for significant amount of time. This could cause a large increase of the inductor current and lead to a long recovery from a transient, over-current condition, or even to a failure especially at high input voltages. To prevent this, the output of the error amplifier is clamped to a fixed value after two clock cycles if severe output voltage excursion is detected, limiting the maximum duty cycle to

$$DC_{MAX} = \frac{V_{OUT}}{V_{IN}} + \frac{2.4}{V_{IN}}$$

This circuit is designed to not interfere with normal PWM operation. When FPWM is grounded, the duty cycle clamp is disabled and the maximum duty cycle is 87%.

Gate Driver section

The Adaptive gate control logic translates the internal PWM control signal into the MOSFET gate drive signals providing necessary amplification, level shifting and shoot-through protection. Also, it has functions that help optimize the IC performance over a wide range of operating conditions. Since MOSFET switching time can vary dramatically from type to type and with the input voltage, the gate control logic provides adaptive dead time by monitoring the gate-to-source voltages of both upper and lower MOSFETs.
The lower MOSFET drive is not turned on until the gate-to-source voltage of the upper MOSFET has decreased to less than approximately 1 volt. Similarly, the upper MOSFET is not turned on until the gate-to-source voltage of the lower MOSFET has decreased to less than approximately 1 volt. This allows a wide variety of upper and lower MOSFETs to be used without a concern for simultaneous conduction, or shoot-through.

There must be a low-resistance, low-inductance path between the driver pin and the MOSFET gate for the adaptive dead-time circuit to work properly. Any delay along that path will subtract from the delay generated by the adaptive dead-time circit and shoot-through may occur.

Frequency Loop Compensation

Due to the implemented current mode control, the modulator has a single pole response with -1 slope at frequency determined by load

$$F_{PO} = \frac{1}{2\pi R_O C_O} \qquad (7)$$

where R_O is load resistance, C_O is load capacitance.

For this type of modulator, Type 2 compensation circuit is usually sufficient. To reduce the number of external components and simplify the design task, the PWM controller has an internally compensated error amplifier. Figure 13 shows a Type 2 amplifier and its response along with the responses of a current mode modulator and of the converter. The Type 2 amplifier, in addition to the pole at the origin, has a zero-pole pair that causes a flat gain region at frequencies between the zero and the pole.

$$F_Z = \frac{1}{2\pi R_2 C_1} = 6\text{kHz} \qquad (8a)$$

$$F_P = \frac{1}{2\pi R_2 C_2} = 600\text{kHz} \qquad (8b)$$

This region is also associated with phase 'bump' or reduced phase shift. The amount of phase shift reduction depends the width of the region of flat gain and has a maximum value of 90°. To further simplify the converter compensation, the modulator gain is kept independent of the input voltage variation by providing feed-forward of VIN to the oscillator ramp.

The zero frequency, the amplifier high frequency gain and the modulator gain are chosen to satisfy most typical applications. The crossover frequency will appear at the point where the modulator attenuation equals the amplifier high frequency gain. The only task that the system designer has to complete is to specify the output filter capacitors to position the load main pole somewhere within one decade lower than the amplifier zero frequency. With this type of compensation plenty of phase margin is easily achieved due to zero-pole pair phase 'boost'.

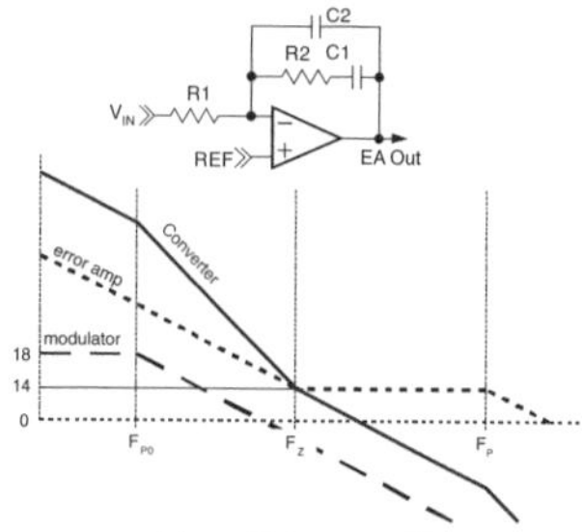

Figure 13. Compensation

Conditional stability may occur only when the main load pole is positioned too much to the left side on the frequency axis due to excessive output filter capacitance. In this case, the ESR zero placed within the 10kHz...50kHz range gives some additional phase 'boost'. Fortunately, there is an opposite trend in mobile applications to keep the output capacitor as small as possible.

If a larger inductor value or low ESR values are called for by the application, additional phase margin can be achieved by putting a zero at the LC crossover frequency. This can be achieved with a capacitor across across the feedback resistor (e.g. R5 from Figure 5) as shown below.

Figure 14. Improving Phase Margin

The optimal value of C(Z) is:

$$C(Z) = \frac{\sqrt{L(OUT) \times C(OUT)}}{R} \qquad (9)$$

Protection

The converter output is monitored and protected against extreme overload, short circuit, over-voltage and under-voltage conditions.

A sustained overload on an output sets the PGx pin low and latches-off the regulator on which the fault occurs. Operation can be restored by cycling the VCC voltage or by toggling the EN pin.

If VOUT drops below the under-voltage threshold, the regulator shuts down immediately.

Over-Current sensing

If the circuit's current limit signal ("ILIM det" as shown in Figure 11) is high at the beginning of a clock cycle, a pulse-skipping circuit is activated and HDRV is inhibited. The circuit continues to pulse skip in this manner for the next 8 clock cycles. If at any time from the 9th to the 16th clock cycle, the "ILIM det" is again reached, the over-current protection latch is set, disabling the regulator. If "ILIM det" does not occur between cycle 9 and 16, normal operation is restored and the over-current circuit resets itself.

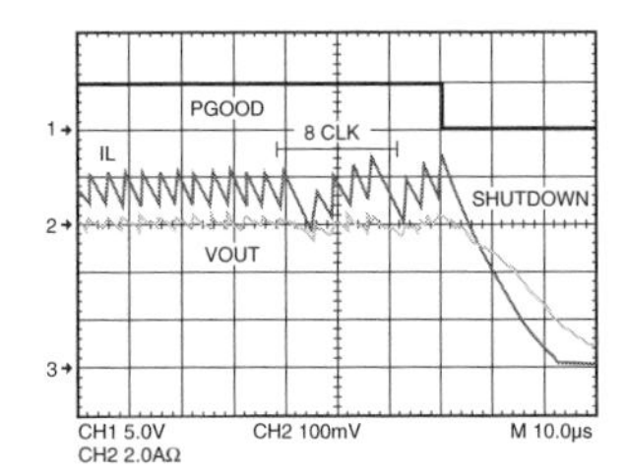

Figure 15. Over-Current protection waveforms

Over-Voltage / Under-voltage Protection

Should the VSNS voltage exceed 120% of VREF (0.9V) due to an upper MOSFET failure, or for other reasons, the over-voltage protection comparator will force LDRV high. This action actively pulls down the output voltage and, in the event of the upper MOSFET failure, will eventually blow the battery fuse. As soon as the output voltage drops below the threshold, the OVP comparator is disengaged.

This OVP scheme provides a 'soft' crowbar function which helps to tackle severe load transients and does not invert the output voltage when activated — a common problem for latched OVP schemes.

Similarly, if an output short-circuit or severe load transient causes the output to droop to less than 75% of its regulation set point, the regulator will shut down.

Over-Temperature Protection

The chip incorporates an over temperature protection circuit that shuts the chip down when a die temperature of about 150°C is reached. Normal operation is restored at die temperature below 125°C with internal Power On Reset asserted, resulting in a full soft-start cycle.

Design and Component Selection Guidelines

As an initial step, define operating input voltage range, output voltage, minimum and maximum load currents for the controller.

Setting the Output Voltage

The interal reference is 0.9V. The output is divided down by a voltage divider to the VSEN pin (for example, R5 and R6 in Figure 4). The output voltage therefore is:

$$\frac{0.9V}{R6} = \frac{V_{OUT} - 0.9V}{R5} \qquad (10a)$$

To minimize noise pickup on this node, keep the resistor to GND (R6) below 2K. We selected R6 at 1.82K. Then choose R5:

$$R5 = \frac{(1.82K)(V_{OUT} - 0.9)}{0.9} = 3.24K \qquad (10b)$$

For DDR applications converting from 3.3V to 2.5V, or other applications requiring high duty cycles, the duty cycle clamp must be disabled by tying the converter's $\overline{\text{FPWM}}$ to GND. When converter's $\overline{\text{FPWM}}$ is GND, the converter's maximum duty cycle will be greater than 90%. When using as a DDR converter with 3.3V input, set up the converter for In-Phase synchronization by tying the VIN pin to +5V.

Output Inductor Selection

The minimum practical output inductor value is the one that keeps inductor current just on the boundary of continuous conduction at some minimum load. The industry standard practice is to choose the minimum current somewhere from 15% to 35% of the nominal current. At light load, the controller can automatically switch to hysteretic mode of operation to sustain high efficiency. The following equations help to choose the proper value of the output filter inductor.

$$\Delta I = 2 \times I_{MIN} = \frac{\Delta V_{OUT}}{ESR} \qquad (11)$$

where ΔI is the inductor ripple current and ΔV_{OUT} is the maximum ripple allowed.

$$L = \frac{V_{IN} - V_{OUT}}{F_{SW} \times \Delta I} \times \frac{V_{OUT}}{V_{IN}} \qquad (12)$$

for this example we'll use:

$V_{IN} = 20V$, $V_{OUT} = 2.5V$
$\Delta I = 20\% * 6A = 1.2A$
$F_{SW} = 300KHz$.

therefore
$L \approx 6\mu H$

Output Capacitor Selection

The output capacitor serves two major functions in a switching power supply. Along with the inductor it filters the sequence of pulses produced by the switcher, and it supplies the load transient currents. The output capacitor requirements are usually dictated by ESR, Inductor ripple current (ΔI) and the allowable ripple voltage (ΔV).

$$ESR < \frac{\Delta V}{\Delta I} \qquad (13)$$

In addition, the capacitor's ESR must be low enough to allow the converter to stay in regulation during a load step. The ripple voltage due to ESR for the converter in Figure 5 is 120mV P-P. Some additional ripple will appear due to the capacitance value itself:

$$\Delta V = \frac{\Delta I}{C_{OUT} \times 8 \times F_{SW}} \qquad (14)$$

which is only about 1.5mV for the converter in Figure 5 and can be ignored.

The capacitor must also be rated to withstand the RMS current which is approximately 0.3 X (ΔI), or about 400mA for the converter in Figure 5. High frequency decoupling capacitors should be placed as close to the loads as physically possible.

Input Capacitor Selection

The input capacitor should be selected by its ripple current rating.

Two-Stage Converter Case

In DDR mode (Figure 4), the VTT power input is powered by the VDDQ output, therefore all of the input capacitor ripple current is produced by the VDDQ converter. A conservative estimate of the output
current required for the 2.5V regulator is:

$$I_{REG1} = I_{VDDQ} + \frac{I_{VTT}}{2}$$

As an example, if average I_{VDDQ} is 3A, and average I_{VTT} is 1A, I_{VDDQ} current will be about 3.5A. If average input voltage is 16V, RMS input ripple current will be:

$$I_{RMS} = I_{OUT(MAX)}\sqrt{D - D^2} \qquad (15)$$

where D is the duty cycle of the PWM1 converter:

$$D < \frac{V_{OUT}}{V_{IN}} = \frac{2.5}{16} \qquad (16)$$

therefore:

$$I_{RMS} = 3.5\sqrt{\frac{2.5}{16} - \left(\frac{2.5}{16}\right)^2} = 1.49A \qquad (17)$$

Dual Converter 180° phased

In Dual mode (Figure 5), both converters contribute to the capacitor input ripple current. With each converter operating 180° out of phase, the RMS currents add in the following fashion:

$$I_{RMS} = \sqrt{I_{RMS(1)}^2 + I_{RMS(2)}^2} \text{ or} \qquad (18a)$$

$$I_{RMS} = \sqrt{(I_1)^2(D_1 - D_1^2) + (I_2)^2(D_2 - D_2^2)} \qquad (18b)$$

which for the dual 3A converters of Figure 5, calculates to:

$$I_{RMS} = 1.4A$$

Power MOSFET Selection

Losses in a MOSFET are the sum of its switching (P_{SW}) and conduction (P_{COND}) losses.

In typical applications, the FAN5236 converter's output voltage is low with respect to its input voltage, therefore the Lower MOSFET (Q2) is conducting the full load current for most of the cycle. Q2 should therefore be selected to minimize conduction losses, thereby selecting a MOSFET with low $R_{DS(ON)}$.

In contrast, the high-side MOSFET (Q1) has a much shorter duty cycle, and it's conduction loss will therefore have less of an impact. Q1, however, sees most of the switching losses, so Q1's primary selection criteria should be gate charge.

High-Side Losses:

Figure 15 shows a MOSFET's switching interval, with the upper graph being the voltage and current on the Drain to Source and the lower graph detailing V_{GS} vs. time with a constant current charging the gate. The x-axis therefore is also representative of gate charge (Q_G) . $C_{ISS} = C_{GD} + C_{GS}$, and it controls t1, t2, and t4 timing. C_{GD} receives the current from the gate driver during t3 (as V_{DS} is falling). The gate charge (Q_G) parameters on the lower graph are either specified or can be derived from MOSFET datasheets.

Assuming switching losses are about the same for both the rising edge and falling edge, Q1's switching losses, occur during the shaded time when the MOSFET has voltage across it and current through it.

These losses are given by:

$$P_{UPPER} = P_{SW} + P_{COND}$$

$$P_{SW} = \left(\frac{V_{DS} \times I_L}{2} \times 2 \times t_S\right) F_{SW} \quad (19a)$$

$$P_{COND} = \frac{V_{OUT}}{V_{IN}} \times I_{OUT}^2 \times R_{DS(ON)} \quad (19b)$$

P_{UPPER} is the upper MOSFET's total losses, and P_{SW} and P_{COND} are the switching and conduction losses for a given MOSFET. $R_{DS(ON)}$ is at the maximum junction temperature (T_J). t_S is the switching period (rise or fall time) and is t2+t3 Figure 15.

The driver's impedance and C_{ISS} determine t2 while t3's period is controlled by the driver's impedance and Q_{GD}. Since most of t_S occurs when $V_{GS} = V_{SP}$ we can use a constant current assumption for the driver to simplify the calculation of t_S:

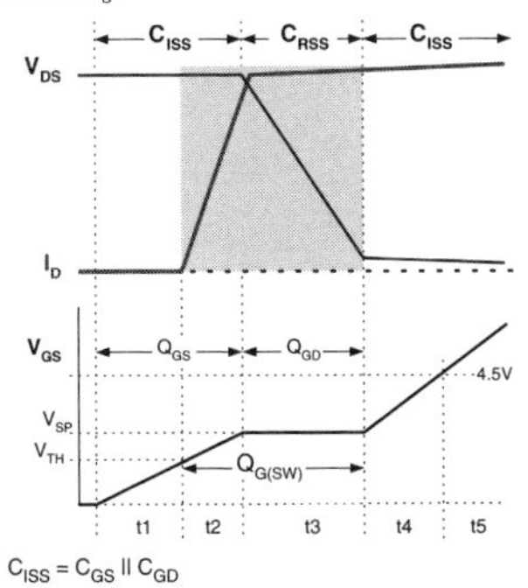

Figure 16. Switching losses and Q_G

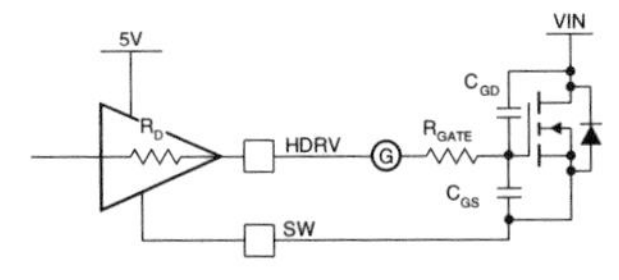

Figure 17. Drive Equivalent Circuit

$$t_S = \frac{Q_{G(SW)}}{I_{DRIVER}} \approx \frac{Q_{G(SW)}}{\left(\frac{VCC - V_{SP}}{R_{DRIVER} + R_{GATE}}\right)} \quad (20)$$

Most MOSFET vendors specify Q_{GD} and Q_{GS}. $Q_{G(SW)}$ can be determined as: $Q_{G(SW)} = Q_{GD} + Q_{GS} - Q_{TH}$ where Q_{TH} is the the gate charge required to get the MOSFET to it's threshold (V_{TH}). For the high-side MOSFET, V_{DS} = VIN, which can be as high as 20V in a typical portable application. Care should also be taken to include the delivery of the MOSFET's gate power (P_{GATE}) in calculating the power dissipation required for the FAN5236:

$$P_{GATE} = Q_G \times VCC \times F_{SW} \quad (21)$$

where Q_G is the total gate charge to reach VCC.

Low-Side Losses

Q2, however, switches on or off with its parallel shottky diode conducting, therefore $V_{DS} \approx 0.5V$. Since P_{SW} is proportional to V_{DS} , Q2's switching losses are negligible and we can select Q2 based on $R_{DS(ON)}$ only.

Conduction losses for Q2 are given by:

$$P_{COND} = (1 - D) \times I_{OUT}^2 \times R_{DS(ON)} \quad (22)$$

where $R_{DS(ON)}$ is the $R_{DS(ON)}$ of the MOSFET at the highest operating junction temperature and

$D = \frac{V_{OUT}}{V_{IN}}$ is the minimum duty cycle for the converter.

Since $D_{MIN} < 20\%$ for portable computers, $(1\text{-}D) \approx 1$ produces a conservative result, further simplifying the calculation.

The maximum power dissipation ($P_{D(MAX)}$) is a function of the maximum allowable die temperature of the low-side MOSFET, the $\theta_{J\text{-}A}$, and the maximum allowable ambient temperature rise:

$$P_{D(MAX)} = \frac{T_{J(MAX)} - T_{A(MAX)}}{\theta_{J-A}} \quad (23)$$

$\theta_{J\text{-}A}$, depends primarily on the amount of PCB area that can be devoted to heat sinking (see FSC app note AN-1029 for SO-8 MOSFET thermal information).

Layout Considerations

Switching converters, even during normal operation, produce short pulses of current which could cause substantial ringing and be a source of EMI if layout constrains are not observed.

There are two sets of critical components in a DC-DC converter. The switching power components process large amounts of energy at high rate and are noise generators. The low power components responsible for bias and feedback functions are sensitive to noise.

A multi-layer printed circuit board is recommended. Dedicate one solid layer for a ground plane. Dedicate another solid layer as a power plane and break this plane into smaller islands of common voltage levels.

Notice all the nodes that are subjected to high dV/dt voltage swing such as SW, HDRV and LDRV, for example. All surrounding circuitry will tend to couple the signals from these nodes through stray capacitance. Do not oversize copper traces connected to these nodes. Do not place traces connected to the feedback components adjacent to these traces. It is not recommended to use High Density Interconnect Systems, or micro-vias on these signals. The use of blind or buried vias should be limited to the low current signals only. The use of normal thermal vias is left to the discretion of the designer.

Keep the wiring traces from the IC to the MOSFET gate and source as short as possible and capable of handling peak currents of 2A. Minimize the area within the gate-source path to reduce stray inductance and eliminate parasitic ringing at the gate.

Locate small critical components like the soft-start capacitor and current sense resistors as close as possible to the respective pins of the IC.

The FAN5236 utilizes advanced packaging technologies with lead pitches of 0.6mm. High performance analog semiconductors utilizing narrow lead spacing may require special considerations in PWB design and manufacturing. It is critical to maintain proper cleanliness of the area surrounding these devices.

Mechanical Dimensions

28-Pin QSOP

Symbol	Inches		Millimeters		Notes
	Min.	Max.	Min.	Max.	
A	0.053	0.069	1.35	1.75	
A1	0.004	0.010	0.10	0.25	
A2	-	0.061	-	1.54	
B	0.008	0.012	0.20	0.30	9
C	0.007	0.010	0.18	0.25	
D	0.386	0.394	9.81	10.00	3
E	0.150	0.157	3.81	3.98	4
e	0.025 BSC		0.635 BSC		
H	0.228	0.244	5.80	6.19	
h	0.0099	0.0196	0.26	0.49	5
L	0.016	0.050	0.41	1.27	6
N	28		28		7
α	0°	8°	0°	8°	

Notes:

1. Symbols are defined in the "MO Series Symbol List" in Section 2.2 of Publication Number 95.
2. Dimensioning and tolerancing per ANSI Y14.5M-1982.
3. Dimension "D" does not include mold flash, protrusions or gate burrs. Mold flash, protrusions shall not exceed 0.25mm (0.010 inch) per side.
4. Dimension "E" does not include interlead flash or protrusions. Interlead flash and protrusions shall not exceed 0.25mm (0.010 inch) per side.
5. The chamber on the body is optional. If it is not present, a visual index feature must be located within the crosshatched area.
6. "L" is the length of terminal for soldering to a substrate.
7. "N" is the maximum number of terminals.
8. Terminal numbers are shown for reference only.
9. Dimension "B" does not include dambar protrusion. Allowable dambar protrusion shall be 0.10mm (0.004 inch) total in excess of "B" dimension at maximum material condition.
10. Controlling dimension: INCHES. Converted millimeter dimensions are not necessarily exact.

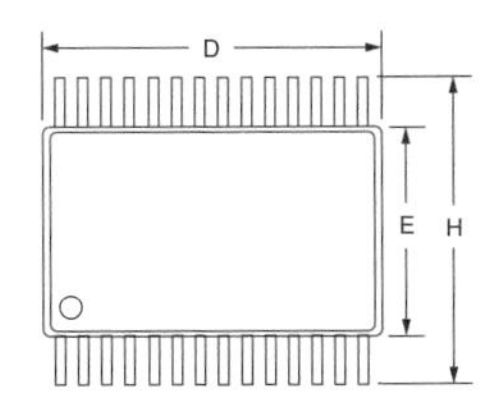

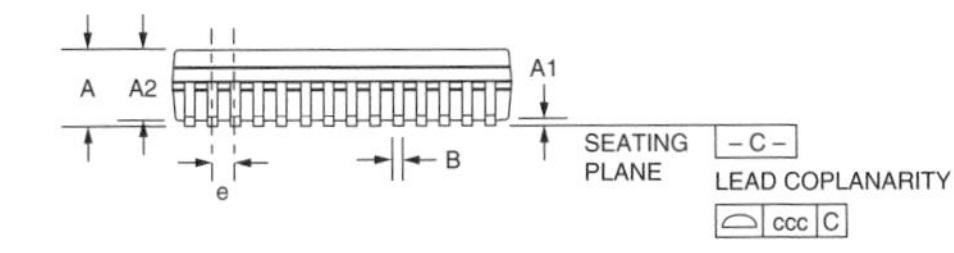

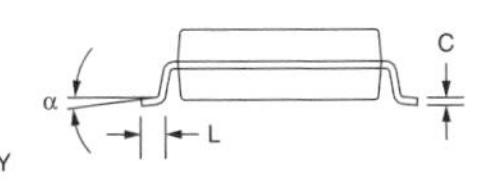

Mechanical Dimensions

28-Pin TSSOP

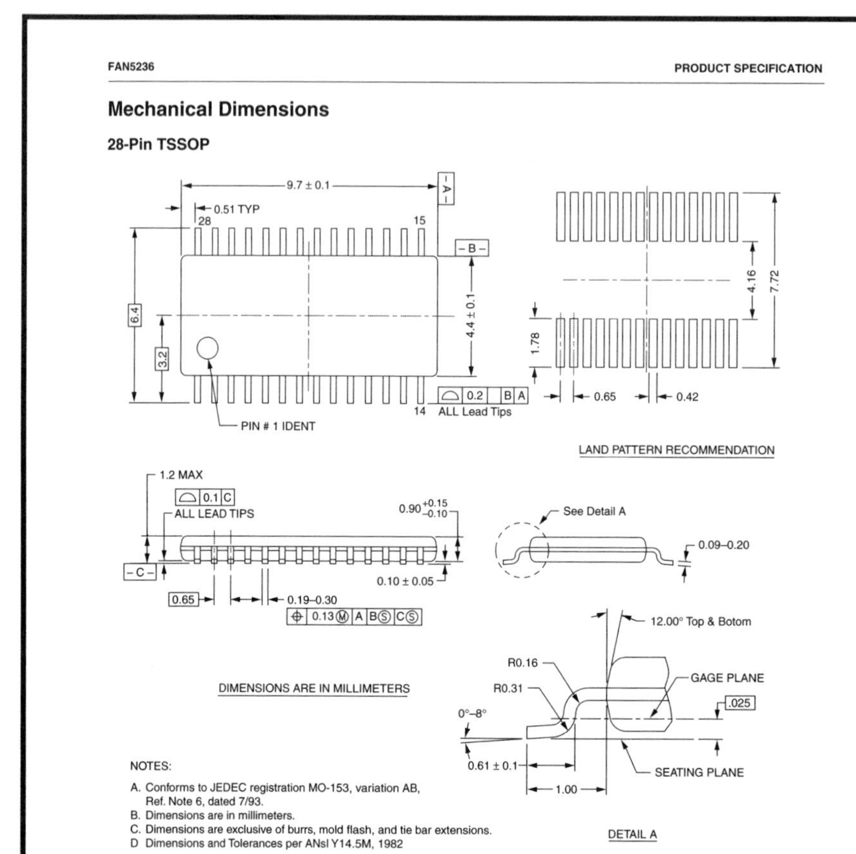

NOTES:

A. Conforms to JEDEC registration MO-153, variation AB, Ref. Note 6, dated 7/93.
B. Dimensions are in millimeters.
C. Dimensions are exclusive of burrs, mold flash, and tie bar extensions.
D Dimensions and Tolerances per ANsI Y14.5M, 1982

Ordering Information

Part Number	Temperature Range	Package	Packing
FAN5236QSC	-10°C to 85°C	QSOP-28	Rails
FAN5236QSCX	-10°C to 85°C	QSOP-28	Tape and Reel
FAN5236MTC	-10°C to 85°C	TSSOP-28	Rails
FAN5236MTCX	-10°C to 85°C	TSSOP-28	Tape and Reel

5/25/05 0.0m 004
Stock#DS30005236

Appendix G

Fairchild Specifications for FAN8702

www.fairchildsemi.com

FAN8702/FAN8702B

6-Channel DSC Motor Driver

Features

- Independent 6-Channel H-Bridge.
- Output Current up to 600mA (Each Channel)
- Constant Current Control on CH5 and CH6.
- Constant Voltage Control on CH1,2,3 and CH4.
- Built in Brake Function on CH3,4 and CH6.
- Built in Short Through Protection.
- Low Saturation Voltage.
- Low Voltage operation.
- Built in Reference Voltage.
- Built in Thermal Shut Down.

Description

The FAN8702 is designed for portable equipments such as DSC and video camera. It consists of 2 constant current and 4 constant voltage drive blocks suitable for shutter, auto-focus, iris and zoom motor drive.

48-LQFP-0707

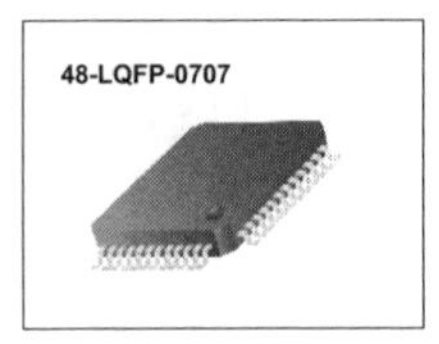

Typical Applications

- DSC One Chip

Ordering Information

Device	Package	Operating Temp.
FAN8702	48-LQFP-0707	-20°C to +80°C
FAN8702B	48-LQFP-0707	-20°C to +80°C
FAN8702_NL	48-LQFP-0707	-20°C to +80°C
FAN8702B_NL	48-LQFP-0707	-20°C to +80°C

Note

NL : Lead Free

Rev.1.0.2

Pin Assignments

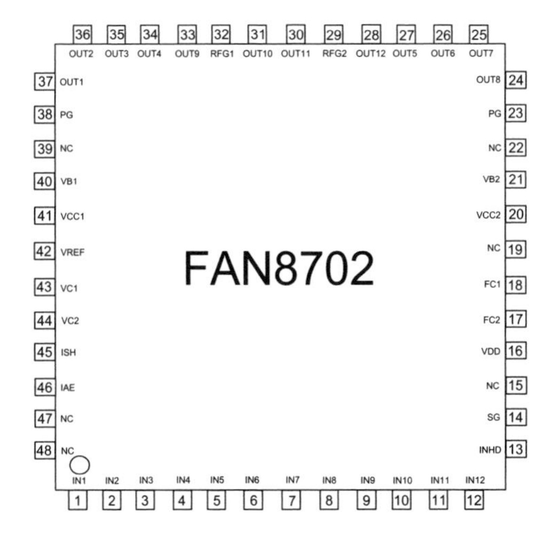

Pin Definitions

Pin Number	Pin Name	I/O	Pin Function Description	Remark
1	IN 1	I	Logic Input 1	-
2	IN 2	I	Logic Input 2	-
3	IN 3	I	Logic Input 3	-
4	IN 4	I	Logic Input 4	-
5	IN 5	I	Logic Input 5	-
6	IN 6	I	Logic Input 6	-
7	IN 7	I	Logic Input 7	-
8	IN 8	I	Logic Input 8	-
9	IN 9	I	Logic Input 9	-
10	IN 10	I	Logic Input 10	-
11	IN 11	I	Logic Input 11	-
12	IN 12	I	Logic Input 12	-
13	INHD	I	Voltage Adjust for Vref	-
14	SGND	P	Signal Ground	-
15	NC	-	Non Connection	-
16	VDD	P	Supply Voltage (Logic Voltage)	-
17	FC2	A	Compensation 2	-
18	FC1	A	Compensation 1	-
19	NC	-	Non Connection	-
20	VCC2	P	Supply Voltage (Current Drive2)	-
21	VB2	P	Supply Voltage (Voltage Drive2)	-
22	NC	-	Non Connection	-
23	PGND	P	Power Ground	-
24	OUT8	A	Voltage Driver OUT8	-
25	OUT7	A	Voltage Driver OUT7	-
26	OUT6	A	Voltage Driver OUT6	-
27	OUT5	A	Voltage Driver OUT5	-
28	OUT12	A	Current Driver OUT12	-
29	RFG2	A	Current Sensing2	-
30	OUT11	A	Current Driver OUT11	-
31	OUT10	A	Current Driver OUT10	-
32	RFG1	A	Current Sensing1	-
33	OUT9	A	Current Driver OUT9	-
34	OUT4	A	Voltage Driver OUT4	-
35	OUT3	A	Voltage Driver OUT3	-
36	OUT2	A	Voltage Driver OUT2	-
37	OUT1	A	Voltage Driver OUT1	-
38	PGND	P	Power Ground	-
39	NC	-	Non Connection	-
40	VB1	P	Supply Voltage (Voltage Drive1)	-
41	VCC1	P	Supply Voltage (Current Drive1)	-

Pin Definitions (Continued)

Pin Number	Pin Name	I/O	Pin Function Description	Remark
42	VREF	A	Reference Voltage Out	-
43	VC1	A	Voltage Adjust for Out 1~4	-
44	VC2	A	Voltage Adjust for Out 5~8	-
45	ISH	A	Voltage Adjust for Shutter(Out9~10)	-
46	IAE	A	Voltage Adjust for IRIS(Out11~12)	-
47	NC	-	Non Connection	-
48	NC	-	Non Connection	-

Internal Block Diagram

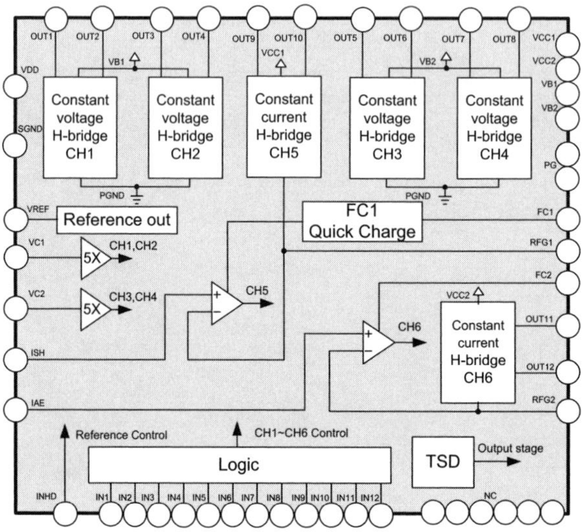

Equivalent Circuits

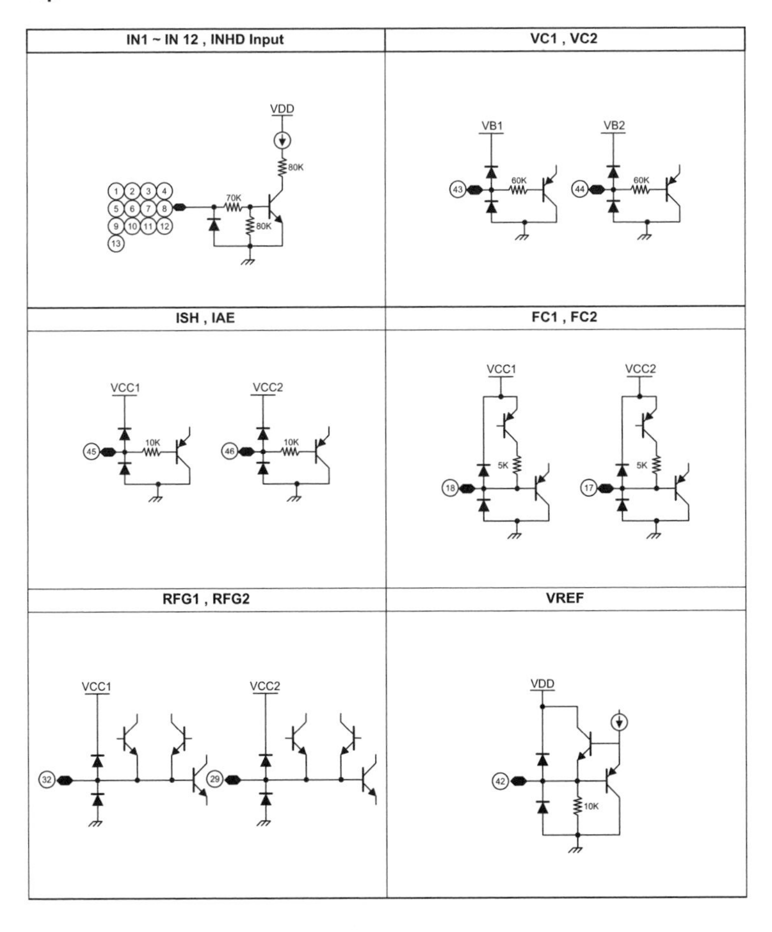

Equivalent Circuits (Continued)

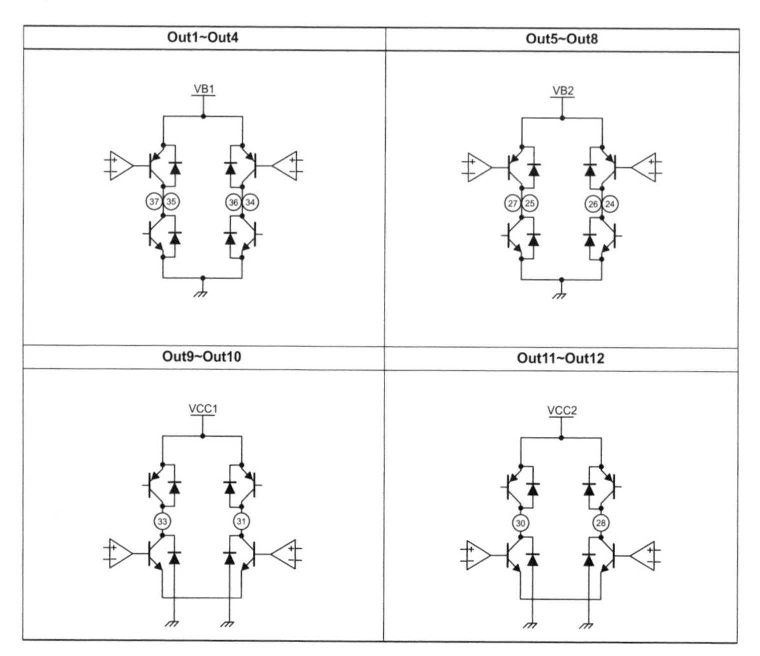

FAN8702/FAN8702B

Absolute Maximum Ratings (Ta = 25°C)

Parameter	Symbol	Value	Unit
Maximum Power Supply Voltage	VBMAX	10.5	V
Maximum Power Supply Voltage	VCCMAX	10.5	V
Maximum Approval Voltage To Input Pin	VINMAX	10.5	V
Maximum Approval Voltage To Output Pin	VOUTMAX	11.5	V
Maximum Output Current	IOUTMAX	600	mA
Maximum Power Dissipation	PdMAX	1000	mW
Operating Temperature	TOPR	-20 ~ +80	°C
Storage Temperature	TSTG	-55 ~ +150	°C

Power Dissipation Curve (Air condition = 0m/s)

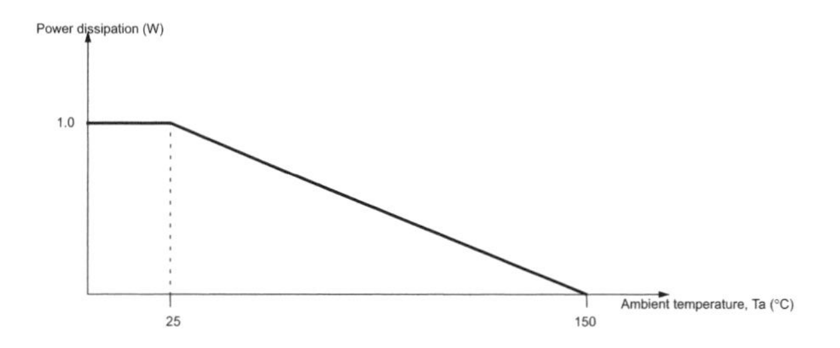

Note:
PCB Condition: Thickness (1.6mm), Dimension (76.2mm * 114.3mm)
Refer: EIA/J SED 51-2 & EIA/J SED 51-3
JESD51-2 : Integrated Circuits Thermal Test Method Environmental Conditions - Natural Convection(Still Air)
JSED51-3 : Low Effective Thermal Conductivity Test Board for Leaded Surface Mount Packages
Should not exceed P_D or ASO value

Recommended Operating Conditions (Ta = 25°C)

Parameter	Symbol	Min.	Typ.	Max.	Unit
Operating Voltage Range	VB1,2	2.2	-	6.5	V
Operating Voltage Range	VCC1,2	2.2	-	6.5	V
Logic Input High Level	VINH	1.8	-	7.0	V
Logic Input Low Level	VINL	-0.3	-	0.4	V

Electrical Characteristics

(Ta = 25°C, VB1=VB2=VCC1=VCC2=VDD=2.4V)

Block	Parameter	Symbol	Conditions	Min.	Typ.	Max.	Unit
Total	Stand-by Current	ISTB	VB=VCC=VDD=7.0V	-	-	1.0	μA
	Operating Consumption Current 1	ICC1	IN1~,IN8 (1Phase) IOV=200mA, Note1	-	8	11	mA
	Operating Consumption Current 2	ICC2	IN1~IN8(2Phase) IOV=400mA, Note1	-	17	25	mA
	Operating Consumption Current 3	ICC3	IN5~IN8(Brake) Note2	-	16	25	mA
	Operating Consumption Current 4	ICC4	IN9~IN12(1 phase) IOI=200mA, Note1	-	6	11	mA
	Operating Consumption Current 5	ICC5	IN11,IN12(Brake)Note2	-	16	25	mA
	Reference Voltage Output Voltage 1	VREF1	IREF=−1mA,INHD=L	0.95	1.0	1.05	V
	Reference Voltage Output Voltage 2	VREF2	IREF=−1mA,INHD=H	0.64	0.67	0.70	V
	Logic Input High Current	IINH	VIN=5.0V	-	60	90	μA
	Logic Input Low Current	IINL	VIN=0.0V	-1	-	1	μA
	Thermal Shutdown	THD	-	-	150	-	°C
Current driver	Output Current 1	IO	RFG=1.0Ω, ISH=0.3V	282	300	318	mA
	Output Saturation Voltage (PNP+NPN)	VSAT1	IO=0.3A	-	0.4	0.6	V
Voltage driver	Output Voltage 1	VO	VC1, 2 =0.4V	1.9	2.0	2.1	V
	Output Saturation Voltage (PNP+NPN)	VSAT2	IO=0.2A	-	0.35	0.50	V

Note :

1. ICC1, ICC2,ICC4 is sum of the current consumption VB1,VB2,VCC1,VCC2 line.
2. ICC3, ICC5 is sum of the current consumption VB1,VB2,VCC1,VCC2 and VDD line.

FAN8702/FAN8702B

Operation Truth Table

Input/ Output Motor Operation	IN 1	IN 2	IN 3	IN 4	IN 5	IN 6	IN 7	IN 8	IN 9	IN 10	IN 11	IN 12	INHD	OUT 1	OT 2	OU T3	OUT 4	OUT 5	OUT 6	OUT 7	OUT 8	OUT 9	OUT 10	OUT 11	OUT 12	Vref
Stand-by	L	L	L	L	L	L	L	L	L	L	L	L	L	Z	Z	Z	Z	Z	Z	Z	Z	Z	Z	Z	Z	Z
	L	L	L	L	L	L	L	L	L	L	L	L	H													0.67
	When one of input of IN1 to IN12 is high												L													1.0
													H													0.67
Voltage driver 1	L	L												Z	Z											
	L	H												L	H											
	H	L												H	L											
	H	H												Z	Z											
			L	L												Z	Z									
			L	H												L	H									
			H	L												H	L									
			H	H												Z	Z									
Voltage driver 2					L	L												Z	Z							
					L	H												L	H							
					H	L												H	L							
					H	H												L	L							
							L	L												Z	Z					
							L	H												L	H					
							H	L												H	L					
							H	H												L	L					
Current driver 1									L	L												Z	Z			
									L	H												L	H			
									H	L												H	L			
									H	H				FAN8702 only								Inhibit				
														FAN8702B only								Z	Z			
Current driver 2											L	L												Z	Z	
											L	H												L	H	
											H	L												H	L	
											H	H												L	L	

Application Information

1. Current Drive Output Current Setting

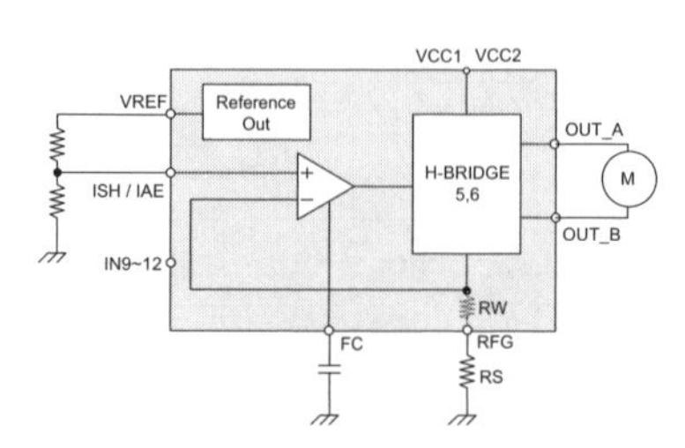

Motor current is determined by ISH/IAE voltage and Rs sensing resistance and calculated by the formula below considering Rw. Generally internal bonding and metal resistance Rw is around 0.05Ω .

$$\text{Motor Current} = \frac{\text{ISH or IAE Input Voltage}}{R_S + R_W}$$

2. Current Drive Block1(CH5)

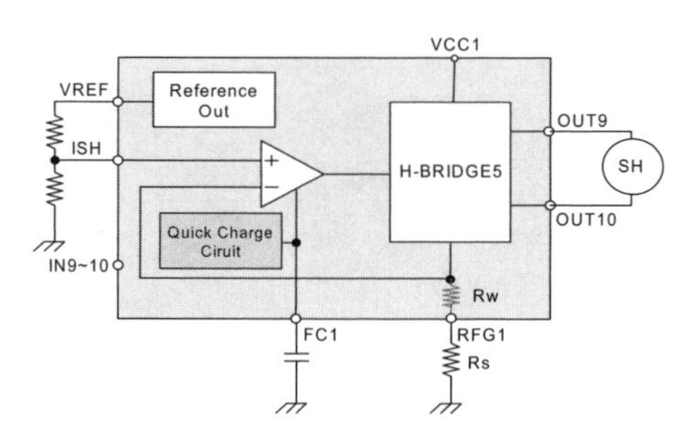

If there is no capacitor on the exterior of the FC terminal or low capacitance is used, it may cause oscillation or overshoot at output terminal. The output stage will not be operating until FC1 terminal voltage reaches around 0.7V (Typical)
The output response time depends on the FC1 capacitance and interval of Input signal. Generally, the quick charging time is 10us~20us. To minimize the delay time difference in the output response between high-speed shutter and bulb shutter a quick charging circuit is built in.

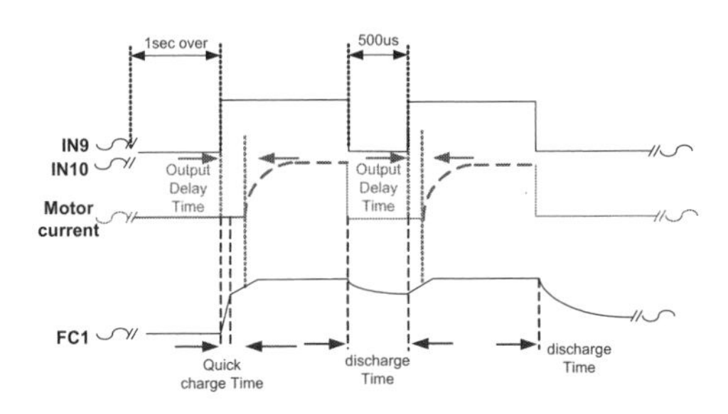

3. Current Drive block2(CH6)

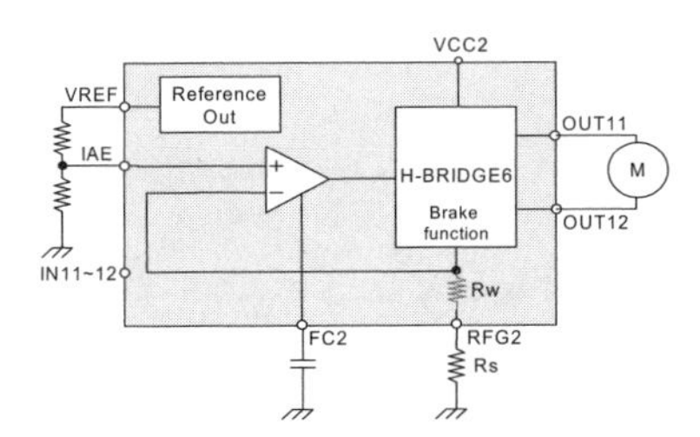

The output response time depends on the FC2 capacitance and interval of Input signal because there is no FC quick charging circuit in current drive2.

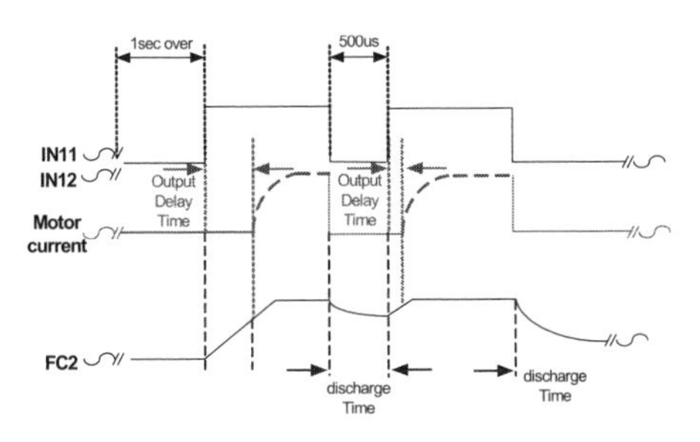

4. Voltage Drive Block

The output voltage as much as 5 times the input voltage VC1,VC2 is produced in the range of motor power VB1/VB2.
If output oscillation occur during constant voltage drive, then 0.01uF~0.1uF capacitor should be installed on the both sides of the output.

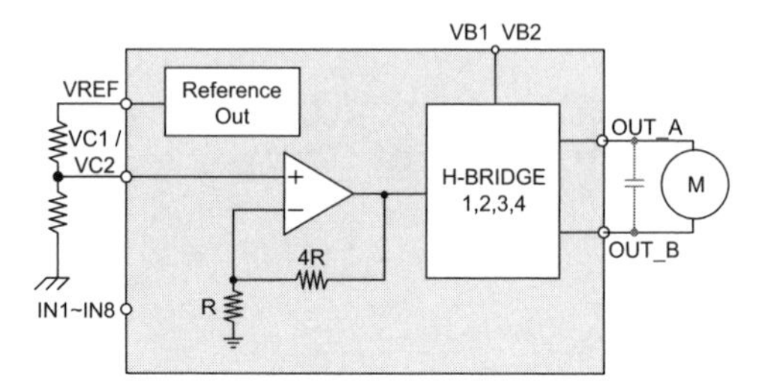

5. H-Bridge Drive Mode

A H-bridge drive mode can be implemented using the current drive block or the voltage drive block.

1) H- bridge drive using current drive block

The current drive block using the H-bridge method can be operated with ISH/AE connected to VREF or supply input, with the FC terminal open and sensing terminal connected to ground.

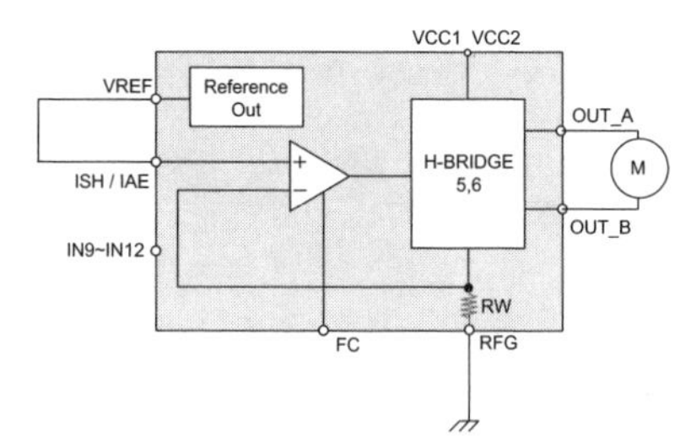

2) H- bridge drive using voltage drive block

When VB1 and VB2 power is less than 5V, VC1/VC2 input should be connected to VREF or motor supply and VB1 and VB2 power is more than 5V, VC1/VC2 input should be connected to motor supply. In H-bridge drive mode, a capacitor to prevent oscillation is not necessary on both sides of the output.

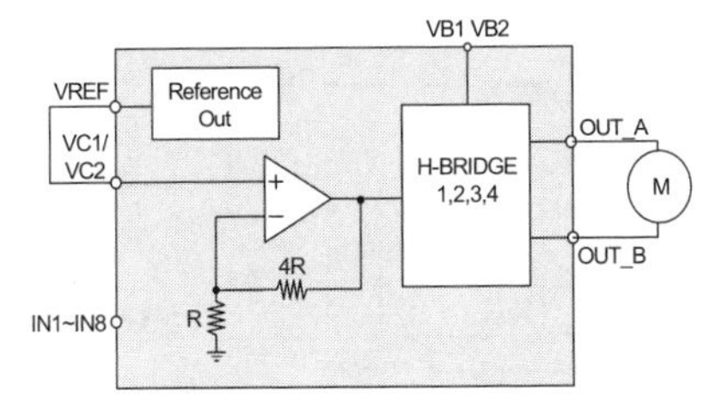

6. Short through protection

When a motor is driven, high/low side TR turn on simulataneously. This range may cause power to be shorted to ground momentarily. To prevent a short through, output is generated with a 1.8us(typical) delay after a high input signal.

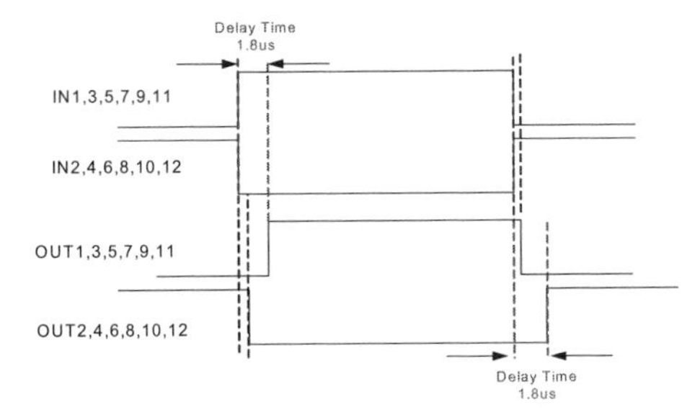

7. Brake function

The brake function is built in Ch3,4 and CH6 . Using the H/H signal on input, it is designed so that a short brake is operated on output.

8. Power Supply

VB1,VB2,VCC1 and VCC2 are separated for motor power of FAN8702 and VDD is used for IC logic power.
VB1,VB2,VCC1 and VCC2 are correspond to H-bridge 1~2, H-bridge 3~4, H-bridge 5 and H-bridge6.

9. Thermal shutdown

When thermal shut down is activated, all the outputs become off except for VREF.

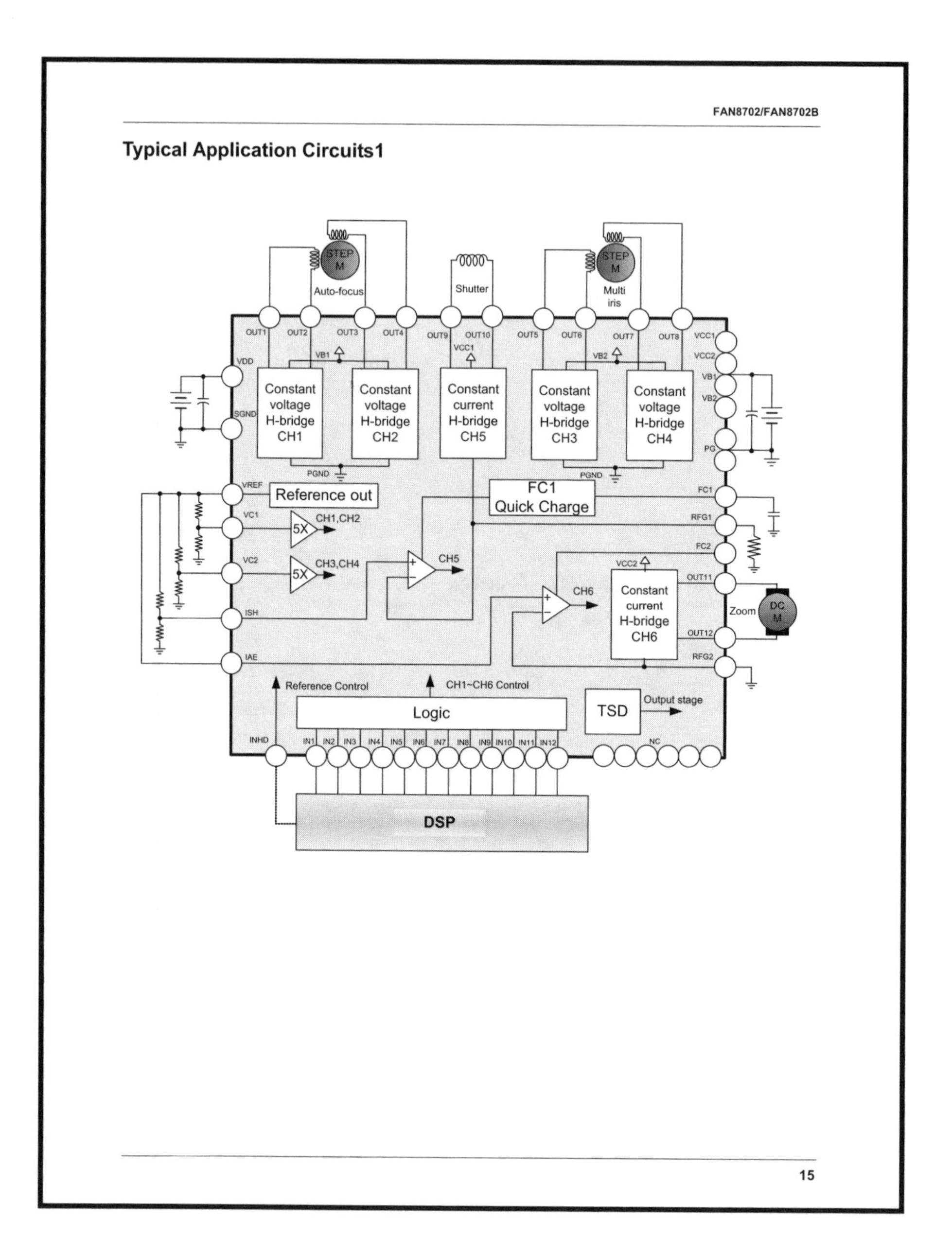

Typical Application Circuits1

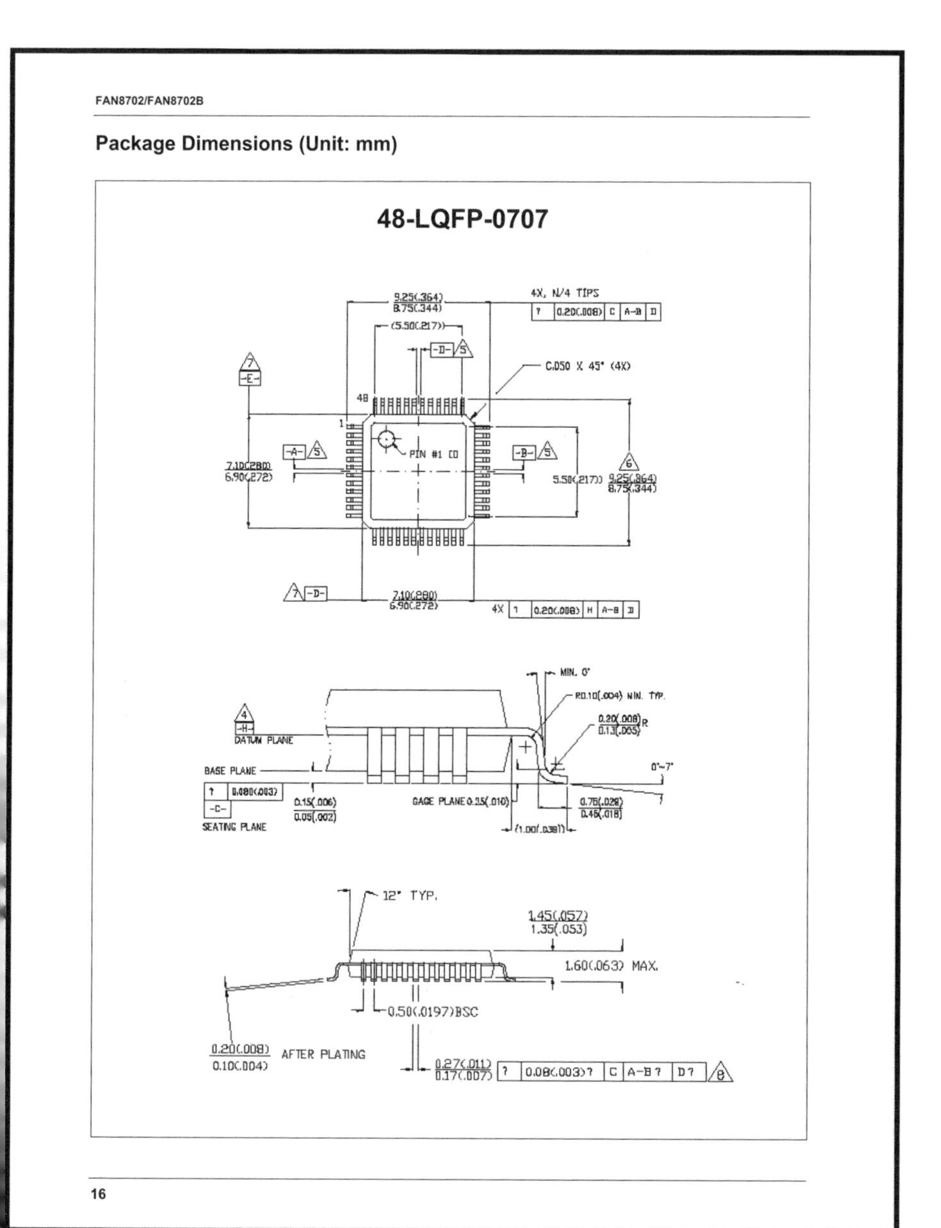
FAN8702/FAN8702B

Package Dimensions (Unit: mm)

48-LQFP-0707

5/25/05 0.0m 001
Stock#DSxxxxxxxx

Glossary

ACMs	Adaptive Computing Machines.
Active components	Transistors like NPN, PNP, NMOS, CMOS etc. The specialty of a transistor is that it can amplify, namely yield an output electrical signal bigger than the signal at its input. For this reason these devices are characterized as "active".
Air gap	The gap in the typically doughnut shaped magnetic core of an inductor. The gap width times the area of the doughnut cross section is the volume in which the inductor's energy is stored.
ANSYS	ANSYS, Inc.—computer-aided engineering technology and engineering design analysis software products and services.
BCD	Bipolar-CMOS-DMOS.
Bipolar	The process technology yielding NPN and PNP transistors.
Bipolar circuit	A circuit built with bipolar transistors such as NPN and PNP transistors.
Bricks	A DC-DC converter built according to industry standard footprint. Examples of form factors are: the popular 1/4 Brick (2.28"×1.45") and the newly defined 1/16 Brick (1.1"×0.9").

BOM Bill Of Materials.

Boost A device or technique that produces an output voltage above the input voltage.

Buck converter A voltage regulator that produces an output lower than its input. Also called step down (and sometimes step-down) converter. The term "buck" appears to be just another reference to the lower value of the output or as the American Heritage Dictionary puts it "of the lowest rank in a category."

Buck xDSP core Step down regulator (buck) for DSP core.

Buffer Transfers voltage transparently from its input to its output while increasing dramatically the current drive.

CDMA Code Division Multiple Access.

Charge pumping The action of biasing the node V_{CHP} above V_{IN} via the capacitor/diode C/D action.

Chip-scale package A package technology that encapsulates the silicon die with minimum overhead, resulting in a final integrated circuit of dimensions virtually identical to the silicon die itself.

CMOS Complimentary Metal Oxide Semiconductor. This is the process technology yielding N-type and P-type MOSFET transistors.

COP Coefficient Of Performance.

CPU Central Processing Unit.

Current mirrors An input/output circuit that produces an output current identical to the input current.

Current source A circuit that produces a constant current invariant to changes of the input voltage.

D flip-flop
A logic memory device that transfers to the output (Q) the logic data (D) present at its input when the positive edge of a clock pulse (positive edge triggered DFF) hits the CLK input. A second output (Q/) reproduces an inverted copy of the data present at the Q output.

Darlington stage
A cascading of two bipolar transistors (emitter on one transistor connected to the base of the second transistor) yielding an amplification gain equal to the product of the gains of the two single transistors.

Digital control architecture
A control architecture in which control parameters (for example control of a motor speed) are immediately digitized and processed digitally thereafter.

Digitally controlled analog system
A control architecture in which control parameters (for example control of a motor speed) are processed analogically while the communication with a host controller happens digitally.

Discrete power MOSFET
A single MOSFET transistor housed in a three terminal package and built to process high levels of power. "Discrete" refers to its 'single' nature and is used as opposed to "Integrated Circuit" which refers to a collection of a high number of transistors on board of a single die.

Disrupting technology
A new technology that enters an established market served by an existing technology and gains acceptance thanks to lower cost. Often disrupting technologies start as low cost and low performance technologies and are ignored by the establishment. After a few learning cycles the disrupting technology is better and cheaper than the existing technology, killing it.

DMOS
Double Diffused MOS. A special type of power MOSFET transistor.

Doped In order to create semiconductors the atomic structure of silicon is altered (doped) by the introduction of atoms of certain other materials (dopants).

DSP Digital Signal Processing.

DSP Core The processing unit in the DSP, as opposed to the periphery or I/O section.

DSTB Digital Set-Top Box.

Duty cycle The proportion of time during which a component, device, or system is operated. The duty cycle can be expressed as a ratio (0 to 1) or as a percentage (0 to 100%).

EMI Electromagnetic Interference.

Energy management The technique of optimizing energy delivery with minimum waste. For example a voltage regulator can produce a well regulated output voltage and waste a lot of energy in the process due to poor efficiency. Typically it costs more to deliver the same performance with less waste.

ESI Equivalent Series Inductance.

ESR Equivalent Series Resistor.

Feature phones High end cellular telephones equipped with features like camera, secondary display etc.

Flip-flop A digital logic circuit tht can be switched back and forth between two states (high and low, or on and off).

Fly-back converter An isolated converter architecture in which the energy is stored in the air gap of a gapped transformer during the first (active) part of the cycle and transferred to the output in the second part. One of the simplest isolated architectures and it is best suited to handle low levels of power.

Forward converter An isolated converter architecture in which the energy is transferred to the output during the first (active) part of the cycle and stored in the output inductor in the second part of the cycle. This architecture is suited for higher levels of power than the flyback converter.

Foundry houses Chip fabrication facilities that can produce chips for companies that are fab-less.

FSB Front System Bus.

Green mode A mode of operation in which the device consumes less than a Watt in stand-by mode.

Green power Power conversion operation in green mode.

GSM Global System Mobile; a communications standard.

Handset Cellular telephone handset.

Heatsink An electronic device in operation produces heat that needs to be disposed of. This is done by tightly connecting the device to heatsinks, highly thermally conductive materials like aluminum or copper molded in shapes—typically fins—that maximize surface for best disposal of heat by contact with circulating air.

Hooks Connections.

Hysteretic comparator A comparator is a device that changes the state of its output (low to high) when a positively sloped voltage signal applied to its positive input equals a reference voltage level present at its negative input. A hysteretic comparator is one in which the reference voltage is lowered after the signal crosses the reference, thereby providing a margin of safety against electrical noise that could induce false triggering of the comparator.

Inherently fast A circuit that is fast because of its nature. For example a sequential circuit can be said to be inherently slow because it will need several clock cycles before executing a command. Conversely, a hard wired circuit can be implemented for maximum speed.

Integrated circuit A collection of transistors connected to perform a given function all fabricated on a single silicon die.

Interleaved multiphase regulation The paralleling and time spacing of multiple regulators.

Interleaving The act of paralleling and time spacing multiple regulators.

I/O Input/Output.

Kirk effect A dramatic increase in the transit time of a bipolar transistor caused by high current densities.

LCD Liquid-Crystal Display.

LDO Low Dropout Regulator. This regulator can work with a small difference between supply and output voltage.

Leading edge An electric pulse is made of a rising or leading edge a flat top and a falling or trailing edge.

Leading edge modulation In leading edge modulation the control loop time-shifts the pulse's leading edge to change the duty cycle. Conversely in trailing edge modulation the control loop time-shifts the pulse's trailing edge to change the duty cycle.

LED Light Emitting Diode.

Li^+ Cell Lithium-Ion cell. The sign "+" is short hand for a (positively charged) Ion.

MCH	Memory Channel Hub.
MCU	Micro Controller Unit.
Mesa	An early type of transistor whose topology resembled that of a broad, flat-topped elevation with one or more clifflike sides, common in the southwest United States.
Microcontroller-based control architecture	A voltage regulator architecture in which the control loop is implemented by a sequential machine like a general purpose microcontroller as opposed to dedicated, hard wired circuitry.
Mixed	As in mixed signal processes. Processes that are capable of integrating on the same piece of silicon MOS and Bipolar transistors, which typically require different processes.
MOS	Metal Oxide Semiconductor
MOSFET	Metal Oxide Semiconductor Field Effect Transistors.
Multiphase interleaved buck converter	A buck converter resulting from the paralleling and time spacing of multiple regulators.
N-MOS	Short for negative-channel metal-oxide semiconductor. A type of semiconductor that is negatively charged so that transistors are turned on or off by the movement of electrons. In contrast, P-MOS (positive-channel MOS) works by moving electron vacancies (holes).
N-type material	A semiconductor material like silicon doped with materials from Column V of the periodic table of elements such as phosphorous that have an excess of negatively charged free electrons.
NAND gate	Negative AND gate. An AND gate performs the 'and' logic function of producing a logic output equal to the product of the logic inputs (like in $1\times1=1$ and $1\times0=0$). The NAND gate takes the result of the AND gate and inverts it.

OEM Original Equipment Manufacturer.

Opamp Operational Amplifier.

P-MOS Short for positive-channel metal oxide semiconductor. A type of semiconductor that is positively charged so that transistors are turned on or off by the movement of holes. In contrast, N-MOS (negative-channel MOS) works by moving electrons.

P-type material A semiconductor material like silicon doped with materials from Column III of the periodic table of elements such as boron that have an excess of positively charged holes.

Passive components Resistors, capacitors, inductors, diodes etc. A network built only with such components will always yield an output electrical signal smaller than the signal at its input. Because of their inability to amplify signals these components are characterized as 'passive'.

Peak current-mode control A type of current-mode control in which the value controlled is the current peak.

PFC Power Factor Correction.

PFM Pulse Frequency Modulation.

Poles The complex frequencies that make the overall gain of the filter transfer function infinite.

POL Point Of Load.

Power management The discipline of powering the circuits in an electronic appliance or transforming raw AC line electricity into finely regulated circuits that can power up delicate circuits like microprocessors and DSPs. Power here is counter posed to Signal, the business of providing computing, amplification, video, or audio capabilities.

PSUs	Power Supply Units.
PWM	Pulse Width Modulator
PWM controller	Pulse Width Modulation controller. A type of switching regulator control technique.
Reference voltage	A stable, known voltage level that is referred to in order to compare any other voltage level in the application.
RF section	Radio Frequency section.
RX block	Receiving block.
SDRAM	Single Data Random Access Memory.
Serial bus	Serial data bus. An I/O port that transfer data serially on one or few wires. As opposed to a parallel data bus.
Set-Reset flip-flop	A flip-flop is an electronic circuit that alternates between two states. When current is applied, it changes to its opposite state (0 to 1 or 1 to 0).) A set-reset is on in which activating the "S" input will switch it to one stable state and activating the "R" input will switch it to the other state.
Silver box	A metallic box containing all the AC-DC power circuitry necessary to power a personal computer motherboard, and found in any PC box.
SIM card	Single In-line Memory cards.
Smart battery	A battery incorporating electronics capable of communicating, via a serial bus, its identy, charge status, and other useful parameters to a host microcontroller.
Smart ICs	The expression refers to chips with intelligence on board, like CPUs and microcontrollers.

Smart phone A device which would have the advanced functionality of a handheld computing device, a digital still camera, a global positioning system, a music player, a portable television set, a mobile phone, and more in one convergent device.

Smooth Slowly varying.

SMPS Switch Mode Power Supply.

Snub network A network typically made of a low value resistor in series with a capacitor and presenting high impedance in normal operation and low impedance to spurious, fast varying signals, thereby short circuiting them and consequently preventing other networks it is connected to from exposure to such fast varying signals.

SOC System On a Chip.

SOI Silicon On Insulator.

SPICE Simulation Program with Integrated Circuit Emphasis.

SPICE deck A sequence of commands in SPICE language. Commands in ancient electronics times were punched on cards, piled on card decks, and fed to a computer for number cruching.

Sweet point An optimum point of operation that is the best compromise between conflicting requirements like high clock frequency versus high efficency.

Switch mode A mode of operation typical of switching regulators in which the energy transfer from input to output is discontinuous or in buckets. This is opposed to a linear mode of operation typical of a linear regulator in which the energy transfer from input to output is continuous.

TDMA	Time Division Multiple Access.
TFT	Thin-Film Transistors.
THD	Total Harmonic Distortion.
Time to market	The time it takes to develop a product and take it to market.
Trailing edge	An electric pulse is made of a rising or leading edge a flat top and a falling or trailing edge.
Trailing edge modulation	In trailing edge modulation the control loop time-shifts the pulse's trailing edge to change the duty cycle. Conversely in leading edge modulation the control loop time-shifts the pulse's leading edge to change the duty cycle.
TX block	Transmitting block.
Valley control	Short for Valley current mode control.
Valley current-mode control	A type of current-mode control in which the value controlled is the ripple current waveform valley.
Virtual prototype	A simulation of a device as opposed to a physical prototype.
VP	Voltage Positioning.
V_{CESAT}	Voltage between collector and emitter of a saturated transistor, namely one forced into heavy conduction.
V_{DD}	Common symbol for a positive power supply voltage.
VLSI	Very Large Scale Integration.
Zeros	The complex frequencies that make the overall gain of the filter transfer function zero.

Further Reading

Analog Design

Gray, P.R., and R.G. Meyer. *Analysis and Design of Analog ICs.* New York: Wiley.

Haskard, M.R., and I.C. May. *Analog VLSI Design NMOS and CMOS.* New Jersey: Prentice Hall.

Banzhaf, Walter. *Computer-Aided Circuit Analysis using SPICE.* New Jersey: Prentice Hall.

Sayre, Cotter W. *Complete Wireless Design.* New York: McGraw-Hill.

Antognetti, P., and G. Massobrio. *Semiconductor Device Modeling with SPICE.* New York: McGraw-Hill.

Control Systems

Melsa, J.L., and D.G. Schultz. *Linear Control Systems.* New York: McGraw-Hill.

Digital Design

Prosser, F.P., and D.E. Winkel. *The Art of Digital Design.* New Jersey: Prentice Hall.

KersHaw, John D. *Digital Electronics.* Boston: PWS-Kent.

Stout, D.F., and M. Kaufman. *Handbook of Microcircuit Design and Applications.* New York: McGraw-Hill.

Roth, Charles H. Jr. *Fundamentals of Logic Design.* Eagan, MN: West Publishing Company.

Taub, H., and D. Schilling. *Digital Integrated Circuits.* New York: McGraw-Hill.

Motors

Sen, P.C. *Principles of Electric Machines.* New York: Wiley.

Robbins and Myers. *DC Motors Speed Controls Servo Systems.* Gallipolis, OH: Electro-craft.

Stan, Sorin G. 1998. *The CD-ROM Drive.* New York: Kluwer Academic Publishers.

Leenhouts, A.C. *The Art and Practice of Step Motor Control.* San Francisco: Intertec Communications.

Power Electronics Devices

Baliga, B.J. *Modern Power Devices.* New York: Wiley

Williams, B.W. *Power Electronics.* New York: Wiley.

Blicher, Adolph. *Field-Effect and Bipolar Power Transistor Physics.* New York: Academic Press.

Smart Power ICs

Antognetti, Paolo. *Power Integrated Circuits.* New York: McGraw-Hill.

Murari, B.B., F. Bertotti, and G.A. Vignola. *Smart Power ICs.* New York: Springer.

Switching regulators

Ang, Simon S. *Power-Switching Converters.* New York: Dekker.

Lenk, Ron. *Practical Design of Power Supplies.* New Jersey: IEEE Press.

Chryssis, George. *High Frequency Switching Power Supplies.* New York: McGraw-Hill.

Severns, Rudolf P., and Gordon E. Bloom. *Modern DC-to-DC Switchmode Power Converter Circuits.* New York: Van Nostrand Reinhold Co.

Lenk, John D. *Simplified Design of Switching Power Supplies.* Burlington, MA: Butterworth-Heinemann.

Basso, Christophe P. *Switch-Mode Power Supply SPICE Cookbook.* New York: McGraw-Hill.

Erickson, R.W., and Dragan Maksimovic. *Fundamentals of Power Electronics.* New York: Kluwer Academic Publishers.

Index

Numerics

A

B

C

D

E

F

G

H

I

J

K

L

M

N

O

Q

R

S

W

X

Z